Development as Freedom
in a Digital Age

Development as Freedom in a Digital Age

Experiences of the Rural Poor in Bolivia

by Björn-Sören Gigler

 WORLD BANK GROUP

Endorsements

Praise for
Development as Freedom in a Digital Age: Experiences of the Rural Poor in Bolivia

"In his first-class study, Gigler provides strong counter-point to a persistent poison: That ICT and development scholarship is all too often atheoretical, methodologically unsound, and marginalizes people by putting all the focus on technology. Gigler first makes an important theoretical contribution by taking the capability approach of Amartya Sen and purpose-adapting it to position humans at the center of their information ecology. Then through a series of remarkably comprehensive survey and focus group studies, he establishes the social, cultural, economic, and informational contexts of the indigenous communities of Bolivia he is working with. The results of this theoretically sound and methodologically rich effort is a fulsome collection of results nuancing many qualities of ICT use: How social context does matter, what role do info-mediaries play, how does access progress to use, and more."

MICHAEL L. BEST
Director, United Nations University Computing and Society

"Gigler's in-depth study is the perfect antidote to the misleading notion of technology as a silver bullet for development. His rigorous analysis of extensive primary data demonstrates that technology access improves people's lives not by replacing existing information channels and community resources, but rather by building on and amplifying them—and illuminates the fascinating mechanisms by which this occurs. In the literature on ICT for development, this book is a true *tour de force*."

MARK WARSCHAUER
Professor of Education and Informatics
Associate Dean, School of Education
University of California, Irvine

"Information and Communications Technologies (ICTs) have changed our lives and they will continue to change our world over the coming years. While ICTs have had a huge impact in wealthy countries they are also tremendously significant in poorer, developing countries. This study provides a careful and in-depth analysis of the effects of ICTs on poor and marginalized people in Bolivia. ICTs appear to promise much as a route to poverty alleviation, but this study provides a balanced assessment of what impacts it can have. Using a notion of "informational capabilities" it concludes that ICTs can have positive impacts on the well-being of poorer people but it highlights that whether and how it does so depends on the role played by various intermediaries in the ICT chain. This excellent study provides a methodological benchmark for how we might assess the contribution of ICTs to the global effort to reduce poverty, inequality, and vulnerability."

J. ALLISTER MCGREGOR
Professor, Institute of Development Studies, Sussex

"This book creatively applies Sen's capability framework to assess the effects of information technologies in human development. Gigler's approach is original, focusing on the key role played by intermediary organizations, which sit between the technological innovations themselves and their use on the ground. He relies on an extraordinary amount of original data, collected via surveys and focus groups, to study the effects of intermediary organizations, and thus understand the conditions under which the spread of information and communication technologies can enhance people's informational capabilities and contribute to social development. The book is a must-read for scholars and students of ICTs and development."

KEN SHADLEN
Professor in Development Studies
Department of International Development
London School of Economics

"Sorting out the role of digital information and communication in international development has been a difficult task for scholars and development practitioners amid facile and false predictions about the power of ICT. Dr. Gigler's closely observed and richly theorized examination of ICT use in Bolivia advances our knowledge of the ways in which "informational capabilities" can be gained even by the poorest of the poor to increase their social and economic well-being. Drawing from and extending Sen's capability framework, systematic analysis of a mountain of carefully drawn empirical data and an understanding of intermediate organizations and influences, Gigler charts a realistic yet promising path for ICT development projects and research. Highly recommended for anyone serious about international development in the digital age."

JANE FOUNTAIN
Distinguished Professor
Director, National Center for Digital Government
University of Massachusetts, Amherst

"Dr. Gigler tackles one of the biggest misconceptions in Information and Communication Technology for Development—that access equals empowerment. As one of the leading scholars of Sen and the "capabilities approach," Gigler deconstructs technology for development projects into information flows and needs, capturing the power relations and politics inherent in any initiative. Through the Alternative Evaluation Framework, Gigler addresses the insufficiency of more commonly employed models and suggests a more-complete treatment of human capability. He adds "information capital" to his six dimensions of human capabilities, updating traditional sustainable livelihood frameworks to recognize information as a critical asset for those in poverty. By calling for explorations of access, communication, indigenous knowledge, and digital literacy, Gigler provides a mechanism for capturing more meaningful metrics and creating more thoughtful interventions in order to actually determine how well we are closing the so-called digital divide."

REVI STERLING
Director, ICTD Graduate Studies
ATLAS Institute
University of Colorado, Boulder

"Björn-Sören Gigler's fascinating book shows that even for the most remote and poor rural people, such as in Boliva, physical access to computers and the Internet is not the only condition for development and well-being. Skills and actual use are also needed. Information capabilities appear to be transformed to human and social capabilities via the poor people's own social network and meaningful practical benefits of information technology for every-day life."

JAN A.G.M. VAN DIJK
Professor of Sociology of the Information Society at the University of Twente, The Netherlands, and author of The Network Society and The Deepening Divide

"Björn-Sören Gigler's important book explains the conditions under which it is possible for Information and Communication Technologies (ICTs) to improve the political, social, cultural, and economic well-being of individuals living within poor communities. Gigler's book is well written and thus accessible to undergraduates, scholars, and development practitioners. Wide-ranging, deeply researched, nuanced, and insightful, *Development as Freedom in a Digital Age* investigates many topics but is particularly powerful in its analysis of (1) what can be measured, what cannot, and how various methodologies best supplement one another; (2) how the activities of women leaders and nonleading women impact their ICT utilization; (3) young people's range of activities regarding ICTs; and (4) the careful, self-conscious roles that development practitioners must undertake if they hope to contribute to improving the well-being of poor people through the use of ICTs."

EVE SANDBERG
Chair, Politics Department, Oberlin College

"Fifteen years after the Nobel-prize crowned work of Amartya Sen, Björn-Sören Gigler takes on the long overdue task of bringing the fundamental idea of Development of Freedom to the digital age. In doing so he goes beyond the much needed discussion of the philosophical aspects of capability deprivation and institution building, and draws his conclusion from a myriad of second- and first-hand statistics. This paints a convincing and enjoyable picture that emerges from the colorful brushes of both theory and hard-fact empirical evidence. The result presents unprecedented insights into a reality that is more complex than it is usually presented: A reality that sometimes seems contradictory at first sight; a reality in which digital tools strengthen horizontal peer networks, but do not eliminate the need for intermediaries; a reality in which the Internet raises high expectations for positive impacts, but is rejected as unfamiliar; a visit to a reality almost as complex as the reality of some the world's most marginalized communities."

MARTIN HILBERT
Department of Communication
University of California, Davis
United Nations Economic Commission for Latin America and the
Caribbean (UN ECLAC / CEPAL)

"This is an interesting and thoughtful book on the impact of ICTs on disadvantaged communities, with extensive empirical evidence from the rural poor in Bolivia. The book criticizes simplistic analyses based solely on ICT access and, instead, develops a more complex ICT impact chain. This involves the need for enhanced information capabilities, including communications capabilities, information literacy, and knowledge sharing abilities. The role of intermediary organizations is emphasized in supporting disadvantaged communities to develop and use these skills. Although the book is based on research in one country only, the content is of relevance to anyone concerned with ICT4D."

GEOFF WALSHAM
Emeritus Professor of Management Studies
Judge Business School
University of Cambridge

"Gigler's book addresses some key questions of the 21st century: Is the new universe of e-communication including or excluding the poor? Under what conditions do the poor have real access and then benefit as a result? The book provides a much needed in-depth exploration of the impacts of ICTs in a low-income country, an exploration that is as carefully socially contextualised, institutionally aware, and attentive to tracing actual impacts on people's lives as were the leading studies a generation ago of the "Green Revolution" in agriculture. Gigler adds to that perspective a refined conceptual framework, from Amartya Sen's capability approach, that helps him to more sensitively investigate ICTs' "impact-chain." And he adds to existing literature on ICTs in low-income countries through the sheer breadth and depth of his study, that draws on 20 months of fieldwork amongst indigenous groups in Bolivia, using surveys, focus groups, participatory observation, and in-depth interviews. A very valuable study."

DES GASPER
International Institute of Social Studies,
The Hague / Erasmus University, Rotterdam

"There is a widespread recognition within the ICTD discourse that "development" has been under-theorized. Gigler's book makes a timely and significant contribution to correcting this imbalance both in terms of its theoretical development and its empirical rigor. A key finding that emerges is that intermediary organizations play a critical role in ensuring that ICTs are locally appropriated by communities—something that needs to be taken seriously by policy makers, managers, local activists, and researchers."

SHIRIN MADON
London School of Economics and Political Science

Contents

PART IV   Conclusions

Box

Figures

Maps

Tables

Foreword

A visitor to a developing country today cannot help but be struck by the appendage attached to every second person's ear on the street, even on otherwise mean or grubby streets. The appendage of course is the mobile phone, and when it is not directly attached to an ear, it is probably linked to a pair of head phones or is being fiddled with to send and receive text messages in a form that even the illiterate can fathom.

What a marvelous technology, to be cheap, attractive, and simple enough to overcome so many of the otherwise grueling handicaps of poverty and under-development. Digital technology today is much more than a conduit for music, exchange of photographs, conversation or the pretense of conversation, for, according to some recent research, many cell phone users in public spaces are in fact pretending to have a conversation when there is no one at the other end, just to avoid the social awkwardness of solitude. It is, importantly, an instrument for trade, exchange and the enhancement of information, and, as such, a contrivance for empowerment.

This still unlocked potential and the ways in which it can be harnessed for the individual and the greater good is the subject of Björn-Sören Gigler's splendid new book. He is interested in much more than mobile phones; he explores the broader technology that makes mobile phones possible and thus asks how and when information and communication technologies (ICTs) in general can be a force for development in its widest sense. The development that this book seeks through ICTs is more than a rise in national or even individual incomes. Gigler is interested in the potential of ICTs to reach the most marginalized communities in poor countries and in helping the people living there to expand their opportunities and realize the potential in their lives.

The central idea is that access to and use of digital technology must now be added to the list of what Amartya Sen has called a person's "human capabilities" — the ability to be and do what that person "values doing or being." The idea of capability goes beyond the functions a person actually undertakes to the larger idea of what the person has the capacity to achieve, and hence captures the critical idea of human empowerment.

To understand how to use ICTs to achieve this end, Gigler begins at the bottom, through rigorous and extensive field data collection from indigenous communities in Bolivia. The mix of qualitative and quantitative methods richly embodies the best traits of good empirical research: Being grounded in theory but not bound by existing theory; laying out triangulated research findings; and giving as much importance to perceptions as to behavior. Gigler deploys this eclectic approach to try to understand what the poor know and think about ICTs, what they would like ICTs to do for them, and what the barriers are to such understanding and use. He is interested in the limits of ICTs for development, especially when their introduction does not take on-the-ground sociocultural realities into account. Only then does he recommend holistic interventions for harnessing ICTs to support development efforts in poor rural communities.

Aside from their policy value, Gigler's findings add to the sociological literature on the role of technology in poor settings by highlighting the importance of vertical (in addition to horizontal) social networks for social and economic change. The results also illustrate the continuing importance of mass media for changing attitudes and behavior.

The study's practical lessons are many and include the caution that, blindly glorified, ICTs can in fact exacerbate existing inequalities and inequities and that those most in need of its potential benefits can actually be left out completely if more thought is not given to the "how" of ICT introduction. Such practicalities must consider the ideal characteristics of strategic intermediaries, the social and cultural heterogeneity of the poor, the complementarity between ICTs and traditional media, and the diverse sector-specific goals.

Today social and economic life is undergoing change at a pace that may well be without parallel in history. If the arrival of the printing press had transformed society, it is likely that we are in the foothills of even more dramatic transformations, as the Internet and digital technology advance and change the conduct of economic and social lives. This has given rise to uncertainties, anxiety, and a multitude of questions.

No single monograph can answer all the questions. But through painstaking research and "thick descriptions" of what is happening to the poor and the most vulnerable, this book sheds light on what new technology is doing for, and also to, us. It harnesses modern welfare economics and moral philosophy, including abstract concepts, such as functionings and capabilities, to conduct grassroots research and data collection in the spirit of old-style anthropology, to address some of the most pressing policy concerns of our age.

Kaushik Basu
Senior Vice President and Chief Economist
The World Bank

Foreword

This important new book by Björn-Sören Gigler explores the opportunities and challenges of adopting the information and communication technologies (ICTs) by indigenous communities in rural Bolivia. The book questions the inevitability of technological determinism for "access" to this information technology, and instead provides a rich in-depth analysis of the dynamic interaction between people, technology, and human development. The assumption that pure access to technology will have a positive impact on people's economic and social development is misguided. Gigler pulls together abundant empirical quantitative and qualitative data to analyze the opportunities and challenges of ICTs in Bolivia's rural setting of poverty and marginality. Gigler provides the favorable and necessary conditions of "institutional embeddedness" where ICT use in Bolivian society can be applied with some degree of success for the rural indigenous despite the complexities and nuances.

Gigler examines a number of governmental and civil society institutions using ICTs, and through case studies from all regions of Bolivia, demonstrates the gap in practice between "access" and "effective use" of ICTs and their impact on community socioeconomic empowerment. Using rich empirical material, including extensive primary data from Bolivia, he demonstrates how this gap can be bridged by the effective use of social capital, namely the "intermediary organizations" of civil society with close ties to, and experience with, the local communities in which they work.

Gigler's richly detailed analysis of the application of ICT to economic and human development within the broader Bolivian development literature has great relevance within a country context where the rural poor have attained significant political and social inclusion under the Evo Morales–led government. However, a new paradigm for an alternative broad-based sustainable development model in rural communities with favorable links both to the market and to indigenous sociocultural values has remained elusive. Innovations in ICTs should be a vital component of any effective strategy for rural development, and Gigler advances our understanding by examining the conditions in which they can be incorporated.

Gigler unpacks the relationship between improved access to information facilitated by ICTs and the improved well-being of the poor in the economic, social, political, and cultural dimensions of people's lives. One of the most innovative aspects of the analysis is that it investigates in great detail, using a multi-disciplinary approach, both the developmental outcomes and the underlying social processes that take place once ICTs are introduced into local communities. And the importance of existing informational capital and indigenous knowledge systems in rural communities is highlighted.

Although Bolivian indigenous Andean communities have been recognized in the development literature for possessing high levels of social capital, this social capital refers more commonly to local democratic governance than economic entrepreneurship and technological innovation. Thus there is a need for civil society organizations, which can provide the skill set for promoting socioeconomic development and empowerment.

In his case studies, both in the Bolivian altiplano region and in the valley of Santa Cruz, Gigler contrasts ICT introduction and its use by municipal governments with the local Aymara indigenous NGO which had deep commitments to, and programmatic activities with, local communities of similar language and cultural background. There are several examples of how ICT can be appropriated by indigenous organizations and made to work within their own normative frameworks for grassroots community action and local development. Gigler's material gives us an approach that demonstrates how traditional knowledge systems can be enhanced by ICTs rather than replaced by them. Another feature of the book is the personal testimonies, and the Aymara claim that ICTs have benefited them both personally and collectively in their grassroots strategies.

Many observers unfamiliar with the importance of social institutional contexts in Bolivia to shape rural development processes often take a more narrow approach to ICT, assuming that its dissemination is simply a function of basic training programs and established developmental designs of local governments. Gigler calls into question this kind of blind faith in ICT technology that claims to work in low-income rural poor communities, irrespective of the rural development institutional framework in which it is placed. To some extent, this technology-as-magic-bullet approach is similar to the Green Revolution's claims for solving the developing world's food problems, which fell short in its primary mission and brought many unanticipated consequences to the rural communities.

I am convinced that the findings presented in this fascinating book will move discussions from the current debate of the potential of new technologies to be transformational for development to a deep reflection about the actual and realized economic, social, political, and cultural impact that new technologies have on poor people's well-being. It provides a refreshing new perspective on a key question for development in today's world: How can poor communities appropriate and own new technologies that fit their local sociocultural context and how they can enact technology to align with their aspirations about a better life for their families and communities?

Kevin Healy

Adjunct Professor, Center for Latin American Studies

Georgetown University

Former Representative for Bolivia

Inter-American Foundation

Author of *Llamas, Weavings, and Organic Chocolate:*

Multicultural Grassroots Development from the Andes and

Amazon of Bolivia, University of Notre Dame Press

Preface

What impact do new information and communication technologies (ICTs) have on people living in poverty in developing countries? Can ICTs make a real difference in the lives of the poor and thus enhance their well-being? Under what conditions can information technology empower poor communities? What are some of the challenges and pitfalls facing local communities in using technology innovations? How can we evaluate the impact that new technologies have on the well-being of poor communities?

It is difficult to imagine a world without cell phones, the Internet, or social media. Our children use smartphones, mobile apps, and social media on a daily basis, and the majority of the poor in the world are gaining access to these technologies for the first time. The real question remains: Does having access to a cell phone, the Internet, or social media have any tangible benefits for the living conditions of the most marginalized among the poor? Is the "digital divide" widening or narrowing the "economic divide"?

In today's world, being digitally literate has become imperative for accessing economic, social, and political opportunities. Technology can act as an enabler for economic, social, and political freedoms. At the same time, most technology innovations, including interactive websites, apps, or social media platforms, were developed for the privileged in the developed world and often do not reflect the socioeconomic and cultural realities of the poor. The Internet played a major role in the Arab Spring as well as the Tunisian uprising. Does this mean that the Internet has the potential to facilitate collective action through "connective action"?

Despite the rapid growth in penetration rates of ICTs in the developing world, the full potential of its impact on the well-being of marginalized communities remains to be seen. What obstacles are stopping ICTs from reaching their full potential? Through which channels can the potential of ICTs be fully realized? The highly successful Gutenberg printing press catered to the information needs of society. Based on a deep understanding of people's information needs, the first printed book was the Bible, followed by instruction manuals such as those for increasing agricultural productivity. In the 1960s, television was considered to hold the potential for improving education among the poorest communities, but it failed to be as transformative as the Gutenberg printing press.

Can the Internet be as transformative for society as the Gutenberg printing press, or will it fail to meet the expectations attached to it? Donor agencies distributed mosquito nets to poor people in Africa in an effort to control the incidence of malaria. Instead of using them as mosquito nets, the recipients turned them into fishing nets. They had access to and were using the product, but they were not using it as intended. Will the people in developing countries use the Internet as a mosquito net or as a fishing net?

The relationship between technology and political economy holds prime importance. How does technology interact with the local political economy? Is technology always neutral, or can it also be disruptive? Under what circumstances does technology remain neutral or threaten existing social and institutional power structures? Is the role of technology in eliminating intermediaries a myth or a reality? Does technology alter the role of intermediaries? Under what conditions can intermediaries help local and marginalized communities to derive maximum benefit from the use of new technologies?

The book tackles these questions by developing a new approach for evaluating the impact of ICT programs on poor people's economic, human, and social development. It looks at the effects of new technologies on well-being in a much more holistic manner than previous studies. Based on the framework and the empirical evidence presented in it, evaluations of ICT programs should take a more holistic approach and address the following issues:

- Measure and assess information and communication needs of the community before implementing a specific technology solution
- Measure "use" of ICTs as opposed to a simple binary variable for "access"
- Measure "purpose of use" for ICTs along with the simple variable for "use"
- Measure and assess the conditions under which effective use can enhance informational capabilities
- Measure and study the link between enhanced informational capabilities and enhanced human and social capabilities

The implementation of this new holistic approach requires the extensive collection and rigorous analysis of primary data. The analysis of the empirical evidence from rural communities in Bolivia discussed throughout the book yields the following key findings.

Marginalized communities think of the Internet as unfamiliar to them. This is primarily due to the fact that it does not speak their local language. Since most programs, applications, and tools on the Internet are designed for the needs of affluent people in developed countries, they often do not satisfy the information and communication needs of poor people living in developing countries. Therefore, the local appropriation of content on the Internet is a necessary condition for ICTs to improve the well-being of marginalized communities.

Technology solutions should build on existing information channels at the community level. New technologies may disrupt local institutions or threaten the bargaining power of the local elite. Case study analysis reveals that technology can only enhance people's well-being if local communities not only have access to it but also use it within their own structure or political economy.

ICTs do not eliminate the need for intermediaries. The role of intermediaries is critical, as they assist marginalized communities in interpreting, evaluating, and understanding the information they receive from ICTs. Intermediaries also help local communities to interpret, appropriate, and adapt ICTs to their local sociocultural context. An important precondition is that the intermediary must have the trust of the community.

While new technologies are strengthening existing horizontal networks among poor people, they are much more limited in enhancing vertical communication channels between policy makers and the poor. For poor people who have strong social networks, strengthening their communication capabilities enhances primarily their horizontal communications with other communities and, to a much lesser degree, their vertical communications with policy makers and the urban population.

Community-level socioeconomic variables are more important determinants of ICT use than individual-level variables in rural communities in Bolivia. This finding is contrary to the literature on the digital divide, which emphasizes the importance of individual variables such as gender, income, or levels of education.

There are significant differences in ICT use among different indigenous groups within Bolivia. The digital inequalities reflect the existing structural inequalities within society, whereby ethnicity is a critical factor influencing whether a person is poor or not and as such lacks the human capabilities and social opportunities to reach the lifestyle that she or he values.

In rural areas of Bolivia, young people do not use technology more than older people in the same communities. Although a gender divide exists, it is not the decisive factor.

The expectations about the positive impact of the Internet are highest among the most marginalized. They are also higher among the least educated.

The analysis of variables generated from perception-based surveys makes a critical contribution to ICT impact evaluations. Perception-based surveys help to identify previously unobserved mechanisms through which new technologies may improve people's well-being. These mechanisms can include the shift in behavior of the actors or agents involved.

The empirical findings also reveal gender differences in the expected impact of the Internet on people's well-being. Women are much more concerned about how the Internet can promote the social good, such as health, child care, education, literacy, and malnutrition, while men are much more interested in how the Internet can expand productive and economic activities.

ICTs by themselves do not provide an escape route from poverty; they need to be paired with complementary factors to have a multiplier effect on well-being. ICTs are so closely embedded in the political economy, which makes it challenging to isolate their effects from others.

Based on the analysis conducted in the book, the study provides the following policy implications and recommendations on the manner in which ICT programs should be designed in order to be most effective in enhancing the well-being of poor communities:

- Integrate ICT programs in the overall objectives of a sectoral program
- Carry out a detailed assessment of existing information flows and information needs of communities
- Analyze the process of how ICTs are being introduced into communities
- Combine different forms of ICTs, as they are most effective when traditional media are combined with newer forms
- Proactively address social, political, and cultural factors during implementation of activities at the community level.

This book goes beyond the traditionally explored research questions about the digital divide and access to ICTs. It instead unpacks the conditions under which as well as the process by which ICTs can have an impact on people's human and collective well-being. Based on the results presented in this book, it is possible to say that, yes, ICTs have an impact on poor people's economic and social development, but only under certain conditions.

Acknowledgments

Along the journey of researching and writing this monograph, I have met numerous people who have provided me with invaluable inputs, have encouraged me to pursue my research, and have helped me in moments when it was difficult to maintain the enthusiasm and energy necessary for completing such as project.

First and most importantly, I would like to thank Karim, my beloved wife, and son Noah who have suffered my efforts for far too long. They have given me love, encouragement, and inspiration to write this book. I am extremely grateful for their moral and emotional support and could not have completed this research without them.

Second, I would like to express my gratitude to my colleagues from the World Bank, from academia, and key partner organizations who have provided me with invaluable intellectual guidance, support, and suggestions on earlier drafts. The work of these colleagues and peers has been a real inspiration, and I am extremely thankful for their unique support throughout the process. A special thanks to Sabina Alkire, Enrica Chiappero-Martinetti, Jean-Paul Faguet, James Foster, Jonathan Fox, Kevin Healy, Shirin Madon, Samia Melhem, Ken Shadlen, and Frances Stewart. Much is owed to Hassane Cisse, Faris Hadad-Zervos, Robert Hunja, Randi Ryterman, Randeep Sudan, and Jeff Thindwa for their support, guidance, and continuous encouragement to work on this research project until its completion. I am especially thankful to Kaushik Basu for the very inspiring conversations about Amartya Sen's capability approach and his comments and suggestions on the entire manuscript.

Third, I would like to give a special thanks to all the peer reviewers and others who have provided me with great advice and comments on how to improve the early drafts. A special thanks to Saher Asad, Michael Best, Stephen Coleman, Jane Fountain, Des Gasper, Kevin Healy, Richard Heeks, Martin Hilbert, Shirin Madon, Robin Mansell, Allister McGregor, Samia Melhem, Eve Sandberg, Janmejay Singh, Revi Sterling, Frances Stewart, Jan van Dijk, Geoff

Walsham, Mark Warschauer, and Ethan Zuckerman for providing feedback, suggestions, and comments.

Fourth, I would like to thank many friends and colleagues who have supported my pursuit of this research and have provided good advice and encouraging words on how best to complete this book. In particular, I would like to thank Norbert Fiess, Marvin Gonzalez, Nick Harrison, Robin King, Erick Langer, Connie Luff, Zoe Marriage, Rosanna Nitti, Denisa Popescu, Victor Quispe, Kenn Rapp, John Rogers, Juan Rubio, Dana Rysankova, Walter Salinas, Marco Scuriatti, Johannes Wiegand, Borge Wietzke, the family Simmons, and James Blackburn whose support has been invaluable.

Fifth, I would like to thank the protagonists of this monograph: The indigenous peoples of Bolivia for their trust and openness as they shared with me their experiences. They offered me the unique opportunity to gain insight into their daily lives, their multiple challenges, and their cultural richness.

I appreciate very much the professional and collegial support from Elizabeth Forsyth, Susan Graham, Patricia Katayama, Carlos Rossel, and Stephen McGroarty of the External and Corporate Relations/Publishing and Knowledge unit and all the services they provided in the process of producing and publishing this book.

I am thankful to Llanco Talamantes for being such a strong creative lead on the design of the cover image and the way it captures the dynamic interaction between indigenous peoples' worldviews and the norms and values embedded in the use of new technologies.

I am very grateful to my parents and brother Torsten who have instilled in me the quest and love for knowledge. My parents have always encouraged me to pursue my dreams and have given me the strong support I needed to complete this research.

Finally, I would like to extend my appreciation to my colleagues from the Open and Collaborative Governance and Innovation Labs teams who have provided their support throughout the process of finalizing this volume. I would like to give a special thanks to Elizabeth Crespo, Nagat Dawaji, Diana Fuentes, Juan Carlos Guzman, Benjamin Holzman, Kaushal Jhalla, Nene Mane, Keith McLean, and Marcos Mendiburu.

About the Author

Björn-Sören Gigler is a Senior Governance Specialist at the World Bank. He is a political economist and his research focuses on information and communication technology (ICT) and how it impacts and enhances the well-being of poor communities. He coordinates the Innovations in Governance program at the Global Governance Practice of the World Bank Group, which leverages innovative ICTs to make development programs more open, inclusive, and effective. Gigler is a Visiting Associate Professor for Political Economy at the Edmund A. Walsh School of Foreign Service at Georgetown University. He has taught and carried out research on the multidimensional aspects of poverty, human development, and ICT for development at the London School of Economics, Georgetown University, the Universidad Católica Boliviana, Bolivia, and the Universidad los Andes, Colombia. He holds a PhD in Development Studies from the London School of Economics; an MS in Economics from the University of Munich; and an MA in International Affairs from the Elliott School, George Washington University. He is the co-author/editor of *Closing the Feedback Loop: Can Technology Bridge the Accountability Gap?* (World Bank 2014) and has authored several articles and book chapters on topics related to information poverty, ICT for development, indigenous people's development, and open government.

Abbreviations and Acronyms

AEF	alternative evaluation framework
APCOB	Apoyo para el Campesino-Indígena del Oriente de Bolivia
CCI	communications channel index
CIDOB	Confederación de Pueblos Indígenas de Bolivia
CIOEC	Coordinadora de Integración de Organizaciones Económicas Campesinas Indígenas y Originarias
CONAMAQ	Consejo Nacional de Ayllus y Markas del Qullasuyu
CV	curriculum vitae
ECADI	Encuesta de Capacidades Informacionales
ETIC	Estrategia Nacional de las Tecnologías de Información y Comunicación
FAM	Federation of Municipal Associations of Bolivia (Federación de Asociaciones Municipales de Bolivia)
GDP	gross domestic product
HDI	human development index
IAI	information availability index
ICMA	International City/County Management Association
ICO	Instituto de Capacitación del Oriente
ICT	information and communication technology
IGI	information gap index
INE	Instituto Nacional de Estadísticas
INI	information needs index
ISI	information source index
IT	information technology
LPP	Law of Popular Participation
MDG	Millennium Development Goal

NGO	nongovernmental organization
OMAK	Organización de Mujeres Aymaras del Kollasuyo (Organization of Aymara Women of the Kollasuyo)
PADEM	Programa de Apoyo a la Democracia Municipal
PII	perceived impact index
UBN	unsatisfied basic needs
UDAPE	Unidad de Análisis de Políticas Económicas y Sociales
UNDP	United Nations Development Programme
USAID	U.S. Agency for International Development
VIPFE	Vice Ministry of Public Investment and External Financing

Part I

ICTs and the Well-Being of Poor Communities

CHAPTER 1

Introduction: People, Technology, and Well-Being

In *Development as Freedom in a Digital Age*, I argue that, for poor people, the knowledge of how to use, manage, and appropriate technology in their own lives is a critical asset for improving their economic, political, and social freedoms. In the information age, the knowledge of how to use information technology is—similar to literacy—a critical human capability that enables a person to realize the "various things he/she values doing or being" in all dimensions of his or her life (Sen 1999, 75). Technology does not stand at the center of this process; at the center is a person's ability to access, process, and act on information facilitated through the use of new technologies. In this book I introduce the concept of "informational capability" and argue that, in the digital age, this new capability becomes central, influencing the extent to which individuals are able to enhance their well-being in the economic, social, political, and cultural dimensions of their lives.

I investigate the impact of information and communication technologies (ICTs) on development from the perspective of poor communities themselves. My approach underscores the notion that communities are not mere "beneficiaries" of development but are "active citizens" who have the capacity and creativity to define their own development priorities, goals, and vision for the future (Chambers 1983, 1995; Cernea 1991; Eade 2003; Max-Neef 1991; Korten and Klauss 1984). Using this concept of people-centered development, I present an alternative framework for evaluating the impact of ICT programs that places people's human development, rather than technology, at the core of its analysis (Madon 2004, 2005) and operationalizes Amartya Sen's capability approach (Sen 1984, 1992, 1999). Specifically, I address two research questions. First, whether and under what conditions can new technologies enhance poor people's human capabilities to achieve the lifestyle they value—thus enhancing their well-being? Second, what role do intermediaries, like poor people's organizations, local nongovernmental organizations (NGOs), or government organizations, play in the process of adapting ICTs to meet the needs of local communities and to what extent do differences in the intermediary process explain variations in the impact of new technologies on poor people's well-being?

At the outset, I argue that improved access to information facilitated by new technologies can improve the well-being of poor people as long as technology programs are fully integrated into the broader sociopolitical realities of local communities (Castells 1997, 1998a, 1998b; Madon 2000; ITDG 2001; O'Farrell, Norrish, and Scott 1999). By placing communities' assets and capabilities at the center, I examine the catalytic role of ICTs in expanding the human capabilities of marginalized groups. Within this framework, I analyze key factors that enable poor communities to make meaningful use of ICTs and to appropriate these technologies as an instrument for their own development.

At the same time, there is the risk that existing inequalities are being perpetuated in the information society and that marginalized groups are being excluded from reaping the benefits from the use of ICTs.

Existing approaches have focused too often on issues related to the digital divide[1] and have overemphasized the role of technology itself in addressing these challenges. A new framework is needed for evaluating the impact of new technologies on economic and social development. We need new ways of thinking to find answers to these questions. We need to go beyond the debate about the "digital divide" and instead focus our attention on issues related to the underlying "capability divide." The central issue is, in fact, how poor communities can own, shape, and enact technology based on their own worldviews and derive real economic, social, and political benefits form the use of ICTs.

Using empirical evidence from indigenous peoples'[2] uses of ICTs in rural Bolivia, I conclude that the most critical factor in determining the impact of ICTs on the well-being of poor communities is the extent to which ICT programs succeed in enhancing people's "informational capabilities." In fact, issues related to the access to ICTs play a secondary role and no direct relationship exists between improved access to ICTs and enhanced well-being. I argue instead for the need to improve our understanding of the complex and dynamic interdependency among people, social institutions, and technology. The manner by which people interact with technologies and the way they adapt them to the local socioeconomic, political, and cultural context of their communities are critical for assessing the impact of new technologies on people's individual and collective well-being.

Emerging Evidence on the Links between ICTs and Development

In recent years, the literature has increasingly articulated the links between ICTs[3] and socioeconomic development (Avgerou 2003, 2008, 2010; Avgerou and La Rovere 2003; Madon 2000; Mansell and Wehn 1998; Heeks 1999, 2010; Braga 1998; Aker 2010; Jensen 2007; Jack and Suri 2014). Proponents of the "ICT for development agenda" claim that these technologies create new opportunities for economic and social development for developing countries and poor communities (Eggleston, Jensen, and Zeckhauser 2002;

Hamelink 1997; Negroponte 1996; Pigato 2001; Pohjola 2002; Beuermann, McKelvey, and Sotelo 2012; Fafchamps and Minten 2012; Goyal 2010).

This literature explicitly or implicitly assumes a direct relationship between ICTs and economic growth, social development, and enhanced democratic participation (Avgerou 2008, 139; Bedi 1999). For instance, Heather Hudson assumes a causal link between ICTs and development: "If information is critical to development, then ICTs, as a means of sharing information, are not simply a connection between people, but a link in the chain of the development process itself" (Hudson 2001). Based on economic growth theories (Rostow 1960), the literature asserts that ICTs can provide developing countries with the necessary mechanisms to "leapfrog" stages of their development by embracing the new "knowledge-based" economy (Davison et al. 2000; Singh 1999; Quah 1997; Soete 1985). This approach emphasizes the physical access to ICTs and assumes that people derive benefits from their use once they have access to the necessary infrastructure (Wresch 1998; Dymond and Oestermann 2004; Navas-Sabater, Dymond, and Juntunen 2002; Hanna 1994).

Researchers and donor agencies, including the World Bank (1995, 1998, 2006), United Nations Development Programme (UNDP 2001), United Nations Conference on Trade and Development (UNCTAD 2013), and International Telecommunication Union (ITU 2012), have emphasized the benefits of ICTs and stressed that developing countries have to embrace these new technologies proactively or suffer the negative consequences of an increasing digital divide, falling further behind in their overall development (Kirkman et al. 2002; Steinmueller 2001; Pohjola 2002; Rodríguez and Wilson 2000).

At the same time, several scholars have argued that the concept of the digital divide is conceptually oversimplified and theoretically underdeveloped (Warschauer 2002, 2003; Selwyn 2004; Katz and Rice 2002; Steyaert 2002a). For instance, according to Warschauer (2002, 5), "This concept attaches overriding importance to the physical availability of computers and connectivity, rather than to issues of content, language, education, literacy, or community and social resources." Moreover, several authors have pointed out that there is no binary divide between the "haves" and "have-nots," but rather a gradation based on degrees of access to information technology (Cisler 2000). Rob Kling concludes,

>> The big problem with the "digital divide" framing is that it tends to connote "digital solutions," i.e., computers and telecommunications, without engaging the important set of complementary resources and complex interventions to support social inclusion, of which informational technology applications may be enabling elements, but are certainly insufficient when simply added to the status quo mix of resources and relationships. (Rob Kling, quoted by Warschauer 2002, 5–6).   <<

More fundamental critiques point out that the ICT for development agenda assumes that improved access to ICTs inevitably leads to enhanced economic and social development—an overly optimistic and unrealistic view that is not based on an adequate conceptual framework or rigorous empirical analysis of the actual impact of ICTs on developing countries and people's well-being (Avgerou 2003; Anderson et al. 1999; Heeks 1999, 2002; Benjamin 2001). Only a few studies have looked at the impact of ICTs on outcomes of well-being in an empirical setting (Jensen 2007; Goyal 2010). The focus of most ICT-related empirical studies is limited to assessing the impact of access to ICTs on economic development. One of the reasons for choosing the ICT access variable over the ICT usage variable is the difficulty of measuring use. Access can be measured much more easily and is also much more readily available in survey data sets. These studies are, however, unable to improve our understanding about the effects of new technologies on people's lives.

Thus more critical studies emphasize that the discourse surrounding this approach hypes ICTs and their potential benefits as the panacea for development without having supporting empirical evidence (Heeks 1999; Kenny 2002; Marker, McNamara, and Wallace 2002; Mansell and Wehn 1998).

At the same time, many studies in the information system literature have found that ICT investments seldom yield the expected outcomes, are of limited value to poor communities, or result in either partial or complete failures (Anderson et al. 1999; Butcher 1998; Heeks 2002; Robey and Boudreau 1999; Avgerou 2000; Madon 1993; Heeks and Kenny 2002). Heeks, for instance, stresses that enthusiasm about ICTs lacks a theoretical framework and is based primarily on scattered anecdotal evidence, which does not justify such optimism (Heeks 1999). Furthermore, Wilson and Heeks emphasize that ICT investments are highly problematic in terms of their opportunity costs; they frequently take away scarce resources from more urgent and direct development priorities, such as improving poor people's access to education, water and sanitation, or health (Wilson and Heeks 2000).

Moreover, scholars have suggested that ICTs actually aggravate existing socioeconomic inequalities and further marginalize the poor (Ciborra 2002; Castells 1998a; Gumucio Dragon 2001; Panos Institute 1998; Wade 2002; Thompson 2004). For instance, Castells emphasizes that ICTs can represent both the cause and the effect of social marginalization and warns that computer-mediated communication is culturally, educationally, and economically restrictive, reinforcing the dominant cultural and social networks and making the poor majority of the development world irrelevant in this new knowledge economy and "network society" (Castells 1997, 1998a).

Finally, a growing number of authors have called for a much deeper and more nuanced understanding of the relationship between ICTs and development (Heeks 2002; Wilson and Heeks 2000; Madon 2000; Burkett 2000; Loader 1998; Barja and Gigler 2005; Gigler 2005a). They point out that a key question for determining whether or not ICTs can have a significant impact on socioeconomic development and people's lives depends on the extent to which these technologies are adaptable to the particular local socioeconomic, political, and cultural context. Such an approach stresses the need for adopting a more holistic approach that fully "integrates ICTs into the overall development objectives of specific programs, rather than being driven solely by technological concerns" (Heeks 2002, 7).

The field of community informatics makes an important contribution to this emerging literature in the sense that it helps us to understand better the conditions under which ICTs can be made more usable by and useful to excluded groups. This literature looks beyond mere access to the effects that ICTs have on local communities within the broader context of existing social systems and cultures (Gurstein 2000; Warschauer 2004). In particular, it draws on the concept of "effective use," developed by Michael Gurstein, which emphasizes that people can derive real benefits from ICTs depending on "the way peoples are making use of ICTs in their daily lives and how well they have integrated ICTs into their social, productive, and cultural activities" (Gurstein 2003, 10).

The proponents of this more nuanced approach have identified a major gap in the literature: the absence of in-depth empirical evidence examining the links between ICTs and socioeconomic development and people's well-being, particularly in developing countries (Avgerou and Walsham 2000; Blattman, Jensen, and Roman 2003; Nulens 2003; Wilson and Heeks 2000). For instance, DiMaggio emphasizes, "We need to move research away from the ideological debate about the relationships between ICTs and development towards robust survey-based and in-depth qualitative work that begins to unpack the complexity of the digital divide" (DiMaggio et al. 2001, 327). Furthermore, to evaluate the impact of ICTs on people's well-being, we need to go beyond assessing outputs that are easily measurable, such as Internet penetration rates, and instead measure outcomes, such as the effects of Internet use on a person's income or access to public services (Gigler 2004).

The Role of Telecenters for Socioeconomic Development

An important approach to using ICTs to enhance the well-being of rural communities is to establish telecenters (Chapman and Slaymaker 2002; Proenza, Bastidas-Buch, and Montero 2001). A telecenter—frequently also called a multipurpose community telecenter, public access point, or information kiosk—is a physical space that provides public access to ICTs for educational, personal, social, and economic development (Gómez, Hunt, and Lamoureux 1999; Roman and Colle 2002). Since the mid-1990s a sizable literature has emerged regarding the use of telecenters as a means for promoting socioeconomic development (Benjamin 2000; Jensen and Esterhuysen 2001; Cecchini and Scott 2003; Keniston 2002; Madon 2005). At first, researchers were optimistic that telecenters could empower poor communities by providing them with direct access to ICTs and information services (Whyte 2000; Harris et al. 2003; Hunt 2001). Increasingly, there is evidence that telecenter projects are more likely to be successful if computer and Internet-based services are complemented by additional services such as capacity building in the use of ICTs, delivery of local content, and access to government services (Kuriyan, Toyama, and Ray 2006; Colle 2005; Conroy 2006; Madon 2004).

The literature on telecenters covers a wide array of topics, including their design (Morelli 2003), access to local and relevant content (Harris 1999;

Roman and Colle 2002), information needs and community participation (Cecchini and Raina 2004; Bailur 2007; Kanungo 2003), and evaluation (Young, Ridley, and Ridley 2001; Reilly and Gómez 2001; Hudson 2001; Whyte 2000). Rothenberg-Aalami and Pal (2005) examine problems that confront evaluations of the effects of telecenters. They mention that, first, telecenters are a combination of entrepreneurial ventures and development projects and that, second, the impact of telecenters can occur on different levels, including the individual, community, regional, and national levels. The issue of telecenter sustainability has emerged in recent years as a key concern, as many telecenter programs have failed or closed due to lack of financial, social, and political sustainability (Bailur 2007; Madon et al. 2007). Bailey (2009) examines operational issues facing telecenters in Jamaica as their roles change to address newer issues that arise. This empirical study, which uses interviews with a range of stakeholders, suggests that there is a need to train telecenter staff to enhance participation by the community. Further findings suggest that there is a need to identify the core capabilities, such as education, training, employment, social network, and socioeconomic factors, grounded in the social context to achieve social sustainability (Bailey 2009).

Goyal (2010) studies the impact of an Internet kiosk program named e-Choupal on the livelihood of farmers in the state of Madhya Pradesh. These kiosks, introduced in 2000, were designed to provide farmers with price information from several markets and their location. They also provided information on the going rate paid by ITC, a large agricultural export business. Therefore, in addition to price information, the project provided marketing information. Regression analysis of the data showed that the project led to an increase in both the price of soybeans and the absolute area under soybean. This change in prices had very important effects on people's well-being since soy is the main cash crop in the local region (Goyal 2010).

Prado (2010) evaluates the impact of a telecenter on 47 individuals in a remote agricultural community in the Caribbean. Analysis of the survey data suggests that this agricultural community has been able to take advantage of Internet access. Respondents to the survey mentioned that they primarily use the telecenter to prepare for work and to learn new things. They reported using the telecenter to search for information on health care, child care, farming, animal care, sanitation, emergency preparedness, and online news. Some of the users also mentioned that they learned how to use the Internet after the telecenter was introduced. In addition, respondents said that they have a positive perception of the telecenter overall and that they trust the information they get from it.

Soriano (2007) employs a case study approach to study the impact of telecenters set up by the United Nations Development Programme (UNDP) and the China Ministry of Science and Technology in selected rural villages in Wu'an, China. Findings suggest that telecenters had a positive impact on economic aspects (such as better earnings or production), e-literacy, farming techniques, and development of knowledge-sharing platforms (Soriano 2007).

Proenza, Bastidas-Buch, and Montero (2001, 2) suggest that telecenters should consider the model of cybercafés, arguing that a private sector model would help to ensure their financial sustainability. However, Simpson, Daws, and Pini (2004) criticize the overemphasis on financial sustainability and recommend defining public access points as essential social community programs, like schools and libraries, deserving of government funding: "This would mean redefining the sustainability of public access only in terms of the outcomes it produces in terms of social and community betterment rather than only in terms of economic gains."

Kuriyan, Toyama, and Ray (2006) investigate the feasibility of using a public-private partnership model as an adequate governance model for implementing telecenter projects in rural India. They examine in detail the Akshaya telecenter project in the state of Kerala and demonstrate that it is problematic to achieve the twin goals of commercial profitability and social development for the poorest and most excluded groups. They argue, "There is a tension inherent between the goals of social development and financial sustainability at the macro level (within the state and political parties) and a micro level (with entrepreneurs and potential consumers) making it difficult to run a financially solvent ICT kiosk that also meets development goals" (Kuriyan, Ray, and Toyama 2008, 2). Furthermore, the Akshaya case study shows that implementing telecenter projects through the public-private partnership model is not simply a technical process of providing ICT services to the poor; it is a highly political process that requires a delicate balance between the various stakeholders of the project (Kuriyan, Ray, and Toyama 2008, 5).

Madon (2005) investigates the social and political sustainability of telecenters based on a longitudinal study of the Akshaya project. Using a "sociology of governance" approach to study implementation of the project, she argues that the extent to which the developmental potential of improved ICT access can have a transformative role for the poor depends on how the interactions between a host of stakeholders, including government agencies, social entrepreneurs, international donors, telecommunications operations, and civil society organizations, are being managed at the local level. The adoption of the "sociology of governance theoretical lens" enables her to analyze "the subtle interaction over time between state and society inviting us to illuminate critical aspects of social reality rather than applying a rigid theoretical governance template within which to 'fit' the reform process" (Madon 2005, 413). Furthermore, she suggests that the focus on "governance as interaction can hopefully provide useful input toward reconceptualizing the nature and scope of telecenter evaluation to move beyond an exclusive focus on good governance indicators."

This overview of the telecenter literature shows that multiple factors influence the extent to which the provision of improved access to ICT and information services can empower poor communities and that access to ICT services alone does not have any transformational effects on the lives of poor communities. It is instead necessary to unpack the factors and processes that result from the introduction of ICTs in local communities. In particular, it is critical to investigate the interactions between the various stakeholders involved in ICT projects and the potential of improved access to ICTs and information to alter existing forms of governance at the community and national levels.

The following section provides a brief overview of the e-governance literature and the opportunities and challenges of e-government programs aiming to improve the well-being of marginalized communities.

The Link between E-Governance, Participation, and Community Development

A growing literature stresses the link between governance, participation, and local community development. However, it often uses the term "governance" in an ambiguous way. In this study, I use the definition of governance commonly used in development studies: governance is "the prevailing patterns by which public power is exercised in a given social context" (Jenkins 2002).

In recent years, the public sector reform literature has stressed that the state is no longer the only provider of social services and has advocated a greater role for the private sector and public-private partnerships in delivering services such as education or health to poor communities. The public sector reform agenda has frequently been linked to the notion of "good governance," which has stressed the importance of increased accountability, efficiency, and responsiveness of government agencies, in particular, in the context of service delivery (World Bank 2004). In addition, these reforms have endorsed the decentralization of the state, assuming that local government would be more efficient and responsive to local communities than the central government (Bardhan 2002).

In this context, development agencies have promoted participatory approaches at the project level in order to "empower" local communities to become active agents in their own development. According to their critics, these public sector reforms and the good governance agenda have led to "depoliticizing development and rule" (Corbridge et al. 2005, 151), weakened the role of the state in favor of the private sector in key areas of the economy (Stiglitz 2007), and led to a "tyranny of participation" that has placed additional burdens on the poor by enlisting them to volunteer their time for the community work that development projects entail (Cooke and Kothari 2001).

Recent research on power relationships within participatory development has highlighted the challenges of this approach, particularly with regard to the ability of different groups within communities to benefit from participatory projects (Agrawal 2001; Eversole 2003). In *Seeing the State*, Corbridge et al. (2005) argue that state-citizen interactions are based on flows of power, money, commodities, and information. In an ethnographic analysis in eastern India, they study the ways in which poor people encounter the state on an everyday basis. They argue that the poor in rural India form their understanding of the state through their direct interaction with local government officials (Corbridge et al. 2005, 29). They note that a poor person interacts most directly with the state when he or she registers a birth certificate, receives a registration form, or processes a financial transaction, such as paying an electricity bill. Such encounters differ by region and public services provided and define people's ability to claim their rights as citizens; they are frequently defined by clientalism and patronage. As Corbridge et al. point out,

>> We learn a great deal about the "state" by examining its changing protocols, for bounding the "poor," and its plans for seeking their development, protection, empowerment, or erasure. For their part, the poor in India learn to see the state through their meetings with particular government officers, with regard for those government conventions and policies with which they gain familiarity. (Corbridge et al. 2005, 19).   <<

Corbridge et al. analyze in depth the power relationship among government officials, local power holders, and the poor, and their analysis outlines the complex ways in which clientalism, patronage, and citizenship are entwined. They acknowledge, "Governmental practices built around participation ... often fail to meet the needs of poorer people" (Corbridge et al. 2005, 261), but they also suggest that there is the potential for citizens to gain strength through encounters with the state: that these practices "can, slowly and unevenly, be instrumental in providing poorer people with a greater sense of self-worth, dignity, and, more rarely, a degree of power over those who would govern them" (Corbridge et al. 2005, 262).

Critical for my research is the literature evaluating the effects of e-government projects on socioeconomic development. E-government is defined "as the use of ICTs to improve the activities of public sector organizations" (Heeks 2003). More specifically, e-government is intended to (a) make government more accessible, effective, and accountable; (b) improve the relationship between government and citizens; (c) strengthen the coordination and cooperation within the public administration; and (d) enhance government's performance in the delivery of public services (Yigitcanlar 2003; Ciborra and Navarra 2005).

The proponents of e-government frequently argue that the application of ICTs in the public sector can play a catalytic role, enhancing democracy, improving government performance, and supporting socioeconomic development more generally (Bhatnagar 2004; Hanna 1994; Krishna and Walsham 2005; Negroponte 1996; Wilson 1999). In particular, the literature frequently points to the potential of e-government programs to promote *good governance* by enhancing the performance and accountability of local governments (Gasco 2003; Heeks 1999; Okot-Uma 2001; OECD 2001).

More critical views of e-government highlight that in praxis many e-government programs in developing countries have failed (Ciborra 2005; Ciborra and Navarra 2005; Heeks and Bhatnagar 1999; Heeks 2002, 2003). In fact, Heeks (2003) ascertains that the empirical evidence shows that 35 percent of e-government programs are complete failures (the program was not implemented or was immediately abandoned after its development) and 50 percent are partial failures (major goals were not attained or the outcomes were undesirable). He introduces the *design-reality gap* framework to analyze the underlying reasons for these failed programs and finds that the major cause is the frequent gap between the local realities within public administrations and the design of e-government projects (Heeks 2003). He concludes that the larger the design-reality gap is in the seven dimensions of information, technology, process, objectives and values, staffing and skills, management systems and structures, and other resources, the greater is the risk that the program will fail.

Based on an interpretive set of case studies in Kerala, India, Madon finds that e-government programs frequently neglect critical context factors, such as adequate resources, skill levels, values, beliefs, and motivations, and that poor people and communities are unable to derive real benefits from such investments (Madon 2004, 2). She concludes that a critical barrier for the successful implementation of e-government programs is that far-reaching back-end administrative reforms need to be implemented in parallel for these programs to have a significant positive effect on local governance (Madon 2004, 9).

Ciborra and Navarra argue that e-government is not conducive to strengthening state capacity per se, but that governments need to take decisive steps to build a strong and efficient state ex ante; otherwise, e-government programs have a high likelihood of failure. They emphasize that developing countries need to address failures due to governance breakdown, corruption, rent seeking, distortions in markets, and absence of democracy before the implementation of e-government programs can succeed (Ciborra and Navarra 2005, 156). Their major criticism, however, is that international donors frequently use e-government as an instrument to construct an electronic, minimally efficient, service delivery–focused state (Ciborra and Navarra 2005). This criticism is particularly relevant in the context of the decentralization agenda in Bolivia.

Furthermore, Heeks and Stanforth (2007) investigate the reasons why some e-government projects succeed while others fail. Drawing on the actor-network theory, they develop a research tool for analyzing e-government trajectories. They stress that studies often pay insufficient attention to "the political in e-government—of the way in which different stakeholders in an e-government project relate to one another via political processes such as coalition and conflict" (Heeks and Stanford 2007, 165). They suggest that e-government research should focus on how stakeholders form networks of influence and exercise power within them. In a case study of the implementation of an integrated financial management system in Sri Lanka, they highlight the importance of investigating the interrelationship between global (international donor) and local (implementing agency) networks of key actors in e-government projects (Heeks and Stanford 2007, 167). Thus the trajectory of projects can be understood by analyzing the power relationships of different actors, using the concept of "power to," which is a dynamic exercise of power, to replace the concept of "power over," which tends to be static and passive (Heeks and Stanford 2007, 172). A key conclusion from their research is that the trajectories of e-government projects should not be characterized in terms of success or failure. Instead, they suggest, "E-government can never completely fail or completely succeed because any given application design or implementation is continuously being reformed" (Heeks and Stanford 2007, 174).

Finally, the literature on e-government and decentralization has stressed that the principal value of ICTs lies in enhancing both the participatory (good governance) and the managerial (improved performance) dimensions of local government (Moon 2002; Musso, Weare, and Hale 2000; Pratchett 1999). Lawrence Pratchett emphasizes that ICT projects have the potential to support local governments in the following three dimensions of their work: (a) enhance local democracy, (b) promote public policy making, and (c) improve the quality of their service delivery (Pratchett 1999). Many scholars argue that local governments use ICTs primarily as instruments to improve the delivery of services and not to enhance good governance (Pratchett 1999; Moon 2002; Musso, Weare, and Hale 2000; Tettey 2002). In fact, an empirical study of the use of ICTs by 270 local governments in California finds that most municipalities do not see ICTs as an instrument for promoting good governance at the local level, but as an additional communications tool—failing to explore their value as change agents and their potential to alter the fundamental relationship between government and citizen (Musso, Weare, and Hale 2000). Evidence from Flanders, Belgium, also shows that local governments primarily use the Internet as a one-way communication stream to reach citizens as customers and neglect its interactive possibilities, which could empower citizens to exercise their rights and responsibilities better (Steyaert 2002b, 201).

Evaluating the Impact of ICTs on the Well-Being of Poor Communities: The Need to Go beyond Conventional Approaches

Based on this literature review, several unanswered questions remain: (a) What is the impact of ICTs on the well-being of poor communities? (b) Under which conditions can new technology transform the lives of the poor? (c) What are some critical factors that influence whether or not ICTs have a positive impact on the human development of marginalized groups? (d) Can poor communities be empowered through the use of new technologies?

The starting point of this study is to investigate these questions from the perspective of poor communities themselves. To this end, I have developed an alternative evaluation framework (AEF) for assessing the impact of ICT interventions based on Sen's capability approach (Gigler 2004). In contrast to most ICT evaluations, I place people's assets and capabilities in the center and examine the extent to which the use of ICTs can act as a catalyst in improving their "informational capabilities," ultimately expanding people's human and social capabilities.

This approach significantly differs from the majority of existing ICT evaluations in the sense that it evaluates ICT programs in terms of their development impacts on people's well-being in the multiple dimensions of their lives (that is, economic, social, and political). It points out that it is necessary to carry out in-depth cases studies to assess the various factors affecting the extent to which ICTs can or cannot result in positive changes in people's lives and enhance the well-being of communities (Harris et al. 2003; Kanungo 2003; Madon 1993). The evaluation literature has stressed that the short time span on which conventional evaluations usually focus is a major shortcoming of such an approach (Cracknell 2000, 357; Fowler 1997).

The AEF stands in contrast to most ICT evaluations, which are predominantly based on conventional evaluation approaches and thus narrowly focus their assessments on measuring the immediate, short-term, and measurable program outputs in terms of enhanced "access" to and "usage" of ICTs (Whyte 2000; Daly 1999; Hudson 1995, 1999; Wilson, Daly, and Griffiths 1998; Ernberg 1998a, 1998b). In the context of evaluating e-government programs in India, Madon has criticized these approaches and stressed that a "more focused approach is needed to ensure that citizens derive true benefits from such investments" (Madon 2004, 2).

These approaches neglect the wider socioeconomic, institutional, and political effects of ICT interventions and instead aim to identify only the direct results of the programs through a narrowly focused analysis. Such an approach to evaluations has been criticized by several scholars, as it is unable to identify the real effects or impact of projects and rather leads to the sole identification of immediate results (Blankenberg 1995; Hopkins 1995; Ruxton 1995).

Furthermore, contrary to conventional approaches that focus on the supply-side benefits of ICT programs, the AEF emphasizes the demand-side effects of ICT interventions on the living conditions of communities at the micro level. In the ICT literature, only a few studies have concentrated on analyzing the demand-side effects of ICT interventions on people's living conditions (Krishna and Madon 2003; Harris 1999; Correa et al. 1997; Delgadillo, Gómez, and Stoll 2002; Hudson 1999). Heeks has stressed the need for more research using a bottom-up approach. He argues that ICT programs should first identify users' actual demands and, through a "wants analysis," then "determine, bottom-up, short term, what users actually demand," before proceeding with a planned ICT program (Heeks 2005).

The AEF furthermore uses poor people's perceptions of the impact of ICTs on their individual and collective well-being as the principal evaluation criteria for the analysis. This notion is based on the work of Chambers, who has called for the "spread of evaluations in which poorer people and communities do their own evaluations" (Chambers 1994, 1438). In the ICT literature, the early work by Menou and his colleagues has stressed that local communities are fully conscious about their own well-being and thus should decide on their own which aspects of the ICT programs contribute to their development (Menou 1993, 2001; Stoll et al. 2004; Menou and Potvin 2000).

The alternative framework does not assume a direct and linear relationship between improved ICT access and enhanced socioeconomic development, but instead develops an impact chain that attempts to unpack the indirect effects of ICTs on people's well-being (Benjamin 2000; Heeks 2005). The approach stresses the need to evaluate both the process of how ICTs are being introduced into communities and the specific outcomes of the interventions on the well-being of the poor (Gigler 2001). It thus draws on the literature on participatory evaluations and incorporates both quantitative and qualitative research methods in its analysis (Oakley, Pratt, and Clayton 1998; Guijt and Gaventa 1998; Roche 1999). In contrast to conventional approaches that base their evaluation solely on quantitative, predetermined evaluation indicators, the approach attempts to be as open and flexible as possible and thus perceives evaluations as a collective, participatory, and continuous learning process among all stakeholders, rather than being driven exclusively by the internal management requirements of funding agencies (Oakley, Pratt, and Clayton 1998).

Finally, the approach emphasizes that ICT programs can have many unintended consequences and stresses the ambivalence of the impact phenomena (Blankenberg 1995; Heeks 2002). Thus the framework ties together both positive and negative effects of ICT interventions. This aspect draws on the work of Richard Heeks, who has demonstrated that the majority of ICT programs have led to either "complete" or "partial" failure, instead of the mostly positive outcomes expected by more conventional ICT evaluations (Heeks 2002).

The Multidimensional Approach to Well-Being Based on the Capability Approach

As the theoretical framework for the AEF, I use the multidimensional approach to well-being developed in Sen's capability approach (Sen 1984, 1992, 1993), which moves away from an income-based perspective of well-being (utilitarianism) and instead emphasizes the nonmaterial (that is, social, cultural, and political) aspects of human well-being. Sen conceives of development as "a process of expanding the real freedoms that people enjoy" and emphasizes the need for the "expansion of 'capabilities' of persons to lead the kinds of lives they value" (Sen 1999, 18).

This view of development places people and human development at its center. What matters, according to Sen, is what people are capable of being or doing with the goods to which they have access. A person's "capability" refers to the alternative combinations of functionings that are feasible for her or him to achieve. Capability is thus a kind of freedom: the substantive freedom to achieve combinations of alternative functionings—or, less formally put, the freedom to achieve various lifestyles (Sen 1999, 75). Capabilities include things that a person actually has done, as well as things that he or she could possibly do. In other words, capabilities refer to the extent of one's positive freedoms (Gasper 2002, 5). The concept of "functionings" "reflects the various things a person may value doing or being" (Sen 1999, 75). In this sense, a person's functionings represent the "various components or aspects of how a person lives," whereby a person's ability to realize these desired and valued functionings depends on her or his capabilities and existing livelihood resources or assets (Gasper 2002, 4).

Recognizing that measuring access to ICTs is not sufficient for evaluating the effects of ICT interventions on poor people's well-being, I apply the capability approach to the study of ICTs, focusing on whether or not, and under which conditions, the *uses* of ICTs can enhance poor people's individual and collective capabilities to achieve the lifestyles they value.

My work draws on previous studies by Garnham (1997), Madon (2004), Mansell (2001), and Zheng (2009), who have stressed the value of using the capability approach to develop an evaluation framework of ICTs. Nicholas

Garnham has pointed out that "thinking in terms of functionings and capabilities allows us to get behind the superficial indices of access and usage that we so often use" (Garnham 1997, 32). Based on a capability perspective, Shirin Madon has developed an evaluation framework that emphasizes human agency rather than structural or institutional variables. She uses this framework to evaluate the development impacts of two e-governance programs in India (Madon 2004). She raises a critical question: Should new options, such as the ability to hold government accountable, the ability to pay bills, or the ability to generate income through e-governance applications, be added to the capability set of individuals, communities, organizations, and states (Madon 2004, 4)?

Zheng emphasizes that the philosophical and conceptual foundation of the capability approach provides a different space for the evaluation of ICT programs and thus can help to avoid some "existing or potential pitfalls in e-development" (Zheng 2009). For instance, the emphasis on human diversity enables researchers to move beyond analyzing only the effects of ICT diffusion and instead to focus on analyzing the "interpersonal variations in conversion factors and decision-making processes in e-development" (Zheng 2009, 76). Furthermore, she points out that the capability approach's emphasis on people's agency enables researchers not only to recognize existing social conditions and cultural values as important factors in the local context for ICT programs, but also to question and critically evaluate those sociocultural factors. She concludes, "Rather than maximizing access to technology, ICT for development should take into account the free flow of valuable information to enhance both well-being and agency freedom of individuals" (Zheng 2009, 79).

Indeed, measuring ICTs in terms of capabilities reveals that there is no linear relationship between having access to ICTs and using them—having Internet access is a necessary, but insufficient, condition for its use. This goes hand in hand with one of the fundamental principles in the conceptual framework of the capability approach, which is that access to a basic good, in this case ICTs, represents an entitlement and key prerequisite for its use; however, differences in people's capabilities determine whether they are indeed able to transform a set of actual opportunities into realized functionings (that is, improved access to information through the use of ICTs). In Sen's words (1999), "People have different ways of transforming the same bundle of goods [ICTs, here] into opportunities for achieving their plans in life."

Thus, when assessing the impact of ICTs on well-being, it is essential not only to evaluate the range of information and communication options made available (the potential use of ICTs), but also to consider people's capabilities—that is, their ability to transform these options into actual or realized functionings (Garnham 1997, 32). Such a process entails examining people's motivations, expectations, and reasons for use as well as the outcomes in relation to their well-being (Mann 2003).

Thus the socioeconomic and cultural milieu is crucial to understanding the potential effects of ICTs on development (Avgerou 2001; Kling 2000; Walsham 1993a, 1995). Technology receives meaning only once it is "enacted" by users, and people can control its use by interpreting and appropriating it to their specific realities (Orlikowski 2000). In essence, human action rather than technology is placed at the center, emphasizing the interdependencies between technology and social context (Orlikowski 2000; Avgerou 2001). In particular, earlier empirical works have used a contextual approach to ICTs (for example, Alampay 2003; Harris et al. 2001; Madon and Sahay 2000; Miller and Slater 2000; Nelson 1996; Tacchi, Slater, and Lewis 2003) and demonstrated that the researcher can discern the specific factors (that is, local culture and social structures) influencing whether or not ICTs have a positive impact on the daily lives of poor communities (Avgerou and Madon 2003).

Focus on Indigenous Peoples and the Impact of New Technologies on Their Well-Being

The empirical evidence presented in this book focuses on the impact of new technologies on the most marginalized segment of Latin American society—indigenous peoples, who, as a group, are particularly deprived of their entitlement to material and nonmaterial goods. Indigenous peoples have historically been the poorest and most excluded populations throughout Latin America. They have faced serious discrimination not only in terms of their basic rights to property, language, culture, and citizenship but also in terms of their access to basic services and essential material conditions for a satisfying life (Partridge, Uquillas, and Johns 1998; Davis 2002). The systematic exploitation of and discrimination against indigenous peoples, which can be traced back to the Spanish colonial period, persist in Bolivia's political, economic, and social life today and are perpetuated through many development policies (Feiring 2003; McNeish 2002; Healy 2001; Ströbele-Gregor 1996; Klein 1982). Thus a high correlation exists between poverty and being indigenous, and socioeconomic conditions and access to basic social services are significantly worse for indigenous peoples than for the nonindigenous population (Psacharopoulos and Patrinos 1994). At the same time, the richness of their cultural diversity and strong customary social and political institutions are at best ignored and at worst undermined by the *mestizo* ruling elites (Healy 2001; Albó 1996).

The rationale to focus on indigenous peoples is based on the following. First, the study is particularly interested in analyzing the effects that ICTs have on marginalized groups and the extent to which these technologies might enable them to overcome some of the underlying causes for their persistent poverty and social exclusion. Second, the research considers indigenous peoples' cultural diversity and their traditional organizational structures as critical assets for human development and deems it critical to take into account the cultural and social factors specific to indigenous peoples. Third, a significant gap exists in the literature about the potential effects of ICTs on socioeconomic development from the perspective of marginalized groups. Fourth, because of their current social exclusion and high levels of poverty, indigenous peoples face enormous barriers to making meaningful use of ICTs to further their economic and social development, while maintaining their cultural identity.

In sum, I believe that taking into account the cultural diversity, social organizations, and continuously high levels of social exclusion and poverty of indigenous peoples adds significant value to the study of the impact of ICTs on their well-being. Doing so allows me to focus on the multiple factors influencing whether or not the meaningful use of ICTs can indeed enhance their individual human and collective social capabilities.

Key Factors for Establishing a Link between ICTs and Poor People's Human Development

Four main factors influence whether, and to what extent, ICT programs can improve access to information, enhance the informational capabilities of poor people, and ultimately improve their well-being:

1. The role of intermediary organizations
2. The role of indigenous knowledge
3. The appropriation process of ICTs
4. The broader sociopolitical context.

The first factor is the role that intermediary organizations play in "brokering" or facilitating improved access to information through ICTs in local communities. This is a key factor influencing the extent to which local communities can use ICTs. The second factor is the role of indigenous knowledge, since the introduction of ICTs should be based on the existing "information ecology"[4] within local communities. Most communities have a rich cultural identity and a wealth of indigenous knowledge that are important aspects of their existing information environment. The third factor is the process by which local communities appropriate ICTs, and the fourth is the broader socioeconomic and cultural milieu. This section describes these four factors in more depth.

The Role of Intermediary Organizations

Recently several authors have highlighted the critical role that intermediary organizations or "brokers" play in the development process (Lewis and Mosse 2006; Mosse 2005; Bierschenk, Chauveau, and de Sardan 2000, 2002). This literature is based on an internationalist approach (Long 2001; Cleaver 1999) suggesting that sociopolitical structures are continuously being redefined by the ongoing and dynamic interaction between social structures and human agencies. Based on this approach, several authors have recently highlighted the critical role that the concept of brokerage plays in analyzing the implementation of development projects. The work by Bierschenk, Chauveau, and de Sardan (2002) has contributed to the concept of "brokers in development" by studying the role of brokerage in the context of international aid, political actions, and community development in the African context. Based on an ethnographic study of the social spaces that exist between donor agencies and donor recipients, they define brokers in the following way:

>> They are supposed to represent the local population, express its "needs" to structures in charge of aid and to external financiers. In fact, from being passive

operators of logic[s] of dependency, development brokers are the key actors in the irresistible hunt for projects carried out in and around African villages. (Bierschenk, Chauveau, and de Sardan 2002, 4)   <<

Furthermore, Lewis and Mosse show that brokers are "a specific group of social actors who specialize in the acquisition, control, and redistribution of development 'revenue'" (Lewis and Mosse 2006, 12). Mosse emphasizes that brokers are critical intermediaries between development institutions and local communities and are unavoidable for development projects since they play a critical role in translating and mediating between two compartmentalized worlds and worldviews—those of international donors and those of local communities (Mosse 2004). He stresses that, at different stages of project implementation,

>> It requires the constant work of translation (of policy goals into practical interests; practical interests back into policy goals), which is the task of skilled brokers (managers, consultants, fieldworkers, community leaders) who read the meaning of a project in the different institutional languages of its stakeholder supporters, ... constantly creating interest and making it real. (Mosse 2004, 647)   <<

According to this concept of "translation," local communities do not confine themselves to being passive victims of development projects; instead, they consider themselves to be local actors with interests and perspectives that contradict, through their everyday practices, the prescription and planned objectives of project designs. As Mosse observes, "During the 'implementation phase,' all the diverse and contradictory interests that were enrolled in the framing of an ambiguous policy model and project design, all the contests and contradictions that are embedded in policy texts, are brought to life and replayed" (Mosse 2004, 664).

He draws the conclusion that the successful implementation of development projects "is never a priori a matter of design or of policy, but instead depends on a complex process of how local actors interact with project staff and interpret the project to fit their local needs" (Mosse 2005, 9). Local intermediaries facilitate or broker this process.

Finally, Anirudh Krishna considers local intermediaries to be key actors in support of local community development (Krishna 2002). Drawing on the work of Putnam, who introduces the concept of "social capital" as "features of social organization such as networks, norms, and social trust that facilitate coordination and cooperation for mutual benefit" (Putnam 1994, 67), Krishna argues that agency is needed for social capital to be beneficial for development. Based on his in-depth study of 69 rural communities in Rajasthan and Madhya Pradesh in India, he demonstrates that local intermediaries—such as village councils or youth groups—act as critical facilitators for ensuring that existing social capital is being channeled in a positive way toward local community development (Krishna 2002).

I draw on this literature and apply the concept of brokerage and intermediation to study the impact of ICTs on poor people's well-being. Furthermore, I draw on studies that specifically highlight the role that intermediaries play in introducing ICTs to local communities (Madon 2000; Heeks 2002; McConnell 2000). For instance, Heeks argues that intermediaries are critical in helping rural communities to overcome some of the barriers to political access, while providing ICT services (Heeks 2002). Building on the work of Madon (2005), Gopakumar (2007) uses a similar approach to study the Akshaya project, but emphasizes the important role that local intermediary organizations play in establishing trust with local communities and providing relevant e-government services to marginalized groups. According to Gopakumar, the Akshaya project succeeded in establishing a high level of trust between the entrepreneurs (intermediary) and people in the local villages because "people had trust in government as an institution and ... the project was spearheaded by the government" (Gopakumar 2007, 29). Furthermore, a local intermediary might be able to gain trust only in a specific area or service, and "although the technology in itself might have the potential for integrating services, the institutional factors like trust associated with the services determine whether the services are actually used or not" (Gopakumar 2007, 31). Therefore, the lack of trust in the intermediary to provide a comprehensive set of e-government services to local communities might be a key factor explaining the failure of so many telecenter projects in developing countries (Gopakumar 2007).

According to Bailur and Maseiro (2012), infomediaries require more than training in information technology skills. They need to consider the communities with whom they are interacting and adjust their role accordingly. Moreover, intermediaries have an active role and can alter their own role or position themselves to create opportunities for development. The role of infomediaries thus needs to be more adaptable and flexible.

However, the literature does not focus on investigating the exact role of intermediaries in the process. Instead, it often takes a more institutional perspective and analyzes the effects of ICTs within organizations (Avgerou 2001; Powell 1999; Meyer 1997). McConnell highlights this shortcoming and draws attention to the "unconnected rural stakeholders," stressing that rural communities due to their social exclusion and geographic remoteness have the most to gain from the expansion of ICTs (McConnell 2000, 1). At the same time, rural communities also require the most support from an "intermediary" in the process of using ICTs to support their local needs. In a study of knowledge and information systems of the urban poor, Schilderman suggests, "Social networks are the foremost source of information of the urban poor" (Schilderman 2002, 4). He stresses, "The poor tend to believe people they trust rather than perhaps better informed contacts who are more distant to them" (Schilderman 2002, 5). He points out that the poor frequently rely on "key informants"—whom he defines as "people inside or sometimes outside a community who are knowledgeable in particular livelihoods aspects and are willing to share that knowledge" to obtain critical information. According to Sey and Fellows (2009), infomediaries are important contributors to the viability and sustainability of telecenters and public access venues.

Drawing on this literature, I use the term *intermediary* to refer to a liaison or broker between individuals or a group of people within a local community and a group or source of information outside the community. I emphasize the role of the intermediary in providing an "enabling environment" for the poor to improve their access to information through the use of ICTs. The notion is that the intermediary plays a critical role in the process of brokering or transferring information (which originates primarily outside the community) in a culturally appropriate manner to the local community (Gould and Gómez 2010).

Based on this overall definition, I investigate the different types and levels of intermediation in ICT programs, which are categorized as (a) ICT or technical intermediaries and (b) social intermediaries. An ICT intermediary or "infomediary" is a person or organization providing "effective" support to local communities in the use and adaptation of technology. Most commonly, an ICT intermediary or infomediary is a specialized organization from outside the community—an NGO, a local government, or an international donor. The main role of this intermediary is to support local communities in all aspects related to their access to and use of ICTs. In particular, the "ICT intermediaries" play a critical role in (a) identifying and providing access to ICT products and services that suit the local community's need for information, (b) supporting the generation of local and relevant content, and (c) providing ongoing support in the areas of training and capacity building (Delgadillo, Gómez, and Stoll 2002).

A "social intermediary" is defined as a local institution, like a community-based organization, that is trusted by the community members and typically has a long-lasting relationship with the community. A social intermediary is critical for facilitating a community engagement process by which all actors within a community are invited to participate fully in the project. In the context of introducing ICTs into communities, the main activities of the social intermediary are related to (a) assessing the existing communications channels and information needs within the community, (b) carrying out a broad-based consultation process about the ICT project, (c) building support for the project among key social organizations and actors within the community (youth or elders) to secure the long-term viability of the project, and (d) supporting the community in adopting ICTs and adapting the information to their local, social, and cultural context.

I also distinguish between high and low levels of intermediation. A high-level intermediation is characterized by a high degree and frequency of direct involvement at the local community level; a low-level intermediation is characterized by centralized management and a lower degree of interaction with the local community.

The Role of Existing Indigenous Knowledge

The importance of "indigenous"[5] or "local" knowledge in development has been increasingly acknowledged in the recent literature. In the current debate, most authors have contrasted "Western scientific" knowledge with indigenous knowledge, emphasizing their separateness in terms of subject matter, methodology, and worldview (Warren 1991; Chambers, Pacey, and Thrupp 1989; Brokensha, Warren, and Werner 1980; Howes and Chambers 1980). Other authors have looked beyond these distinctions and instead focused on how the community's broader social and political context affects knowledge (Agrawal 1995; Scoones and Thompson 1994; Bebbington 1993; Altieri 1989).

In contrast to a more conventional approach to ICTs, this study highlights the key role that indigenous information systems—based on indigenous knowledge and communication practices—play in improving the well-being of indigenous peoples (Agrawal 1995; Chambers, Pacey, and Thrupp 1989; Brokensha, Warren, and Werner 1980; Harris 2001). These traditional information systems are embedded in the existing social and organizational structures at the community level. It is crucial to analyze this "information ecology" and its communication channels before exploring the impact of ICTs (Madon 2004; ITDG 2001; O'Farrell, Norrish, and Scott 1999; Slater, Tacchi, and Lewis 2002; Wang and Dissanayake 1994). A common reason that ICT programs fail is that key community members (for example, elders and other "information brokers") perceive the new technologies as undermining the information systems that are embedded in the social and organizational structures of the community (Robinson 1998; Long and Villareal 1994).

Agrawal notes that this recent emphasis on indigenous knowledge shifts the perspective from a preoccupation with centralized, technically oriented development to an embrace of "development from below" (Agrawal 1995, 414). According to Brokensha, Warren, and Werner, "Indigenous knowledge makes the adaptation of technology to local needs possible, encourages community self-diagnosis, heightens awareness, and involves users in feedback systems, thus constituting a strong case for incorporating indigenous knowledge in development programs" (Brokensha, Warren, and Werner 1980, 7–8).

Such a "technical" approach to indigenous knowledge suffers from several contradictions and weaknesses. First, indigenous peoples view indigenous knowledge and information systems as related to their culture, identity, and spiritual and religious beliefs, while development professionals recognize only the instrumental value of indigenous knowledge as a means to resolve certain problems of development. Jorge Terena, an indigenous leader from Brazil, expressed this view in the following way: "It is on this knowledge that our community depends for its living. The important thing is our community has a traditional belief, a spiritual belief that controls this knowledge. This goes far beyond just thinking of the economic value that the knowledge has …. It is something sacred to us" (Davis and Ebbe 1993, 7). Agrawal sums up the technical approach, stating that it "divorces the knowledge from the source that presumably provides it with its vigor—the people and their needs" (Agrawal 1995, 429).

Second, many poor communities are aware of the crucial importance of indigenous knowledge to the survival of their communities and are, consequently, wary of sharing this knowledge with outsiders (Davis and Ebbe 1993).

Third, the concept of indigenous knowledge neglects the political dimension of knowledge. Agrawal stresses that significant shifts in power relationships are crucial to overcoming sociopolitical and cultural barriers and that only when power has been realigned are poor communities able to improve their access to information and knowledge. In fact, development programs have frequently led to conflict between Western and traditional knowledge by attempting to impose Western values onto non-Western communities (Salas 1994). However, information and knowledge gaps are often mutual: that is, poor communities frequently lack access to information and knowledge, while policy makers in capital cities lack knowledge about the local economic, political, social, and cultural realities of poor and marginalized communities (Scoones and Thompson 1994; Long and Villareal 1994).

The Process of Technology Appropriation

Building on this contextualized approach to ICTs (Avgerou 2001; Madon 2000; Walsham 1993a, 1993b, 1998), my work emphasizes that the local appropriation of technologies by communities and the contextualization of information provided through ICTs are the key factors determining whether poor communities derive real benefits from ICT use. This is also reflected in Sen's thinking related to the contextual aspects of ICTs:

>> The easy way these communicational technologies have been absorbed by people shows that, despite their modernity and, I suppose, "globality," they are not things that are completely alien to the local culture. The important issue is what we can do with all the technologies that are available. The right way of seeing IT [information technology] is also not to cast it in terms of what we can do on the basis of our own culture, unaided, because we do not have any unaided culture. IT has become an interactive culture across the world, and the important question is how we can make people more functionally efficient, not just with their own things, but with everything—the global, as well as the local. (Sen 2010, 2)   <<

Pure access to ICTs does not translate into improved well-being for the poor (O'Farrell 2001). A major challenge for the local appropriation of ICTs is that the content on the Internet often does not reflect the realities of local communities (Ballantyne 2002). Much of the Internet's content is written in an academic or business style, making it difficult to understand and apply locally, and most is in English, limiting access in poor communities across most developing countries (Gurstein 2000).

In this study, I draw on the work of Sabine Michiels and Loy Van Crowder and use their definition of local appropriation of ICTs:

>> Local appropriation of ICTs is about communities and groups selecting and adopting communication tools according to the different information and com- munication needs identified by them and then adapting the technologies so that they become rooted in their own social, economic, and cultural processes. (Michiels and Van Crowder 2001, 1)   <<

According to ethnographic research in the field of cultural studies, "Local peoples being confronted by technologies, rather than being 'duped' and 'controlled' by mass media, initiated to integrate media within their lives and made sense out of it in locally specific ways" (Michiels and Van Crowder 2001, 1).

In their early work, Cochrane and Atherton (1980) analyze the conditions needed for communities with distinct cultural backgrounds to overcome information poverty. They stress that the following steps need to be taken so that information can be successfully incorporated into a different cultural context: (a) contextualization—fitting the material into the cultural environment; (b) incrementalism—deciding how much can be done at each step; (c) motivation—assessing the receptivity to information; and (d) the absorptive process—determining how best to acquire information.

The Broader Enabling Sociopolitical Context

The literature on the social impact of new technologies on development argues that existing barriers to the use of ICTs are not simply lack of access to technology or inadequate access to relevant and timely information; in fact, they are rooted much more deeply in the underlying social and institutional structures that perpetuate existing socioeconomic inequalities between the elites and marginalized groups of society, ultimately reproducing themselves in the uneven benefits that accrue from ICTs (Castells 1998a, 1998b; Hamelink 1994; Hewitt de Alcántara 2001; Mansell 2004; Skuse 2000). The analysis investigates the barriers that prevent poor communities from attaining improved access to information and knowledge. It analyzes the extent to which existing political, economic, social, and cultural barriers between the urban elites and marginalized communities prevent them from having equal access to information and making their "voices" heard on the national stage, thus barring or limiting their participation in the dominant society's political and economic system.

The Theoretical Framework: Placing Human Well-Being at the Center of Impact Evaluations

This section describes the theoretical underpinnings of the AEF in more depth. It is divided into five subsections: (a) the need to expand the capability approach, (b) poor people's well-being and the capability approach, (c) capabilities and the sustainable livelihood approach, (d) the move from information literacy to informational capabilities, and (e) the interlinkages between capital, agency, and capabilities.

Expanding the Capability Approach

The capability approach is theoretically underspecified when it comes to groups and the role that collective action plays in expanding not only the individual but also the social capabilities of the poor. Thus it needs to be expanded when analyzing the impact of development interventions on groups (Stewart 2001). This point is particularly relevant for indigenous communities, since their strong collective identity and worldview incline them to define their well-being in collective, not individual, terms (see appendix C for a description of social capabilities).

Furthermore, while Sen emphasizes the importance of political freedom for the development process itself, he does not analyze the political processes involved with the expansion of such freedoms (Sen 1999). Corbridge has raised this point, arguing that Sen's vision has only limited force as a political tool because it fails to address the presence of entrenched power and symbolic violence within cultures (Corbridge 2001). It thus fails to recognize the significance of *social struggles and contestational politics* that have been historically necessary for people in most countries to achieve a redistribution of resources—a crucial ingredient for freedom.

Poor People's Well-Being and the Capability Approach

The capability approach is well suited to conceptualizing the well-being of marginalized communities. In particular, it reflects the holistic view of the development of indigenous communities—well-being in the social, economic, political, and cultural dimensions (Gigler 2005b).

Like Sen, I take a holistic view and stress the multiple dimensions of development as the expansion of people's well-being, including social, political, cultural, and spiritual elements. Sen's (1999) criticism of both utilitarianism (assumed by welfare economics) and the primary-good approach (demanded by Rawls 1971)—that they overemphasize the material aspects of development—has significant similarities with indigenous peoples' notions of development based on their cultural diversities, traditional knowledge, and worldviews.

Poor people frequently stress their community's rich cultural identity, traditional knowledge, and strong customary social and political institutions. In particular, indigenous communities consider themselves to be not poor but rich in knowledge, social networks, and cultural diversity:

>> The basic question is, what is poverty in economic and in spiritual terms? Many people may be economically poor but at the same time they can be qualitatively rich in knowledge and values. The contributions of these people to humanity, in the areas of medicine, music, textiles, architecture, oral tradition etc., should also be taken into consideration. Establishing indicators of poverty should not be a unilateral process but a dialogue between cultures.[6]   <<

The development paradigm needs to change fundamentally: it needs to stop relying on economic and material aspects and to recognize the relationships among the social, political, economic, and ecological components of communities. Both the capability approach and the local communities' perspective on development emphasize the nonincome dimensions of well-being and recognize the multiple dimensions in which a poor person's well-being can be enhanced.

Furthermore, the capability approach evaluates people's well-being directly (indicators of quality of life) rather than indirectly through measures of inputs or means, such as higher incomes or consumption (Sen 1999). This goes hand in hand with the perspective of poor communities, and in particular indigenous peoples, who believe that conventional approaches to development overemphasize their material and other forms of deprivation (lack of income and low levels of literacy) without adequately taking into account their rich cultural identity and traditional knowledge.

Third, Sen strongly believes that value judgments about the priorities of development have (a) to be made by individuals and society themselves and (b) to be made explicit by the theoretical framework. On the one hand, he criticizes conventional economic theories for assuming that people's behavior is individualistic and based on rational profit maximization. Furthermore, he criticizes the focus of conventional welfare economics on the efficient allocation of resources at the expense of equity.[7] On the other hand, he proposes an "evaluative exercise" in which individuals and society hold public discussions and engage in democratic processes. This is the reason that Sen, in contrast to Nussbaum (2000) and Alkire (2002), never defines a specific list of "basic capabilities," deliberately leaving this open to be defined by the local context and people's priorities. The capability approach has been criticized for its theoretical incompleteness, and this has certainly led to difficulties in operationalizing the approach (Comim 2001). However, this particular *openness of approach* makes it especially suitable for evaluating the well-being of marginalized communities.

Fourth, the capability approach stresses the crucial role that process itself plays in development. Therefore, Sen sees capabilities as representing the *potentials* or *opportunities* that a person has to achieve his or her valued functionings. Capabilities are a kind of freedom enabling a person to reach the life he or she values. Sen underscores the importance of the *process* of expanding a person's capabilities as itself essential to development and considers the analysis of the actual outcomes (functionings) as secondary.

Operationalizing the Capability Approach

In the last couple of years, there has been lively debate in the literature on the operationalization of Sen's capability approach and its applicability to empirical research (Alkire 2002; Comim 2001; Corbridge 2001; Gasper 1997, 2002; Stewart and Deneulin 2002). Comim has suggested that the capability framework is well suited to "evaluating and assessing social arrangements, standard of living, inequality, poverty, justice, quality of life, or well-being" (Comim 2001, 4). Other scholars note the difficulties of operationalizing it (Corbridge 2001; Alkire 2002). Comim points out that these difficulties derive from the capability approach's "theoretical underspecification and inclusive view of operationalization which contest not only the evaluative but also the practical foundations of utilitarianism" (Comim 2001, 2). A key challenge has been to define a priori a set of basic capabilities—a baseline from which to start specific evaluations (Nussbaum 2000; Alkire 2002).

Comim suggests that the capability approach is particularly suited to micro-level studies, since it focuses on nonincome variables (Comim 2001). Such an approach, he argues, reveals more interesting findings at the micro than at the macro level, as research at this level can better analyze people's ability to choose what to do or be.

Moving from Digital Literacy to Informational Capabilities

I also draw on existing information science literature on "digital" or "infomation literacy" in order to conceptualize "ICT capabilities" and "informational capabilities" (Brevik 1992; Eisenberg and Berkowitz 1990; Horton 1983; McClure 1994; Menou 2002; Ochs et al. 1991; Zurkowski 1974). Paul Zurkowski first used the term "information literacy" in 1974, pointing out that individuals need the ability to find, evaluate, and use various sources of information. This ability should include five capabilities: (a) knowing what kind of information is helpful, (b) knowing where to get that information, (c) knowing how to inspect the information, (d) evaluating and organizing the information, and (d) immediately transmitting the information. While a standard definition of information literacy has yet to appear, I use the following commonly quoted definition provided by the American Library Association's Presidential Committee on Information Literacy: "Information literacy is a set of abilities enabling individuals to recognize when information is needed and have the ability to locate, evaluate, and use effectively the needed information" (ALA 1989, 2). The concept of information literacy stresses people's capability to use information to solve problems. According to McClure (1994), information literacy should include the following four components: (a) traditional literacy—the basic capability of reading and writing; (b) media literacy—the ability to use multimedia (compact discs, microfilms) to solve information problems; (c) computer literacy—the capability to operate a computer; and (d) network literacy—the ability to identify, access, and use electronic information from the network.

Applying the capability perspective to ICTs, I begin with the concept of "ICT capabilities," which constitute the extent to which a person is proficient in the use of a computer, software, and the Internet. Specifically, ICT capabilities are defined as "a set of skills and understandings required by people to enable meaningful use of ICT appropriate to their needs" (Oliver and Tower 2000, 384). This functional view of ICT literacy stresses the ability of people to appropriate technology to meet their local and cultural needs.

Furthermore, I differentiate this concept from the notion of "ICT use," which encapsulates the simple use of ICTs, such as the Internet, regardless of one's proficiency or ability to make meaningful, efficient use of the technology. As such, the "ICT capabilities" variable is important for evaluating a person's proficiency of use, while the "ICT use" variable defines whether or not a person is using ICTs. For studying the effects of ICTs on people's human well-being, it is critical to understand their ICT capabilities; nevertheless, the "ICT use" variable is important, since "actual use" is a prerequisite for realizing the potential benefits of ICTs for people's lives. In sum, use is a necessary, but insufficient, condition for affecting people's well-being.

In a second step, I introduce the concept of "informational capabilities," which deemphasizes the role of technology and instead focuses on an information-centric approach and analyzes people's abilities to apply its uses in a meaningful manner to the social, economic, and political aspects of their lives.

The differences between ICT capabilities and informational capabilities are significant. The concept of ICT capabilities encapsulates a person's ability to use computer hardware, software, and ICT tools; the concept of informational capabilities is much broader and relates to the role of information itself and the ability to analyze and place information into one's own sociocultural context (Horton 1983; Castells 1995). Furthermore, it argues that strengthening a person's ability to process and share local knowledge with others constitutes an important aspect of a person's overall informational capability.

Figure 1.1 visualizes the four components of the "informational capabilities" concept, which refers to a person's capability (a) to use ICTs in an effective manner (ICT capability); (b) to find, process, evaluate, and use information (information literacy); (c) to communicate effectively with family members, friends, and professional contacts (communication capability); and (d) to produce and share local content with others through the network (content capability).

An important factor determining the extent to which ICTs and the stronger informational capabilities can expand poor people's human capabilities is related to the "agency role" of information. As the review of the literature has demonstrated, only the combination of stronger resources and stronger agency can enhance individual and collective capabilities. An important aspect of agency in the context of human capabilities is the ability of the poor to negotiate and bargain with the formal institutions of the state, market, and civil society. Within this negotiation, information frequently plays a key role.

According to Sen (1997), human capabilities can be agents of change. They enhance people's ability to question, challenge, propose, and ultimately usher in new ways of doing things, which, in turn, enhances people's capability to represent their interests by engaging and negotiating with formal institutions to receive better services and to defend their basic rights. This notion can be applied to "informational capabilities." The ability to access information better, to process and evaluate it, and to communicate and share it with family members and friends can play a vital role in supporting the poor's ability to act and to improve their livelihoods.

FIGURE 1.1   The concept of informational capabilities

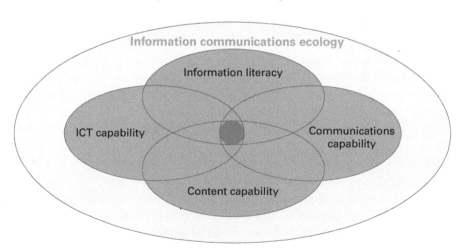

Source: Gigler 2004, based on McClure 1994.

In Sen's words, informational capabilities are a person's "capability" to transform access to ICTs (entitlement) into real opportunities through meaningful use and to achieve the things she or he values doing or being (Garnham 1997, 32). In other words, informational capabilities refer to a person's positive freedom to use ICTs.

Thus it is critical to understand (a) the process by which a person's use of ICT enhances her or his informational capabilities and (b) how these enhanced informational capabilities expand a person's human and social capabilities.

Incorporating the Sustainable Livelihood Approach

Bebbington (1999) suggests a particularly interesting way to operationalize the capability approach. His analysis starts with the livelihood approach and underscores the importance of the capital or assets to which people have access. From here onward, I use the definition of sustainable livelihoods developed mainly by Chambers and Conway (1992), as quoted by Scoones:

>> A livelihood compromises the capabilities, assets (including both material and social resources), and activities required for a means of living. A livelihood is sustainable when it can cope with and recover from stress and shocks, maintain or enhance its capabilities and assets, while not undermining the natural resource base. (Scoones 1998, 5)   <<

Key principles of the livelihood approach include shifting the focus from resources to people and from livelihood constraints to people's assets or capital. Furthermore, the approach evaluates livelihood outcomes rather than outputs and emphasizes the institutional and organizational setting and processes. It asks, What combination of livelihood resources (different types of capital) enables the poor to combine various livelihood strategies (that is, livelihood diversification), and what are the outcomes for their well-being (Scoones 1998, 3)?

Based on the livelihood and capability approaches, Bebbington develops a framework that highlights the importance of combining capital with capabilities. He argues, "Assets (or capital) are not simply resources that people use in building livelihoods: they are assets that give them the capability to be and act" (Bebbington 1999, 5). He refers to Sen's discussion of the significance of human capital to strengthen the capabilities of the poor. According to Sen, the possession of human capital means not only that people produce more, and more efficiently, but also that they can engage more fruitfully and meaningfully with the world and, ultimately and most importantly, can change the world (Sen 1997, 1960).

Bebbington argues that this is the case for all of the other types of capital. He understands these assets not only as things that allow survival, adaptation, and poverty alleviation: "They are also the basis of agents' power to act and to reproduce, challenge, or change the rules that govern the control, use, and transformation of resources" (Bebbington 1999, 5).

This analysis is of particular interest to my research for three reasons. First, it stresses the importance of assets or capital as the basis for the "agency" role of individuals, which is crucial to the notion of empowerment (Kabeer 1999b). Second, Bebbington's framework was developed in order to analyze rural live-lihoods and poverty in the Andes, which is the same geographic focus as my own research. Third, at the core of my investigation is the question of whether and under which conditions the access to improved information and ICTs can act as an "agent" for change, similar to "human capital," and thus empower poor communities.

Interlinkages between Capital, Agency, and Capabilities

According to Naila Kabeer, the notion of empowerment entails a process of change in existing power relationships and the ability to make strategic life choices (Kabeer 1999a, 437). To be disempowered, therefore, implies to be denied choices. There exists a "logical association between poverty and dis-empowerment because an insufficiency of the means for meeting one's basic needs often rules out the ability to exercise meaningful choice" (Kabeer 1999a, 439). Consequently, understanding the process of change is crucial for under-standing the concept of human capabilities. For change to happen, however, it is critical to discuss the dimension of power that relates to agency. Kabeer defines agency "as the ability to define one's own goals and act upon them" (Kabeer 1999a, 438). She points out that agency is usually operationalized as "decision making"; however, in the context of people's human capabilities, it is more important to see it within the context of the poor's ability to negoti-ate or bargain with the formal institutions of the market, civil society, and the state. Within this context and in the positive sense of the "power to," agency refers to the poor's ability to define their own strategic life choices and to pur-sue their own goals, even in the face of opposition from others (Nelson and Wright 1995).

What significance does this notion of agency and its interlinkages with the poor's capital have for operationalizing the capability approach? As Kabeer has demonstrated, the combination of resources (or capital) and agency con-stitutes what Sen refers to as capabilities. In this sense, improving the access to resources for the poor—for instance, providing access to girl's education or access to ICTs—represents only a potential for enhancing their capabili-ties; it does not automatically lead to empowerment. There does not seem to be a direct and automatic causal relationship between improving access to resources (such as, for example, access to ICTs) and improving people's well-being. Kabeer instead emphasizes the notion of agency and the role it plays in determining whether or not the increase in resources can expand the poor's realized functionings.

When designing indicators of livelihood outcome, it is important to make reasonable assumptions about the potential "agency" embodied in the resource being provided through the intervention. Furthermore, a deep knowledge is needed of the development priorities and goals of the marginalized group itself (in Kabeer's case, women). Otherwise, the interventions run the danger of pre-scribing the process of empowerment, which would violate its essence.

This last point addresses the issue of who defines the desirable and val-ued livelihood outcomes. Robert Chambers argues that within the capability (or well-being) approach to poverty and livelihoods, the analysis may allow people themselves to define the criteria that they deem important (Chambers 1997). This may result in a range of sustainable livelihood outcomes, including self-esteem, security, happiness, stress, vulnerability, power, and exclusion, as well as more conventionally measured material concerns.

Based on these theoretical underpinnings, the following section applies the AEF to analyzing the development impacts of ICTs on the well-being of marginalized groups.

Toward an Alternative Evaluation Framework of ICT Programs

Based on the theoretical framework described above, this section develops the AEF in more detail (Gigler 2004, 2005a). Richard Heeks (2002) has argued that information, instead of technologies, must be at the center of the analysis; he has developed an "integral and systematic model of ICTs," in which ICTs form an integral part of an overall institutional environment. This information-centered model stresses that ICTs can be fully integrated into projects only if their use is driven by the program's development objectives and the potential contribution of ICTs rather than by technological concerns (Heeks 2002, 7).

The evaluation model presented in figure 1.2 places human and social capabilities at the center; information and ICTs occupy the outermost circle, indicating that they can play a catalytic role in development, but are not the central aim of development. In fact, ICTs alone are not a means to an end: only under certain conditions can they expand the capabilities of the poor and help them to realize improved economic, social, political, and cultural opportunities.

Although the right to information and knowledge is an important asset and its absence can contribute to poverty, this notion needs to be balanced against the broader context of social and economic inequalities, which may be reinforced through the technology (Hewitt de Alcántara 2001; Castells 1997). Consequently, the sustainable livelihood framework is integrated into the framework to attempt a more holistic socioeconomic analysis of the possible effects of ICTs.

This approach argues for including the informational dimension as an additional asset, or capital, in the sustainable livelihoods framework. The view that information represents an important asset for the poor is the rationale for including the concept of "information capital" in the livelihoods approach. In this model, I highlight the existing information and communication ecologies within communities compounded by (a) existing "access to information capital"—the extent to which poor communities have access to information from the formal institutions of the market, state, and civil society; (b) "communications capital"—the strength of traditional communication practices and channels within communities and social networks (oral traditions, use of community meetings, and traditional forms of ICTs such as community and amateur radio); (c) "indigenous knowledge capital"— the extent to which communities produce, share, and transmit traditional

FIGURE 1.2   Alternative evaluation framework for the impact of ICTs on well-being

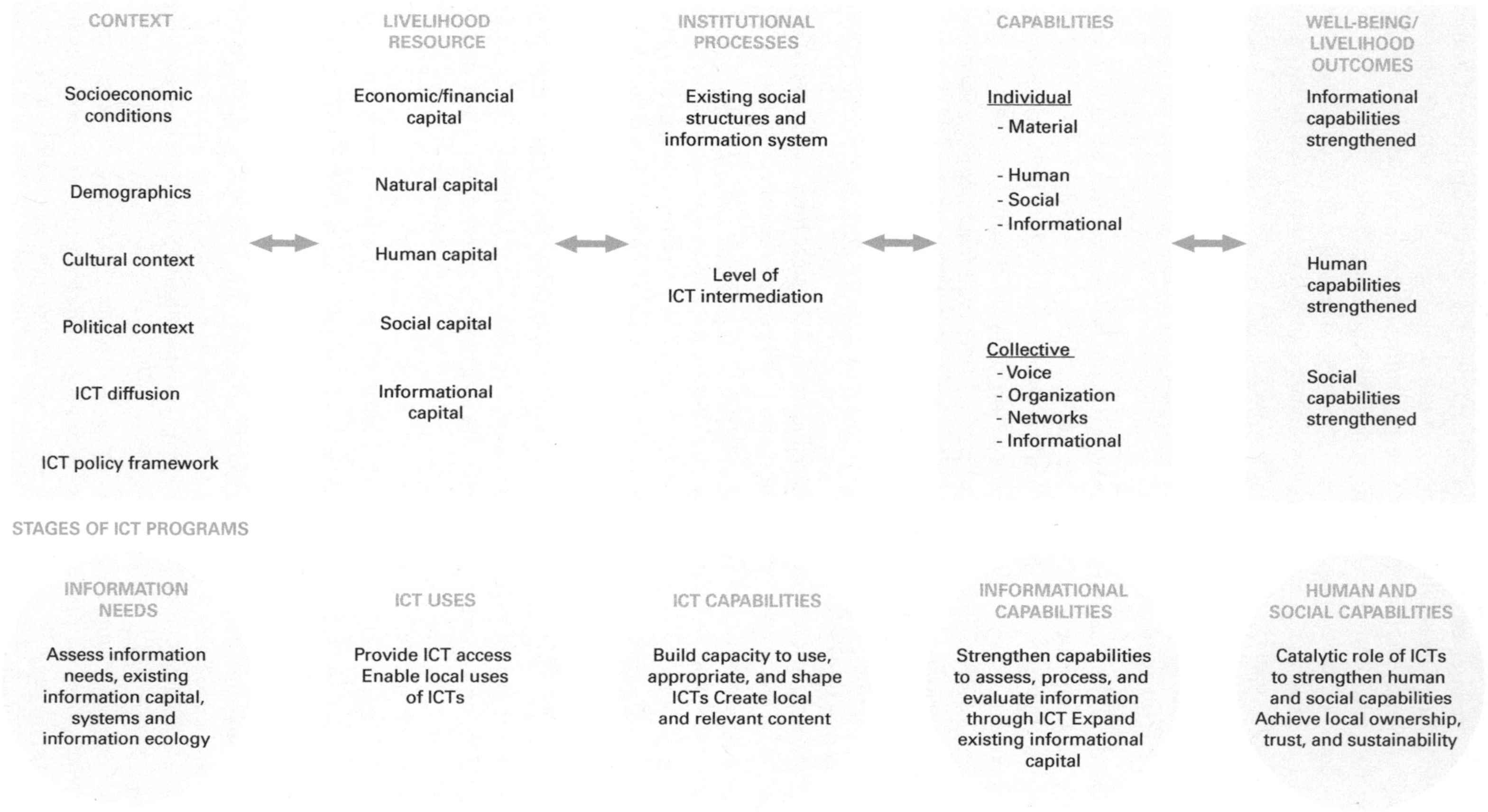

CONTEXT	LIVELIHOOD RESOURCE	INSTITUTIONAL PROCESSES	CAPABILITIES	WELL-BEING/ LIVELIHOOD OUTCOMES
Socioeconomic conditions	Economic/financial capital	Existing social structures and information system	Individual - Material	Informational capabilities strengthened
Demographics	Natural capital		- Human - Social - Informational	
Cultural context	Human capital	Level of ICT intermediation		Human capabilities strengthened
Political context	Social capital		Collective - Voice	Social capabilities strengthened
ICT diffusion	Informational capital		- Organization - Networks - Informational	
ICT policy framework				

STAGES OF ICT PROGRAMS

INFORMATION NEEDS	ICT USES	ICT CAPABILITIES	INFORMATIONAL CAPABILITIES	HUMAN AND SOCIAL CAPABILITIES
Assess information needs, existing information capital, systems and information ecology	Provide ICT access Enable local uses of ICTs	Build capacity to use, appropriate, and shape ICTs Create local and relevant content	Strengthen capabilities to assess, process, and evaluate information through ICT Expand existing informational capital	Catalytic role of ICTs to strengthen human and social capabilities Achieve local ownership, trust, and sustainability

knowledge within their community and networks; and (d) information literacy—the existing skills to process, evaluate, and solve information-related problems.

These four types of capital are derived from the earlier discussion on applying the capability approach to ICTs. In particular, the four types of capital describe the existing informational ecologies within communities in terms of people's ability to obtain information from outside the community, communicate with others, use their own traditional knowledge, and interpret and apply information to their own context. As described above, information can play an important "agency role" in expanding poor people's human and social capabilities. Thus it is critical for the AEF to assess the existing informational assets or (capital) of a community *before* investigating the effects of enhancing a community's access to information through ICTs.

As figure 1.2 shows, the informational capital dimension has been added—in the second column of the model—as an additional asset to the set of livelihood resources of the poor. Thus the framework underscores that the interlinkages between informational capital and all the other types of capital, such as economic, human, or social capital, are crucial for evaluating the role of information and ICTs in the livelihoods of the poor. At the same time, it is argued that information itself is an important asset for the poor to use to improve or secure their livelihoods. Integrating "informational capital" into this approach allows me to study the role that information plays in the context of an asset-based livelihoods approach.

The approach underscores that the capability of individuals and social groups to transform valued functionings into realized functionings depends largely on their livelihood resources with regard to the different types of capital (Bebbington 1999). The expansion of capabilities is understood as the strengthening of people's capital. The strengthening of existing informational capital thus translates into enhanced informational capabilities.

What is the role that information plays in this context, and what justifies broadening the livelihood approach with the addition of the "informational capital" dimension? The main argument for including this dimension in the framework is that information and ICTs can play an important role not only in their own right, but also as "agents" that strengthen the poor's capital in multiple areas. As the review of the literature (Kabeer 2000; Bebbington 1999) has demonstrated, only the combination of strengthened resources and agency can enhance individual and collective capabilities.

Under what conditions can the expansion of informational capabilities have a positive "multiplier effect" on the other capabilities? In other words, does the expansion of the poor's capability to make meaningful use of ICTs strengthen their capability to achieve valued functionings in multiple areas? This notion stems from Sen's concept of the role that human capital plays not only in enhancing a person's ability to generate income, but also in expanding her or his capabilities to lead a freer and more fulfilled life and to reach her or his valued functionings (Sen 1997, 1960). In this sense, the focus is on the agency role of individuals to expand their human capabilities and thus to bring about social change. I apply this concept to the field of ICTs, arguing that better access to information and enhanced informational capabilities act like improved literacy in increasing people's capabilities to make choices in their lives in various areas, including the economic, social, and political spheres. Thus, as a result of expanded informational capabilities, individuals will be able to increase their control over important life choices; in this sense, information and ICTs play a critical role in strengthening poor people's human capabilities.

For instance, a critical aspect of improving the "agency" of the poor is to enhance their ability to negotiate and bargain with the formal institutions of the state, market, and civil society. Information, of course, is critical to negotiation. In this context, ICTs could be critical in improving the access of local communities to market prices, thus potentially reducing the existing "information asymmetries" between the poor and potential buyers in local and national markets. This process might ultimately strengthen people's human capabilities in the economic dimension, since the market price information strengthens their power to negotiate prices in the marketplace. In this example, ICTs can strengthen the informational assets and agency of the poor, thus enhancing people's informational and human capabilities. Improved access to ICTs has indirect, rather than direct, effects on the well-being of the poor.

The framework suggests that a quite complex process needs to take place for ICTs to have an impact on the lives of poor communities. The five stages of ICT projects at the bottom of figure 1.2 describe the different steps in this process. A direct and causal relationship does not exist between ICTs, information, and human well-being; the relationship between these variables is much more multidimensional and needs to be seen within the broader context of sustainable human development.

Based on the importance of the local socioeconomic context for ICT projects, the AEF places the context in the first column of figure 1.2. As described in the literature review, a critical factor that determines whether ICT programs can significantly enhance people's well-being depends on the extent to which the technologies and approaches used are amenable to the particular local socioeconomic, demographic, political, and cultural context in which ICTs are being inserted. These contextual factors, which are external to people's own capabilities, are frequently critical if ICT programs are to enhance poor communities' well-being. The framework also considers the enabling environment—the existing ICT policy framework and level of ICT diffusion—within a country.

The second column shows the livelihood resources within a community, including the economic, natural, human, social, and information capital. As mentioned, the informational dimension has been fully integrated into the model.

The AEF also stresses the need to understand the institutional structures and processes that mediate the process of transforming livelihood resources into the expansion of human and social capabilities, thus enhancing people's well-being. Similar to the livelihood approach itself, the AEF emphasizes the study of institutions and organizations. This is also apparent in figure 1.2, since the institutional processes are placed at the center of the model. It is important to analyze the interrelationship between existing social structures and the ICT intermediation. The framework contends that a successful mediation process by an effective local intermediary is essential to transforming the sheer access of ICTs into meaningful uses that can, in turn, enhance the human and social capabilities of the poor.

The AEF also stresses that the lives of poor people are embedded in a broader social, economic, political, and institutional context and thus form a critical part of the overall institutional setting. For instance, in the case of indigenous peoples, belonging to a specific marginalized ethnic group can significantly constrain their access to key resources, such as education or electricity, independent from their individual characteristics (Gigler 2001, 2005b). Sen argues, "Social and economic factors such as basic education, elementary health care, and secure employment are important, not only on their own, but also for the role that they can play in giving people the opportunity to approach the world with courage and freedom" (Sen 1999, 63). However, structural barriers, such as the systematic lack of access to education services, often prevent even people who recognize the significance and value of ICTs from using them. In relation to the capability approach, such a situation is described as "unrealized functionings," and it is also captured in Sen's concept of "unfreedom" (Sen 1999). As such, the AEF analyzes the barriers that prevent people from deriving benefits from the effective use of ICTs.

The fourth column of the model describes the individual and collective capabilities that can result from the use of ICTs. The framework mentions material, human, social, and informational capabilities. Literacy skills are an example of an individual human capability, since the knowledge of how to read and write is directly associated with a person. The AEF categorizes the areas of voice, organization, and networks as collective capabilities. For instance, the effects of ICT programs on strengthening community organizations or the networks among communities are good examples of collective capabilities that are realized only for the community as a whole. The framework includes both individual and collective informational capabilities. This categorization emphasizes that enhanced "informational capabilities"—for instance, the ability to obtain more information about government programs—can enhance a person's human capabilities as well as increase the capabilities of the community as a whole. In other words, the informational dimension has an important public goods aspect, in the sense that it is a nonrivalrous and nonexcludable good.

The last column in the AEF summarizes the various outcomes of enhanced well-being, in terms of strengthened informational, human, and social capabilities. The indicators used to assess the outcomes of ICT projects for three different capabilities and for the well-being of individuals and communities are described below.

Dimensions of Human Capabilities

The AEF distinguishes the following six dimensions of *human capabilities*: informational, psychological, social, economic, political, and cultural. In order to go beyond studying the effects of ICT projects on people's individual capabilities, it also analyzes the impact of ICTs on social capabilities, which refer to the collective capabilities of a community. As Sen (2010) states, ICTs do not just have an impact on individuals but also have positive externalities for the wider community:

>> A telephone owned by a person helps others to call the person up, as well as to receive calls from him or her, and so the increased freedom of the phone owner adds to the freedom of others. (Sen 2010, 1)   <<

As described in more depth below, it also includes six dimensions of *social capabilities*: informational, organizational, social development, economic development, political participation, and cultural identity.

Finally, the AEF describes the five stages of ICT projects, as visualized in the line of boxes at the bottom of figure 1.2. At the outset of an ICT program, the framework highlights the need to assess the existing information environment and information systems within communities in order to determine the strengths and weaknesses of the existing information capital and to identify the key stakeholders and their interests in the information system. Such an assessment makes explicit the role of information in the community and the traditional channels of information and communications, such as oral tradition and community radio (Brown 1991; Robinson 1998).

The AEF thus provides an overview of the most important information needs in the community, regarding which type of information people value the most and have the least access to. Furthermore, it shows the level of existing information capital and ICT uses, defining the extent to which communities have access to information from the formal institutions of the market, state, and civil society prior to intervention of the ICT program.

As a third step, the AEF highlights the central role that the ICT intermediary and social intermediary play in enhancing people's ICT capabilities. The main role of the ICT intermediary is to facilitate the access to ICTs, to provide training in the use of ICTs to local communities, and to support program participants in identifying local and relevant content. Another critical factor is the extent to which the community's social institutions and community-based organizations (social intermediaries) play a proactive role in "brokering" the improved access to information of community members and in supporting the ICT program in achieving its main objective.

The AEF also stresses the local appropriation of technologies and the contextualization of information provided through ICTs. Simply providing access to ICTs will not allow the poor to derive meaningful and lasting benefits from their use—only if poor communities appropriate the technology to their local needs and realities can they enhance their informational capabilities and thus derive real benefits from their use.

Finally, successful ICT projects will reach the final stage in which local communities have gained ownership of the program and the long-term social and financial sustainability of the program is secured. Only if this final step is reached can the enhancement of peoples' informational capabilities be translated into improved human and social capabilities and can the program contribute to enhancing people's well-being.

Key Hypotheses for Evaluating the Impact of ICT Projects

As the theoretical framework has shown, improved access to ICTs does not have a direct positive effect on poor people's well-being. Instead, the framework suggests that a complex process needs to take place to translate enhanced access to and use of ICTs into improved well-being. The framework therefore emphasizes the critical role that an "intermediary" plays in facilitating the necessary "enabling environment." Thus the following hypothesis is instrumental in analyzing the conditions under which ICT projects can enhance the well-being of poor communities:

>> *Hypothesis 1.* The access to and use of ICTs have to be facilitated by an effective and local intermediary organization in order to improve poor people's well-being.   <<

The second hypothesis is derived from the importance of providing ICT services to local communities in a culturally appropriate manner. This topic is particularly important to indigenous communities based on their strong cultural identity and strong local institutions. As has been pointed out, the appropriation of ICTs to local realities is critical for ensuring the long-term viability and sustainability of an ICT program. Consequently, the following hypothesis is also critical for the analysis:

>> *Hypothesis 2.* ICTs have to be appropriated by local communities in order to enhance their well-being.   <<

Operationalizing the AEF for Research in Bolivia

The research presented in this book is based on poor people's own priorities, as identified during 46 focus groups with rural communities in Bolivia (summarized in appendix B). These results formed part of the national Indigenous Peoples Profile developed by the Bolivian government with the technical assistance of the World Bank (VAIPO 1999). I worked for two years on this project and was responsible for coordinating the consultations with indigenous peoples at the community level. For this reason, I am thoroughly familiar with both the process and the results of these consultations. I use this extensive qualitative study on poor communities' own visions of human development as the baseline for this study. The main contribution of the VAIPO study is that it describes in great detail poor people's own definition and vision of human development, based on their local social, cultural, and political context. Indigenous communities in that study identified five key dimensions for their individual and collective well-being: social development, economic development, political participation, cultural identity, and organizational development. The AEF uses this classification scheme to assess the impact of ICTs on the well-being of indigenous peoples.

The research uses three types of indicators. First, it uses quantitative indicators, including sociodemographic (age, income) and socioeconomic (poverty levels, illiteracy rates) data to analyze the multiple and complex factors that influence whether and to what extent marginalized groups are able to make use of ICTs. It also includes a set of indicators about people's uses and levels of ICT (type and frequency of use) and studies the extent to which individual and external socioeconomic factors influence people's ICT capabilities. Finally, it studies the influence of ethnicity (the belonging to a specific marginalized group) and institutional factors (the role of intermediary organizations) on people's use of, and facility with, ICTs.

Second, the research includes a set of qualitative indicators to assess whether or not the identified uses of ICTs have indeed enhanced people's human and social capabilities and thus improved their well-being. It uses qualitative research methods (focus groups, in-depth interviews) to understand people's expectations and perceptions of the benefits and risks involved in using ICTs (Madon 2004; O'Farrell 2001; ActionAid 1999).

Third, it uses a set of process indicators to understand *how* ICTs are introduced into communities. The intent is to examine whether the ICT program is based on real needs as expressed by local communities or on an externally driven agenda of a donor agency, NGO, or government agency. An important process indicator is whether or not an information needs assessment was conducted at the community level before proceeding with the ICT program. Frequently, the implementation of ICT programs fails to identify the community's communication and information needs, leading to resistance within the community or even refusal to accept the new technologies (Robinson 1998). Subsequently, the study assesses whether or not the project has allowed for a convergence between new ICTs and traditional ones, building on, rather than replacing, existing communication and information channels (O'Farrell 2001). Considering that the core of the research aims to assess the impact of ICTs on well-being, the following section goes into more detail about the key outcome indicators of enhanced human, informational, and social capabilities.

Indicators of Enhanced Human Capabilities

The analysis distinguishes from among the following six dimensions: informational, psychological, social, economic, political, and cultural (table 1.1). The framework uses specific indicators for each of these dimensions, stressing the interdependencies among the various dimensions of well-being and investigating whether or not they reinforce each other. Because of the instrumental role of ICTs in enhancing people's human capabilities, the analysis evaluates

TABLE 1.1   Indicators of enhanced human capabilities

Dimension	Objective	Outcome indicators
Informational	To improve the access to information and informational capabilities	• Improved capacity to use different forms of ICTs • Enhanced information literacy • Enhanced capacity to produce and publish local content • Improved ability to communicate with family members and friends abroad
Psychological	To support a process of self-reflection (critical conscientization) and problem-solving capacity	• Stronger self-esteem • Improved ability to analyze one's own situation and solve problems • Stronger ability to influence strategic life choices • Sense of inclusion in the "modern" world
Social (human capital)	To strengthen people's human capital (skills, knowledge, ability to work, and good health)	• Enhanced ICT literacy and technology skills (for example, computer repair) • Enhanced leadership skills • Improved program management skills
Economic	To enhance people's capacity to interact with the market	• Improved access to markets • Alternative sources of income • Stronger productive assets • Improved employment opportunities • Improved income through (a) lower transaction costs (fewer time constraints), (b) reduced transport needs, (c) increased timeliness of sales, and (d) increased remittances from family members living abroad
Political	To improve people's participation in decision-making processes at the community level and the political system	• Improved access to government information and services (e-government) • Improved awareness about political issues • Improved capabilities to interact with local governments
Cultural	To strengthen people's cultural identity	• Use of ICTs as a form of cultural expression (design of computer graphics, websites) • Increased awareness of one's own cultural identity

people's actual ICT uses and their level of use. For instance, in order to assess whether or not ICTs have improved people's capability to access markets, it is critical to understand their patterns of ICT use, the extent to which use has improved their informational capabilities, and the positive effect this use has had on their human capabilities.

The following section specifies several outcome indicators to determine whether and to what extent ICT programs have improved people's informational capabilities.

Indicators of Enhanced Informational Capabilities

Four outcome indicators are relevant for analyzing the informational dimension of an ICT program. For instance, in order to assess whether or not ICTs have improved people's capability to access market prices or to improve their income, it is critical to understand people's motivation to use ICTs and to assess which types of ICTs—for instance, community radio or the Internet—and which specific applications they are using.

The second is related to "communication capabilities" and the extent to which an ICT program has improved poor people's ability to communicate with family members, friends, and professional contacts.

The third is related to whether or not an ICT program has enhanced people's *information literacy* and thus their ability to find, evaluate, and use information for problem solving. Specifically, it assesses a person's capacity to access the needed information effectively and efficiently, to evaluate information, to incorporate the selected information into his or her own knowledge base, and to process information effectively (table 1.2).

TABLE 1.2   Indicators of enhanced informational capabilities

Dimension	Objective	Outcome Indicators
ICT capabilities	To strengthen people's human capital in terms of their ICT uses	*Enhanced ICT capabilities* • Improved ICT skills to use computer hardware and software • Improved use of a variety of ICTs
Communications capabilities	To strengthen people's capabilities to communicate with family members and friends	*Enhanced communications capabilities* • Improved ability to communicate with family members and friends abroad • Knowledge on how to use Internet telephony services
Information literacy	To enhance people's capabilities to find, use, evaluate, and process information	*Enhanced information literacy* • Enhanced capacity to access information and to use multiple methods to retrieve information (search engine, e-mail, websites) • Enhanced capacity to evaluate information, to assess different information sources, and to analyze and contextualize information • Enhanced capacity to process information and to extract, record, and manage information
Content capabilities	To improve people's capabilities to produce and share information based on their local context and existing base of indigenous knowledge	*Enhanced content capabilities* • Improved capacity to produce and disseminate local content • Improved capacity to place content into local context • Enhanced ability to share information effectively

The final outcome indicator for enhanced informational capabilities is related to whether or not an ICT program has enhanced people's *content capabilities*—that is, has it enhanced people's capacity to meet their information needs by improving their access to local and relevant content and strengthening their ability not only to use, but also to produce, information based on their local and indigenous knowledge?

Indicators of Enhanced Social Capabilities

The analysis of social capabilities emphasizes a process of community empowerment encompassing a dynamic continuum that includes small groups, community organizations, partnerships, and political action (Rissel 1994; Labtone 1990). This view illustrates the concept of "power to" and "power with," which coincides with two of the most important assets of poor communities: their ability to form groups and organizations at the community level and their ability to pursue goals collectively based on a shared vision (Rowlands 1997; Menchú 1984).

However, local communities are not a homogeneous entity, and many development projects have underestimated the role of existing social norms and social institutions within them. This may lead to power struggles and reinforce the marginalization and exclusion of certain groups (Agrawal and Gibson 1999). The AEF specifically addresses issues related to the distribution of power within the community (Oakley 2001). Power relations define the basic patterns of social and economic relations and therefore should be included in any analysis examining the potential impact of ICTs on community empowerment.

As previously mentioned, the AEF considers six key dimensions of social capabilities, identified by local communities themselves: informational, organizational, social development, economic development, political participation, and cultural identity (summarized in table 1.3).

TABLE 1.3 Indicators of enhanced social capabilities

Dimension	Objective	Outcome indicator
Informational	To improve access to information and informational capabilities	• Stronger traditional information system (that is, information flows within community improved)
		• Stronger horizontal knowledge exchanges with other communities
		• Stronger vertical knowledge exchanges with the state, donors, and NGOs
Organizational	To strengthen organizational capabilities	• Increased efficiency and transparency of poor people's organizations (that is, more transparent selection of leaders)
		• Better coordination among different organizations
		• Stronger networks with other community-based organizations through improved information flows

(continued)

TABLE 1.3   Indicators of enhanced social capabilities *(continued)*

Dimension	Objective	Outcome indicator
Social development	To improve access to and quality of basic social services	• Improved access to education through bilingual and intercultural education programs (that is, curricula available through e-learning)
		• Improved access to more culturally appropriate health services
Economic development	To promote economic opportunities for the communities	• Improved system of collective land titling (that is, through digitalization of land records)
		• Improved productive activities through enhanced knowledge (that is, better knowledge about agricultural practices)
		• Improved transparency markets through reduced information asymmetries (that is, agricultural market prices disseminated)
		• Enhanced opportunities for the commercialization of products (that is, fair trade through e-commerce)
		• Enhanced capacity to mobilize resources from outside donors for community-based programs
Political participation	To improve participation in the political system and strengthen poor people's networks	• Enhanced decision-making power in political process
		• Improved "voice" and participation in development process
		• Improved transparency of political institutions (e-government)
		• Better coordination of political activities between different indigenous organizations
		• Enhanced transparency of information flows between poor people's organizations and communities
		• Direct participation in international policy dialogue (United Nations permanent forum)
Cultural identity	To strengthen the community's cultural identity	• Stronger indigenous languages
		• Stronger indigenous knowledge

The individual and collective processes are often interdependent. However, this separation adds significantly to the analysis, since it provides a clear, logical framework for breaking up the empowerment processes into smaller, more comprehensive, and manageable units.

Structure of the Book

This section provides an overview of the structure of the book. The study consists of 10 chapters, including this introduction and the conclusion. Figure 1.3 summarizes the main structure of the study and visualizes how the different chapters are related to the AEF.

Chapter 2 presents the research methodology of the case study approach (Orlikowski and Icono 2001; Walsham 1995). It describes the rationale for adopting this research methodology and shows how the case study approach enables me (a) to study the specific factors that influence whether or not ICTs can contribute to the well-being of indigenous peoples and (b) to unpack the

FIGURE 1.3   Structure of the study

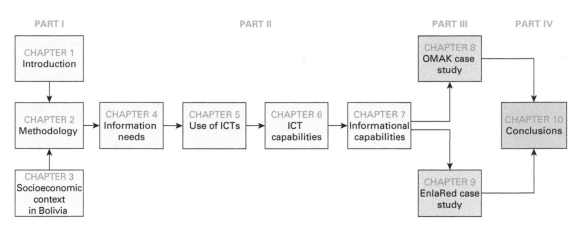

process of how ICTs are introduced in local communities and which individuals or institutions support or resist the new technologies (Yin 1994). Finally, it outlines the data collection methods used and illustrates the importance of combining quantitative with qualitative research methods in order to analyze both the key factors and the processes that influence whether and to what extent poor people can enhance their well-being through the use of ICTs.

Chapter 3 presents an overview of the demographic characteristics and the socioeconomic situation of indigenous peoples in Bolivia. It examines the persistent social and economic gaps between the indigenous and nonindigenous populations and analyzes the key factors responsible for their continuous economic and social exclusion.

Chapter 4 analyzes indigenous peoples' local information needs, their existing patterns of communication, uses of ICT, and their perceptions and expectations of ICTs' ability to enhance their well-being based on the empirical findings from the non–ICT user survey. The analysis establishes a baseline of "information ecologies"[8] in communities, identifies people's information needs, and analyzes the effects of ICTs (Tacchi, Slater, and Hearn 2003; O'Farrell, Norrish, and Scott 1999). This approach is particularly relevant to the study of indigenous communities in rural Bolivia, where communication and information flows are thoroughly embedded in the traditional social institutions of the communities. The introduction of new forms of ICTs has to be based on these existing structures. The chapter concludes by presenting indigenous peoples' perspectives on the value of the Internet for promoting their development.

Chapter 5 looks beyond the concept of "ICT access" to the multiple factors that influence whether or not indigenous peoples use ICTs. The first section presents the empirical evidence from two user surveys carried out in indigenous communities. The second section analyzes the data from a national ICT survey administered by the UNDP in 2004 (ECADI household survey) in order to (a) cross-validate the findings from the user surveys with the results from a national data set; (b) generalize the findings about indigenous peoples' uses of ICTs to a broader national scale; and (c) determine the nature of the relationship between the dependent variables (ICT uses) and the independent variables (gender and ethnicity), as well as the magnitude of the interrelationship. The chapter uncovers some of the current barriers preventing indigenous peoples from using ICTs and provides insight into the role that intermediary organizations can play in promoting their ICT use. The final section draws conclusions from both the statistical analysis and the regression models and places the findings into the broader context of the study.

Chapter 6 investigates the magnitude of the effects of multiple factors on people's ICT capabilities by assessing their capacity to make meaningful use of ICTs. In the first section, it presents the empirical evidence of indigenous peoples' ICT capabilities based on micro-level data gathered through the two user surveys described in chapter 5. The second section cross-validates the survey results by applying a series of regression models to the ECADI data.

Chapter 7 investigates the extent to which indigenous peoples' use of ICTs translates into enhanced human and social capabilities and thus positively affects their well-being. It is divided into two parts. The first analyzes indigenous peoples' perceptions of the Internet's impact on their well-being, as expressed in their responses to an impact user survey ($n = 91$) administered to them during the fieldwork. The empirical results describe their perceptions of how Internet use has affected their well-being in the political, economic, social, personal, organizational, and cultural dimensions of their lives. The second part discerns the factors that influence their perceptions of the impact of the Internet on their well-being.

Chapter 8 presents the findings from an in-depth case study evaluating the impact of the grassroots-level program of the Organización de Mujeres Aymaras del Kollasuyo (Organization of Aymara Women of the Kollasuyo, OMAK) on the well-being of indigenous women. The program focused on the innovative uses of ICTs as a conduit for empowering indigenous women and supporting them in their daily struggle to overcome racial discrimination and social exclusion. The case study, set in Tiahuanaco—a small town located in the Andean highlands in one of the poorest regions of Bolivia—explores whether and to what extent ICTs can enhance the human and social capabilities of indigenous women, even in such a challenging socioeconomic context, and thus can contribute significantly to improving their well-being. It also investigates the extent to which the success of the program can be attributed to OMAK and its grassroots-based approach to promoting the use of ICTs among indigenous communities.

Chapter 9 contrasts the OMAK case study with a case study of EnlaRed Municipal—a major ICT program coordinated by the Federation of Municipal Associations of Bolivia intended to support local governments. This program was implemented with a low level of intermediation, centralized management, and a low degree of interaction with the local community. The chapter discusses the program's ambitious aims to institutionalize major reforms in local governments by (a) strengthening their institutional and technical capacity, (b) enhancing their accountability and transparency, and (c) improving their delivery of services. The chapter uses empirical evidence of impact and interviews with local government officials and community leaders.

Chapter 10 offers a brief conclusion. It discusses first the findings of the study regarding its principal research question: Whether and under which conditions can ICTs enhance indigenous peoples' human and social capabilities to achieve the lifestyle they value and thus enhance their well-being? It then discusses the extent to which the study's two main hypotheses are confirmed and some of the theoretical implications. Finally, based on the theoretical framework presented in this introduction, it develops an ICT impact chain that summarizes the principal findings of the study and illustrates the process by which ICTs can significantly expand indigenous peoples' human and social capabilities and thus enhance their individual and collective well-being.

Notes

1. The digital divide is often defined as "the differential extent to which rich countries and poor countries benefit from various forms of ICT" (James 2007, 284). An important shortcoming of this definition is that it neglects the divides that exist in access to and use of ICTs within countries and between socioeconomic groups.
2. No single definition captures the diversity of indigenous peoples, and due to the history of political repression, discrimination, and assimilation policies by states, indigenous peoples usually reject being defined by external agencies. In order to identify rather than "define" indigenous identities, I use the working definitions provided by the International Labour Organization Convention 169 and by the United Nations special rapporteur, José Martínez-Cobo, which both emphasize that self-identification as indigenous should be the main criterion used to identify indigenous peoples (Martínez-Cobo 1986).
3. For this research I use Hamelink's definition of ICTs: "Information and Communication Technologies (ICTs) encompass all of those technologies that enable the handling of information and facilitate different forms of communication among human actors, between human beings and electronic systems, and among electronic systems" (Hamelink 1997:3). This functional definition of ICTs includes both the new (i.e., Internet, e-mail) and traditional (i.e., community radio) forms of ICT in its definition.
4. This term comes from "communicative ecology," which describes using ethnographic research methods to examine all of the social processes, relations, and technologies through which information and communications flow at the community level (Tacchi, Slater, and Hearn 2003).
5. The literature does not provide a single definition of "indigenous knowledge." Instead, it emphasizes the concept's main characteristics as being (a) rooted in a particular local community and situated within broader cultural traditions; (b) experimental, meaning that it is being tied to action and based on experiences from trial and error; (c) implicit or "tacit," meaning that it is often difficult for people to express this

knowledge unambiguously and to find words to express what they know (Giddens 1984); (d) transmitted orally or through imitation and demonstration; and (e) a dynamic mix of past traditions and present innovations (Bicker, Sillitoe, and Pottier 2004; Ellen and Harris 1996; Warren 1991).

6. Marcelo Fernandez Osco from the Taller de la Historia Oral Andina, an indigenous peoples' NGO.

7. A good example for this is the Pareto optimum, which leads to the efficient allocation of resources, but is not sensitive to the issue of equity. It makes an implicit value judgment by confirming the "status quo," even if this leads to a socially unacceptable disparity in the distribution of resources and wealth.

8. See Tacchi, Slater, and Hearn (2003) for a more detailed description of the term "information ecologies."

References

ActionAid. 1999. *Fighting Poverty Together: ActionAid's Strategy 1999–2003*. London: ActionAid.

Agrawal, Arun. 1995. "Dismantling and Divide between Indigenous and Scientific Knowledge." *Development and Change* 26 (3): 413–39.

———. 2001. "Common Property Institutions and Sustainable Governance of Resources." *World Development* 29 (10): 1649–72.

Agrawal, Arun, and Clark Gibson. 1999. "Enchantment and Disenchantment: The Role of Community in Natural Resource Conservation." *World Development* 27 (4): 629–49.

Aker, Jenny. 2010. "Information from Markets Near and Far: Mobile Phones and Agricultural Markets in Niger." *American Economic Journal: Applied Economics* 2 (3): 45–69.

ALA (American Library Association). 1989. *Presidential Committee on Information Literacy Final Report*. Chicago: ALA.

Alampay, Edwin. 2003. *Using the Capability Approach to Analyze Access to Information and Communication Technologies (ICTs) by the Poor*. University of the Philippines, National College of Public Administration and Governance.

Albó, Xavier. 1996. "Poverty, Development, and Indigenous Identity." In *Indigenous Development: Poverty, Democracy, and Sustainability*, 33–40. Washington, DC: Inter-American Development Bank.

Alkire, Sabina. 2002. *Valuing Freedoms: Sen's Capability Approach and Poverty Reduction*. Oxford: Oxford University Press.

Altieri, Miguel. 1989. "Rethinking Crop Genetic Resource Conservation: A View from the South." *Conservation Biology* 3 (1): 77–79.

Anderson, Jon, L. V. Crowder, D. Dion, and W. Truelove. 1999. "Applying the Lessons of Participatory Communication and Training to Rural Telecentres." FAO, Sustainable Development Department, Rome, July.

Avgerou, Chrisanthi. 2000. "IT and Organizational Change: An Institutionalist Perspective." *Information Technology and People* 13 (4): 234–62.

———. 2001. "The Significance of Context in Information Systems and Organizational Change." *Information Systems Journal* 11 (1): 43–63.

———. 2003. "The Link between ICT and Economic Growth in the Discourse of Development." In *Organizational Information Systems in the Context of Globalization*, edited by Mikko Korpela, Ramiro Montealegro, and Angeliki Poulymenakou, 373–86. Dordrecht: Kluwer.

———. 2008. "Information Systems in Developing Countries: A Critical Research Review." *Journal of Information Technology* 23 (3): 133–46.

———. 2010. "Discourses on ICT and Development." *Information Technologies and International Development* 6 (3): 1–18.

Avgerou, Chrisanthi, and R. L. La Rovere, eds. 2003. *Information Systems and the Economics of Innovation*. Cheltenham: Edward Elgar.

Avgerou, Chrisanthi, and Shirin Madon. 2003. "Framing IS Studies." Working Paper, London School of Economics, Department of Information Systems, London.

Avgerou, Chrisanthi, and Geoff Walsham. 2000. "Introduction: IT in Developing Countries." In *Information Technology in Context: Studies from the Perspective of Developing Countries,* edited by Chrisanthi Avgerou and Geoff Walsham. London: Ashgate Publishing.

Bailey, Arlene. 2009. "Issues Affecting the Social Sustainability of Telecentres in Developing Contexts: A Field Study of Sixteen Telecentres in Jamaica." *Electronic Journal of Information Systems in Developing Countries* 36 (4): 1–18.

Bailur, Savita. 2007. "Using Stakeholder Theory to Analyze Telecenter Projects." *Information Technology for Development* 3 (3): 61–80.

Bailur, Savita, and Silvia Masiero. 2012. "The Complex Position of the Intermediary in Telecenters and Community Multimedia Centers." *Information Technologies and International Development* 8 (1): 27–42.

Ballantyne, Peter. 2002. "Collecting and Propagating Local Development Content." *INASP Newsletter* 20 (June): n.p.

Bardhan, Pranab. 2002. "Decentralization of Governance and Development." *Journal of Economic Perspectives* 16 (4): 185–206.

Barja, Gover, and Björn-Sören Gigler. 2005. "The Concept of Information Poverty and How to Measure It in the Latin American Context." In *Digital Poverty: Latin American and Caribbean Perspectives*, edited by Hernan Galperin and Judith Mariscal, 11–28. Ottawa: IDRC.

Bebbington, Anthony. 1993. "Modernization from Below: An Alternative Indigenous Development?" *Economic Geography* 69 (3): 274–92.

———. 1999. "Capitals and Capabilities: A Framework for Analysing Peasant Viability, Rural Livelihoods, and Poverty in the Andes." *World Development* 27 (12): 2021–44.

Bedi, Arjun S. 1999. "The Role of Information and Communication Technologies in Economic Development: A Partial Survey." Development Policy Discussion Paper 7, Universität Bonn, ZEF, Bonn.

Benjamin, Peter. 2000. "African Experience with Telecenters." *On the Internet* (November-December).

———. 2001. "Telecentres in South Africa." *Journal of Development Communication Telecenters and ICT for Development: Critical Perspectives and Visions for the Future* 12 (2): 32–38.

Beuermann, Diether W., Christopher McKelvey, and Carlos Sotelo. 2012. "The Effects of Mobile Phone Infrastructure: Evidence from Rural Peru." University of Wisconsin. http://www.aae.wisc.edu/mwiedc/papers/2011/Beuermann_Diether.pdf.

Bhatnagar, Subhash C. 2004. *E-Government from Vision to Implementation: A Practical Guide with Case Studies*. New Delhi: Sage Publications.

Bicker, Alan, Paul Sillitoe, and Johan Pottier, eds. 2004. *Investigating Local Knowledge: New Directions, New Approaches*. Aldershot: Ashgate Publishing.

Bierschenk, Thomas, Jean-Pierre Chauveau, and Jean-Pierre Olivier de Sardan, eds. 2000. *Courtiers en développement: Les villages africains en quête de projets*. Paris: Karthala.

———. 2002. "Local Development Brokers in Africa: The Rise of a New Social Category." Working Paper 13, Johannes Gutenberg University, Department of Anthropology and African Studies, Mainz, Germany.

Blankenberg, Floris. 1995. *Methods of Impact Assessment Research Programme: Resource Pack and Discussion*. The Hague: Oxfam UK/I and Novib.

Blattman, Christopher, Robert Jensen, and Raul Roman. 2003. "Assessing the Need and Potential of Community Networking for Development in Rural India." *Information Society* 19 (5): 349–65.

Braga, Carlos. 1998. "Inclusion or Exclusion." Information for Development (InfoDev), World Bank, Washington, DC.

Brevik, Patricia Senn. 1992. "Information Literacy: An Agenda for Lifelong Learning." *AAHE Bulletin* (March): 6–9.

Brokensha, David, Dennis M. Warren, and Oswald Werner, eds. 1980. *Indigenous Knowledge Systems and Development*. Lanham, MD: University Press of America.

Brown, David. 1991. "Methodological Considerations in the Evaluation of Social Development Programmes: An Alternative Approach." *Community Development Journal* 26 (4): 259–65.

Burkett, Ingrid. 2000. "Beyond the 'Information Rich and Poor': Futures Understandings of Inequality in Globalising Informational Economies." *Futures* 32 (7): 679–94.

Butcher, Neil. 1998. "The Possibilities and Pitfalls of Harnessing ICTs to Accelerate Social Development: A South African Perspective." South African Institute for Distance Education, Johannesburg, June.

Castells, Manuel. 1995. "Information Technology, Cities, and Development." *Urban Age* 3 (1): 15.

———. 1997. *The Information Age: Economy, Society, and Culture.* Vol. 2: *The Power of Identity.* Oxford: Blackwell.

———. 1998a. *The End of Millennium. The Information Age: Economy, Society, and Culture.* Vol. 3. Oxford: Blackwell.

———. 1998b. "Information Technology, Globalisation, and Social Development." Paper prepared for the Conference on Information Technologies and Social Development, United Nations Research Institute for Social Development (UNRISD), Geneva, June.

Cecchini, Simone, and Monica Raina. 2004. "Electronic Government and the Rural Poor: The Case of Gyandoot." *Information Technologies and International Development* 2 (2): 65–75.

Cecchini, Simone, and Christopher Scott. 2003. "Can Information and Communications Technology Applications Contribute to Poverty Reduction? Lessons from Rural India." *Information Technology for Development* 10 (2): 73–84.

Cernea, Michael, ed. 1991. *Putting People First: Sociological Variables in Rural Development*, 2d ed. Oxford: Oxford University Press.

Chambers, Robert. 1983. *Rural Development: Putting the Last First.* Harlow, U.K.: Pearson Education.

———. 1994. "Participatory Rural Appraisal (PRA): Challenges, Potentials, and Paradigm." *World Development* 22 (10): 1437–54.

———. 1995. "Poverty and Livelihoods: Whose Reality Counts?" *Environment and Urbanization* 7 (1): 173–204.

———. 1997. "Editorial: Responsible Well-Being; A Personal Agenda for Development." *World Development* 25 (11): 1743–54.

Chambers, Robert, and Gordon Conway. 1992. *Sustainable Rural Livelihoods: Practical Concepts for the 21st Century.* Brighton, U.K.: Institute of Development Studies.

Chambers, Robert, Arnold Pacey, and Lori Ann Thrupp, eds. 1989. *Farmer First: Farmer Innovation and Agricultural Research.* London: Intermediate Technology Publication.

Chapman, Robert, and Tom Slaymaker. 2002. "ICTs and Rural Development: Review of the Literature, Current Interventions, and Opportunities for Action." ODI Working Paper 192, Overseas Development Institute, London.

Ciborra, Claudio. 2002. "Unveiling E-Government and Development: Governing at a Distance in the New War." Public lecture at the London School of Economics.

———. 2005. "Interpreting E-Government and Development: Efficiency, Transparency, or Governance at a Distance?" *Information Technology and People* 18 (3): 260–79.

Ciborra, Claudio, and Diego Navarra. 2005. "Good Governance, Development Theory, and Aid Policy: Risks and Challenges of E-Government in Jordan." *Journal of Information Technology for International Development* 11 (2): 141–59.

Cisler, Steve. 2000. "Subtract the Digital Divide: Online Essay." *San Jose Mercury News*, January 16. http://home.inreach.com/cisler/divide.htm.

Cleaver, Frances. 1999. "Paradoxes of Participation: Questioning Participatory Approaches to Development." *Journal of International Development* 11 (4): 597–612.

Cochrane, Glynn, and Pauline Atherton. 1980. "The Cultural Appraisal of Efforts to Alleviate Information Inequality." *Journal of American Society for Information Services* 31 (2): 282–92.

Colle, Royal. 2005. "Memo to Telecentre Planners." *Electronic Journal of Information Systems in Developing Countries* 21 (1): 1–13.

Comim, Flavio. 2001. "Operationalizing Sen's Capabilities Approach." Paper prepared for the conference "Justice and Poverty: Examining Sen's Capability Approach," St. Edmund's College, Von Hugel Institute, Cambridge, U.K., June 5–7.

Conroy, Czech. 2006. "Telecentre Initiatives in Rural India: Failed Fad or the Way Forward?" TeleSupport Working Paper 4, TeleSupport Initiatives, Natural Resources Institute, Greenwich, U.K.

Cooke, Bill, and Uma Kothari, eds. 2001. *Participation: The New Tyranny?* London: Zed Books.

Corbridge, Stuart. 2001. "Development as Freedom: The Spaces of Amartya Sen." *Progress in Development Studies* 2 (3): 183–217.

Corbridge, Stuart, Glyn Williams, Manoj Srivastava, and René Véron. 2005. *Seeing the State: Governance and Governmentality in India*. Cambridge, U.K.: Cambridge University Press.

Correa, Antoinette F., Djibril Ndiaye, Kingo J. Mchombu, Gloria M. Rodríguez, Diana Rosenberg, and N. U. Yapa. 1997. "Rural Information Provision in Developing Countries: Measuring Performance and Impact." Report prepared for UNESCO on behalf of IFLA, UNESCO, General Information Programme and UNSIST, Paris.

Cracknell, Basil E. 2000. *Evaluating Development Aid: Issues, Problems, and Solutions*. New Delhi: Sage.

Daly, John. 1999. "Measuring Impacts of the Internet in the Developing World." *iMP Magazine*, May 21.

Davis, Shelton. 2002. "Indigenous Peoples, Poverty, and Participatory Development: The Experience of the World Bank in Latin America." In *Multiculturalism in Latin America*, edited by Rachel Sieder. London: Palgrave Macmillan.

Davis, Shelton, and Katrina Ebbe, eds. 1993. "Traditional Knowledge and Sustainable Development." Environmentally Sustainable Development Proceedings Series 4, World Bank, Washington, DC.

Davison, Robert, Doug Vogel, Roger Harris, and Noel Jones. 2000. "Technology Leapfrogging in Developing Countries: An Inevitable Luxury?" *Electronic Journal of Information Systems in Developing Countries* 1 (5): 1–10.

Delgadillo, Karen, Roberto Gómez, and Klaus Stoll. 2002. "Community Telecentres for Development: Lessons from Community Telecentres in Latin America and the Caribbean." IDRC, Ottawa.

DiMaggio, Paul, Eszter Hargittai, W. Russell Neuman, and John P. Robinson. 2001. "Social Implications of the Internet." *Annual Review of Sociology* 27 (August): 307–36.

Dymond, Andre, and Sonja Oestermann. 2004. "Rural ICT Toolkit for Africa." Information for Development Programme (infoDev), World Bank, Washington, DC, December.

Eade, Deborah. 2003. *Capacity-Building: An Approach to People-Centred Development*. Oxford: Oxfam.

Eggleston, Karen, Robert Jensen, and Richard Zeckhauser. 2002. "Information and Communication Technologies, Markets, and Economic Development." In *The Global Information Technology Report 2001–2002: Readiness for the Networked World*, edited by Geoffrey S. Kirkman, Peter G. Cornelius, Jeffrey D. Sachs, and Klaus Schwab, 62–75. New York: Oxford University Press.

Eisenberg, Michael E., and Robert Berkowitz. 1990. *Information Problem-Solving: The Big Six Skills Approach to Library and Information Skills Instruction*. Norwood, NJ: Linworth.

Ellen, Roy, and Holly Harris. 1996. "Concepts of Indigenous Environmental Knowledge in Scientific and Development Studies Literature: A Critical Assessment." Draft paper for the East-West Environmental Linkages Network Workshop 3, Canterbury.

Ernberg, Johan. 1998a. "Integrated Rural Development and Universal Access: Towards a Framework for Evaluation of Multipurpose Community Telecentre Projects Implemented by ITU and Its Partners." International Telecommunication Union, Geneva.

———. 1998b. "Review of the ITU MCT Pilot Project Programme: Towards a Framework for Impact Evaluation." International Telecommunication Union, Geneva.

Eversole, Robyn. 2003. "Managing the Pitfalls of Participatory Development: Some Insights from Australia." *World Development* 31 (5): 781–95.

Fafchamps, Marcel, and Bart Minten. 2012. "Impact of SMS-Based Agricultural Information on Indian Farmers." *World Bank Economic Review* 28 (1): 1–32.

Feiring, Birgitte. 2003. *Indigenous Peoples and Poverty: The Cases of Bolivia, Guatemala, Honduras, and Nicaragua*. London: Minority Rights Group International. www.minorityrights.org/download.php?id=77.

Fowler, Alan. 1997. *Striking a Balance: A Guide to Enhancing the Effectiveness of Nongovernmental Organisations in International Development*. London: Earthscan.

Garnham, Nicholas. 1997. "Amartya Sen's 'Capabilities' Approach to the Evaluation of Welfare: Its Application to Communication." *Javnost: The Public Journal of the European Institute for Communication and Culture* 4 (4): 25–34.

Gasco, Mila. 2003. "New Technologies and Institutional Change in Public Administration." *Social Science Computer Review* 21 (1): 6–14.

Gasper, Des. 1997. "Sen's Capability Approach and Nussbaum's Capability Ethics." *Journal of International Development* 9 (2): 281–302.

———. 2002. "Is Sen's Capability Approach an Adequate Basis for Considering Human Development?" Working Paper Series 360, Institute of Social Studies, The Hague.

Giddens, Anthony. 1984. *The Constitution of Society: Outline of the Theory of Structuration.* Cambridge, U.K.: Polity.

Gigler, Björn-Sören. 2001. "Empowerment through the Internet: Opportunities and Challenges for Indigenous Peoples." *TechKnowLogia* (July-August): 33–37. http://www.techknowlogia.org/tkl_articles/pdf/300.pdf.

———. 2004. "Including the Excluded: Can ICTs Empower Poor Communities?" In *Proceedings of the 4th International Conference on the Capability Approach.* Pavia, Italy, September 5–7.

———. 2005a. "Enacting and Interpreting Technology: From Usage to Well-Being; Experience of Indigenous Peoples with ICTs." In *Empowering Marginal Communities with Information Networking*, edited by Hakikur Rahman, 124–64. London: Idea Group Publishing.

———. 2005b. "Indigenous Peoples, Human Development and the Capability Approach." In *Proceedings of the 5th International Conference on the Capability Approach.* Paris: UNESCO.

Gómez, Ricardo, Patrik Hunt, and Emmanuelle Lamoureux. 1999. *Telecentre Evaluation: A Global Perspective.* Ottawa: IDRC.

Gopakumar, Kiran. 2007. "E-Governance Services through Telecenters: The Role of Human Intermediary and Issues of Trust." *Information Technologies and International Development* 4 (1): 19–35.

Gould, Elizabeth, and Ricardo Gómez. 2010. "Community Engagement and Infomediaries: Challenges Facing Libraries, Telecentres, and Cybercafés in Developing Countries." Paper presented at iConference 2010, Urbana Champaign, IL.

Goyal, Aparjita. 2010. "Information, Direct Access to Farmers, and Rural Market Performance in Central India." *American Economic Journal: Applied Economics* 2 (3): 22–45.

Guijt, Irene, and John Gaventa. 1998. "Participatory Monitoring and Evaluation: Learning from Change." IDS Policy Briefing 12, Institute for Development Studies, Brighton, November.

Gumucio Dragon, Alfonso. 2001. *Making Waves: Stories of Participatory Communication for Social Change.* New York: Rockefeller Foundation.

Gurstein, Michael. 2000. *Community Informatics: Enabling Communities with Information and Communications Technologies.* Hershey, PA: Idea Group Publishing.

———. 2003. "Effective Use: A Community Informatics Strategy beyond the Digital Divide." *First Monday* 8 (12): 1–27.

Hamelink, Cees J. 1994. *Trends in World Communication, on Disempowerment and Self-Empowerment.* Penang, Malaysia: Southbound, Third World Network.

———. 1997. "New Information and Communication Technologies, Social Development, and Cultural Change." UNRISD Discussion Paper 86, UNRISD, Geneva.

Hanna, Nagy. 1994. *Exploiting Information Technology for Development.* Discussion Paper 246. Washington, DC: World Bank.

Harris, Roger W. 1999. "Evaluating Telecentres within National Policies for ICTs in Developing Countries." In *Telecentre Evaluation: A Global Perspective; Report on an International Meeting on Telecentre Evaluation.* Ottawa: IDRC, September.

———. 2001. "Information and Communication Technologies for Rural Development in Asia: Methodologies for Systems Design and Evaluation." In *Proceedings, Conference on Information Technology in Regional Areas,* edited by Stewart Marshall, Wallace Taylor, and Xing Huo Yu. Central Queensland University, September.

Harris, Roger W., Poline Bala, Peter Songan, Elaine G. L. Kho, and Tingang Trang. 2001. "Challenges and Opportunities in Introducing Information and Communication Technologies to the Kelabit Community of North Central Borneo." *New Media and Society* 3 (3): 271–96.

Harris, Roger W., et al. 2003. "Sustainable Telecentres? Two Cases from India." In *The Digital Challenge: Information Technology in the Development Context,* edited by Santos Krishna and Shirin Madon, 124–35. Aldershot, U.K.: Ashgate.

Healy, Kevin. 2001. *Llamas, Weavings, and Organic Chocolate: Multicultural Grassroots Development in the Andes and Amazon of Bolivia.* South Bend, IN: Notre Dame University Press.

Heeks, Richard. 1999. "Information and Communication Technologies, Poverty, and Development." Development Informatics Working Paper 5, University of Manchester, Institute for Development Policy and Management, Manchester.

———. 2002. "I-Development not E-Development: Special Issue on ICTs and Development." *Journal of International Development* 14 (1): 1–11.

———. 2003. "Most E-Government-for-Development Projects Fail: How Can Risks Be Reduced?" Manchester University, Institute for Development Policy and Management, Manchester.

———. 2005. "Foundation of ICTs in Development: The Information Chain." e-Development Briefing 3, Manchester University, Institute for Development Policy and Management, Manchester.

————. 2010. "Do Information and Communication Technologies (ICTs) Contribute to Development?" *Journal of International Development* 22 (5): 625–40.

Heeks, Richard, and Subhash Bhatnagar. 1999. "Understanding Success and Failure in Information Age Reform." In *Reinventing Government in the Information Age,* edited by Richard Heeks, 49–74. London: Routledge.

Heeks, Richard, and Charles J. Kenny. 2002. "The Economics of ICTs and Global Inequality: Convergence or Divergence for Developing Countries?" Development Informatics Working Paper 10a/2002. Manchester University, Institute for Development Policy and Management, Manchester.

Heeks, Richard, and Carolyne Stanforth. 2007. "Understanding E-Government Project Trajectories from an Actor-Network Perspective." *European Journal of Information Systems* 16 (2): 165–77.

Hewitt de Alcántara, Cynthia. 2001. "The Development Divide in a Digital Age." Technology, Business, and Society Programme Issue Paper 4, UNRISD, Geneva.

Hopkins, Raul. 1995. *Impact Assessment: Overview and Methods.* Oxford: Oxfam/Novib.

Horton, Forest Woody. 1983. "Information Literacy vs. Computer Literacy." *Bulletin of the American Society for Information Science* 9 (4): 6.

Howes, Michael, and Robert Chambers. 1980. "Indigenous Technical Knowledge: Analysis, Implications, and Issues." In *Indigenous Knowledge Systems and Development,* edited by David Brokensha, D. M. Warren, and Oswald Werner. Lanham, MD: University Press of America.

Hudson, Heather. 1995. "Economic and Social Benefits of Rural Telecommunications: A Report to the World Bank." University of San Francisco. http://usf.usfca.edu/fac_staff/hudson/papers/Benefits%20 of %20Rural%20Communication.pdf.

————. 1999. "Designing Research for Telecentre Evaluation." In *Telecentre Evaluation: A Global Perspective*, edited by Ricardo Gómez and Patrik Hunt. Report of an international meeting on telecentre evaluation, Far Hills Inn, Quebec, September 28–30.

————. 2001. "Telecentre Evaluation: Issues and Strategies." In *Telecentres: Case Studies and Key Issues,* edited by Colin Latchem and David Walker, 169–82. Vancouver: Commonwealth of Learning.

Hunt, Patrik. 2001. "True Stories: Telecentres in Latin America and the Caribbean." *Electronic Journal of Information Systems in Developing Countries* 4 (5): 1–17.

ITDG (Intermediate Technology Development Group). 2001. "Enable People to Make Technologies Work for Them: ITDG's Response to the Human Development Report 2001." ITDG, London.

ITU (International Telecommunication Union). 2012. *Measuring the Information Society 2012.* Geneva: ITU.

Jack, William, and Tavneet Suri. 2014. "Risk Sharing and Transactions Costs: Evidence from Kenya's Mobile Money Revolution." *American Economic Review* 104 (1): 183–223.

James, Jeffrey. 2007. "From Origins to Implications: Key Aspects in the Debate over the Digital Divide." *Journal of Information Technology* 22 (3): 284–95.

Jenkins, Rob. 2002. "The Emergence of the Governance Agenda: Sovereignty, Neoliberal Bias, and the Politics of International Development." In *The Companion to Development Studies,* edited by Vandana Desai and Robert Potter. London: Edward Arnold.

Jensen, Mike, and Anriette Esterhuysen. 2001. "The Community Telecentre Cookbook for Africa: Recipes for Self-Sustainability." CI-2001/WS /2, UNESCO, Paris. http://unesdoc.unesco.org/images/0012/001230 /123004e.pdf.

Jensen, Robert. 2007. "The Digital Provide: Information (Technology), Market Performance, and Welfare in the South Indian Fisheries Sector." *Quarterly Journal of Economics* 122 (3): 879–924.

Kabeer, Naila. 1999a. "The Conditions and Consequences of Choice: Reflection on the Measurement of Women's Empowerment." UNRISD Discussion Paper 108, UNRISD, Geneva.

———. 1999b. "Resources, Agency, Achievement: Reflections on the Measurement of Women's Empowerment." *Development and Change* 30 (3): 261–302.

———. 2000. "Social Exclusion, Poverty, and Discrimination towards an Analytical Framework." *IDS Bulletin* 31 (4): 83–97.

Kanungo, Shivraj. 2003. "Information Village: Bridging the Digital Divide in Rural India." In *The Digital Challenge: Information Technology in the Development Context,* edited by Santos Krishna and Shirin Madon, 104–23. Aldershot, UK: Ashgate.

Katz, James Everett, and Ronald E. Rice. 2002. *Social Consequences of Internet Use: Access, Involvement, and Interaction.* Cambridge, MA: MIT Press.

Keniston, Kenneth. 2002. "Grassroots ICT Projects in India: Some Preliminary Hypotheses." *ASCI Journal of Management* 31 (1-2): 1–9.

Kenny, Charles. 2002. "The Costs and Benefits of ICTs for Direct Poverty Alleviation." World Bank, Washington, DC.

Kirkman, Geoffrey S., Peter G. Cornelius, Jeffrey D. Sachs, and Klaus Schwab, eds. 2002. *The Global Information Technology Report 2001–2002: Readiness for the Networked World.* New York: Oxford University Press.

Klein, Herbert S. 1982. *Bolivia: The Evolution of a Multi-Ethnic Society.* Oxford: Oxford University Press.

Kling, Rob. 2000. "Learning about Information Technologies and Social Change: The Contribution of Social Informatics." *Information Society* 16 (3): 217–32.

Korten, David, and Rudi Klauss, eds. 1984. *People-Centered Development: Contributions toward a Theory and Planning Framework.* West Hartford, CT: Kumarian Press.

Krishna, Anirudh. 2002. *Active Social Capital: Tracing the Roots of Development and Democracy.* New York: Columbia University Press.

Krishna, Santos, and Shirin Madon, eds. 2003. *The Digital Challenge: Information Technology in the Development Context.* Aldershot, U.K.: Ashgate.

Krishna, Santos, and Geoffrey Walsham. 2005. "Implementing Public Information Systems in Developing Countries: Learning from a Success Story." *Information Technology for Development* 11 (2): 1–18.

Kuriyan, Renee, Isha Ray, and Kentaro Toyama. 2008. "Information and Communication Technologies for Development: The Bottom of the Pyramid Model in Practice." *Information Society* 24 (2): 93–104.

Kuriyan, Renee, Kentaro Toyama, and Isha Ray. 2006. "Integrating Social Development and Financial Sustainability: The Challenges of Rural Computer Kiosks in Kerala." Paper presented at the ICTD'06, International Conference on Information and Communication Technologies and Development.

Labtone, Ronald. 1990. "Empowerment: Notes on Professional and Community Dimensions." *Canadian Review of Social Policy* 26: 64–75.

Lewis, David, and David Mosse, eds. 2006. *Development Brokers and Translators: The Ethnography of Aid and Agency.* Bloomfield, CT: Kumarian Press.

Loader, Brian D. 1998. "Cyberspace Divide: Equality, Agency, and Policy in the Information Society." In *Cyberspace Divide: Equality, Agency, and Policy in the Information Society,* edited by B. D. Loader, 1–15. London: Routledge.

Long, Norman. 2001. *Development Sociology: Actor Perspectives.* London: Routledge.

Long, Norman, and N. M. Villareal. 1994. "The Interweaving of Knowledge and Power in Development Interfaces." In *Beyond Farmer First: Rural People's Knowledge, Agricultural Research and Extension Practice,* edited by Ian Scoones and John Thompson. London: Institute of Development Studies.

Madon, Shirin. 1993. "Introducing Administrative Reform through the Application of Computer-Based Information Systems: A Case Study in India." *Public Administration and Development* 13 (1): 37–48.

———. 2000. "The Internet and Socio-Economic Development: Exploring the Interaction." *Information Technology and People* 13 (2): 85–101.

———. 2004. "Evaluating the Developmental Impact of E-Governance Initiatives: An Exploratory Framework." *Electronic Journal of Information Systems in Developing Countries* 20 (5): 1–13.

———. 2005. "Governance Lessons from the Experience of Telecentres in Kerala." *European Journal of Information Systems* 14 (4): 401–16.

Madon, Shirin, Nicolau Reinhard, Dewald Roode, and Geoff Walsham. 2007. "Digital Inclusion Projects in Developing Countries: Processes of Institutionalisation." IFIP WG9.4, ninth international conference "Taking Stock of E-Development," São Paulo.

Madon, Shirin, and Sundeep Sahay. 2000. "Democratic Governance and Information Flows: A Case Study in Bangalore." *Information, Communication, and Society* 3 (2): 173–91.

Mann, Catherine L. 2003. "Information Technologies and International Development: Conceptual Clarity in the Search for Commonality and Diversity." *Information Technologies and International Development* 1 (2): 67–79.

Mansell, Robin. 2001. "New Media and the Power of Networks." First Dixons Public Lecture and Inaugural Professorial Lecture, London School of Economics, London.

———. 2004. "Political Economy, Power, and New Media." *New Media and Society* 6 (1): 74–83.

Mansell, Robin, and Uta Wehn. 1998. *Knowledge Societies: Information Technology for Sustainable Development.* Oxford: Oxford University Press.

Marker, Phil, Kerry McNamara, and Lindsay Wallace. 2002. "The Significance of Information and Communication Technologies for Reducing Poverty." DFID, London.

Martínez-Cobo, José. 1986. *The Study of the Problem of Discrimination against Indigenous Populations.* Vols. 1–5. UN Document E/CN.4/Sub.2/1986/7. New York: United Nations.

Max-Neef, Manfred. 1991. *Human-Scale Development: Conception, Application, and Further Reflections.* New York: Apex Press.

McClure, C. R. 1994. "Network Literacy: A Role for Libraries." *Information Technology and Libraries* 13 (2): 116–17.

McConnell, Scott. 1998. "Connecting with the Unconnected: Proposing an Evaluation of the Impacts of the Internet." In *The First Mile of Connectivity: Advancing Telecommunications for Rural Development through a Participatory Communication Approach,* edited by Don Richardson and Lynnita Paisley. Rome: FAO.

McNeish, John Andrew. 2002. "Globalization and the Reinvention of History and Community in Highland Bolivia." *Journal of Peasant Studies, Special Issue on Latin American Peasantry* 29 (3-4, April-July): n.p.

Menchú, Rigoberta. 1984. *I, Rigoberta Menchú, an Indian Woman in Guatemala,* edited by Elisabeth Burgoes-Debray. London: Verso.

Menou, Michel. 1993. *Measuring the Impact of Information on Development.* Ottawa: International Development Research Centre.

———. 2001. "Digital and Social Equity? Opportunities and Threats on the Road to Empowerment." Paper presented at the LIDA (Libraries in the Digital Age) 2001 annual course and conference, Dubrovnik, Croatia, May.

————. 2002. "Information Literacy in National Information and Communications Technology (ICT) Policies: The Missed Dimension, Information Culture." White Paper prepared for UNESCO, Information Literacy Meeting of Experts, Prague, Czech Republic.

Menou, Michel, and Joseph Potvin. 2000. "Toward a Conceptual Framework for Learning about ICTs and Knowledge in the Process of Development." Background document, Global Knowledge Learning and Evaluation Action Program, Global Knowledge Initiative, Washington, DC.

Meyer, Carrie A. 1997. "The Political Economy of NGOs and Information Sharing." *World Development* 25 (7): 1127–40.

Michiels, Sabine I., and Loy Van Crowder. 2001. "Discovering the 'Magic Box': Local Appropriation of Information and Communication Technologies (ICTs)." SD Dimensions, FAO, Sustainable Development Department, Rome, June.

Miller, Daniel, and Don Slater. 2000. *The Internet, an Ethnographic Approach.* Oxford: Berg.

Moon, Jae. 2002. "The Evolution of E-Government among Municipalities: Rhetoric or Reality." *Public Administration Review* 62 (4): 424–33.

Morelli, Nicola. 2003. "Product-Service Systems, a Perspective Shift for Designers: A Case Study; the Design of a Telecentre." *European Journal of Information Systems* 24 (1): 73–99.

Mosse, David. 2004. "Is Good Policy Unimplementable? Reflections on the Ethnography of Aid Policy and Practice." *Development and Change* 35 (4): 639–71.

————. 2005. *Cultivating Development: An Ethnography of Aid Policy and Participation.* London: Pluto Press.

Musso, Juliet, Christopher Weare, and Matt Hale. 2000. "Designing Web Technologies for Local Governance Reform: Good Management or Good Democracy?" *Political Communication* 17 (1): 1–19.

Navas-Sabater, Juan, Andrew Dymond, and Niina Juntunen. 2002. "Telecommunications and Information Services for the Poor." Discussion Paper 432, World Bank, Washington, DC.

Negroponte, Nicholas. 1996. *Being Digital.* New York: Vintage.

Nelson, Diana. 1996. "Maya Hackers and the Cyberspatialised Nation State: Modernity, Ethnostalgia, and a Lizard Queen in Guatemala." *Cultural Anthropology* 11 (3): 287–308.

Nelson, Nici, and Susan Wright. 1995. "Participation and Power." *Power and Participatory Practice*, edited by Nici Nelson and Susan Wright. London: Practical Action.

Nulens, Gert. 2003. "The Digital Divide and Development Communication Theory." *Communication: South African Journal of Communication Theory* 29 (1–2): 68–78.

Nussbaum, Martha. 2000. *Women and Human Development: The Capabilities Approach.* Cambridge: Cambridge University Press.

Oakley, Peter, ed. 2001. *Evaluating Empowerment: Reviewing the Concept and Practice.* Oxford: INTRAC.

Oakley, Peter, Brian Pratt, and Andrew Clayton. 1998. *Outcomes and Impact: Evaluating Change in Social Development.* NGO Management and Policy Series 6. Oxford: INTRAC.

Ochs, Mary, Bill Coon, Darla Van Ostrand, and Susan Barnes. 1991. "Assessing the Value of an Information Literacy Program." Cornell University, Mann Library, Ithaca, NY.

OECD (Organisation for Economic Co-operation and Development). 2001. "Engaging Citizens in Policy-Making: Information, Consultation, and Policy Participation." Puma Policy Brief 10, OECD, Paris.

O'Farrell, Clare. 2001. "Information Flows in Rural and Urban Communities: Access, Processes, and People." University of Reading, IRDD, Reading, U.K.

O'Farrell, Clare, Patricia Norrish, and Andrew Scott. 1999. "Information and Communication Technologies (ICTs) for Sustainable Livelihoods." Preliminary study, ITDG, AERDD, London.

Okot-Uma, Rogers. 2001. "Electronic Governance: Re-inventing Good Governance." World Bank, Washington, DC.

Oliver, Ron, and Stephen Tower. 2000. "Benchmarking ICT Literacy in Tertiary Learning Settings." In *Learning to Choose: Choosing to Learn,* edited by Rod Sims, Meg O'Reilly, and Sue Sawkins. Proceedings of the 17th annual conference Australian Society for Computers in Learning in Tertiary Education, Southern Cross University, Coffs Harbour, December 9–14.

Orlikowski, Wanda J. 2000. "Using Technology and Constituting Structures: A Practice Lens for Studying Technology in Organizations." *Organizational Science* 11 (4): 404–28.

Orlikowski, Wanda J., and Suzanne C. Icono. 2001. "Research Commentary: Desperately Seeking the IT Artifact in IT Research: A Call to Theorizing the IT Artifact." *Information Systems Research* 12 (2): 121–34.

Panos Institute. 1998. "The Internet and Poverty." Panos Media Briefing 28, Panos Institute, London.

Partridge, William, Jorge Uquillas, and Kathryn Johns. 1998. "Including the Excluded: Ethnodevelopment in Latin America." In *Poverty and Inequality: Proceedings of the Annual World Bank Conference on Development in Latin America and the Caribbean,* 229–50. Washington, DC: World Bank.

Pigato, Miria. 2001. "Information and Communication Technology, Poverty, and Development in Sub-Saharan Africa and South Asia." Africa Region Working Paper 20, World Bank, Washington, DC.

Pohjola, Matti. 2002. "The New Economy in Growth and Development." *Oxford Review of Economic Policy* 18 (3): 380–96.

Powell, Mike. 1999. *Information Management for Development Organisations.* Oxford: Oxfam.

Prado, Paola. 2010. "Lighting up the Dark: Telecenter Adoption in a Caribbean Agricultural Community." *Journal of Community Informatics* 6 (3): n.p. http://ci-journal.net/index.php/ciej/article/view/727/604.

Pratchett, Lawrence. 1999. "New Technologies and the Modernization of Local Government: An Analysis of Biases and Constraints." *Public Administration* 77 (4): 731–50.

Proenza, Francisco, Roberto Bastidas-Buch, and Guillermo Montero. 2001. "Telecenters for Socioeconomic and Rural Development in Latin America and the Caribbean: Investment Opportunities and Design Recommendations with Special Reference to Central America." FAO/ITU/IADB Paper, FAO, Rome.

Psacharopoulos, George, and Harry Patrinos. 1994. *Indigenous Peoples and Poverty in Latin America: An Empirical Analysis.* Washington, DC: World Bank.

Putnam, Robert D. 1994. *Making Democracy Work: Civic Traditions in Modern Italy.* Princeton, NJ: Princeton University Press.

Quah, Danny T. 1997. "The Weightless Economy: The Mass of Evidence." Centre Piece, London School of Economics, Centre for Economic Policy, London, June.

Rawls, John. 1971. *A Theory of Justice.* Oxford: Oxford University Press.

Reilly, Katherine, and Ricardo Gómez. 2001. "Comparing Approaches: Telecentre Evaluation Experiences in Asia and Latin America." *Electronic Journal of Information Systems in Developing Countries* 4 (3): 1–17.

Rissel, Christopher. 1994. "Empowerment: The Holy Grail of Health Promotion?" *Health Promotion International* 9 (1): 39–47.

Robey, Daniel, and Marie Claude Boudreau. 1999. "Accounting for the Contradictory Organizational Consequences of Information Technology: Theoretical Directions and Methodological Implications." *Information Systems Research* 10 (2): 167–85.

Robinson, Scott. 1998. "Telecenters in Mexico: Learning the Hard Way." Presented at the conference "Partnerships and Participation in Telecommunications for Rural Development: Exploring What Works and Why," University of Guelph, Guelph.

Roche, Chris. 1999. *Impact Assessment for Development Agencies: Learning to Value Change.* Oxford: Oxfam.

Rodríguez, Francisco, and Ernest Wilson. 2000. *Are Poor Countries Losing the Information Revolution?* Washington, DC: World Bank.

Roman, Paul, and Royal D. Colle. 2002. "Themes and Issues in Telecentre Sustainability." Development Informatics Working Paper 10, University of Manchester, IDPM, January.

Rostow, Walt W. 1960. *The Stages of Economic Growth: A Non-Communistic Manifesto.* Cambridge, U.K.: Cambridge University.

Rothenberg-Aalami, Jessica, and Joyojeet Pal. 2005. "Rural Telecenter Impact Assessments and the Political Economy of ICT for Development (ICT4D)."

BRIE Working Paper 164, University of California, Berkeley. http://brie
.berkeley.edu/publications/WP%20164revised.pdf.

Rowlands, Jo. 1997. *Questioning Empowerment.* Oxford: Oxfam.

Ruxton, Robert. 1995. *Participation in Impact Assessment.* Oxford: Oxfam.

Salas, María A. 1994. "The Technicians Only Believe in Science and Cannot
Read the Sky: The Cultural Dimension of the Knowledge Conflict in the
Andes." In *Beyond Farmer First: Rural People's Knowledge, Agricultural
Research and Extension Practice,* edited by Ian Scoones and John Thompson.
London: Intermediate Technology Publications.

Schilderman, Timo. 2002. *Strengthening the Knowledge and Information
Systems of the Urban Poor.* London: DFID, ITDG. http://practicalaction
.org/docs/ia3/kis-urban-poor-report-2002.pdf.

Scoones, Ian. 1998. "Sustainable Rural Livelihoods: A Framework for
Analysis." IDS Working Paper 72, IDS, Brighton.

Scoones, Ian, and John Thompson, eds. 1994. *Beyond Farmer First: Rural
People's Knowledge, Agricultural Research, and Extension Practice.*
London: Intermediate Technology Publications.

Selwyn, Neil. 2004. "Reconsidering Political and Popular Understandings of
the Digital Divide." *New Media and Society* 6 (3): 341–62.

Sen, Amartya. 1984. "Well-Being, Agency, and Freedom: The Dewey Lectures
1984." *Journal of Philosophy* 82 (4): 169–221.

———. 1992. *Inequality Reexamined.* Cambridge, MA: Harvard University
Press.

———. 1993. "Capability and Well-Being." In *The Quality of Life,* edited by
Martha Nussbaum and Amartya Sen, 30–53. Oxford: Clarendon Press.

———. 1997. "Editorial: Human Capital and Human Capability." *World
Development* 25 (12): 1959–61.

———. 1999. *Development as Freedom.* New York: Knopf Press.

———. 2010. "The Mobile and the World." *Information Technologies and
International Development* 6 (special issue): 1–3.

Sey, Araba, and Michelle Fellows. 2009. "Literature Review on the Impact of
Public Access to Information and Communication Technologies." University
of Washington, Center for Information and Society, Seattle, WA.

Simpson, Lyn, Leonie Daws, and Barbara Pini. 2004. "Public Internet Access
Revisited." *Telecommunications Policy* 28 (3-4): 323–37.

Singh, J. P. 1999. *Leapfrogging Development? The Political Economy of
Telecommunications Restructuring.* Albany, NY: SUNY Press.

Skuse, Andrew. 2000. "Information Communications Technology, Poverty,
and Empowerment." Dissemination Note 3, DFID, London. http://www
.imfundo.org/knowledge/skuse.htm.

Slater, Don, Jo A. Tacchi, and Peter Lewis. 2002. "Ethnographic Monitoring
and Evaluation of Community Multimedia Centres: A Study of Kothmale
Community Radio Internet Project, Sri Lanka." DFID, London.

Soete, Luc. 1985. "International Diffusion of Technology, Industrial Development, and Technological Leapfrogging." *World Development* 13 (3): 409–22.

Soriano, Cheryl Ruth. 2007. "Exploring the ICT and Rural Poverty Reduction Link: Community Telecenters and Rural Livelihoods in Wu'an, China." *Electronic Journal of Information Systems in Developing Countries* 32 (1): 1–15.

Steinmueller, W. Edward. 2001. "ICTs and the Possibilities of Leapfrogging by Developing Countries." *International Labour Review* 140 (2): 193–210.

Stewart, Frances. 2001. "Horizontal Inequalities: A Neglected Dimension of Development." WIDER Annual Lecture 5, UNU-WIDER, Helsinki.

Stewart, Frances, and Severine Deneulin. 2002. "Amartya Sen's Contribution to Development Thinking." *Studies in Comparative International Development* 37 (2): 61–70.

Steyaert, Jo. 2002a. "Inequality and the Digital Divide: Myths and Realities." In *Advocacy, Activism, and the Internet,* edited by Steven Hick and John McNutt, 199–211. Chicago, IL: Lyceum Press.

———. 2002b. "Local Governments Online and the Role of the Resident: Government Shop versus Electronic Community." *Social Science Computer Review* 18 (1): 3–16.

Stiglitz, Joseph. 2007. *Making Globalization Work.* London: Penguin.

Stoll, Klaus, Michel J. Menou, K. Camacho, and Y. Khellady. 2004. "Learning about ICT's Role in Development: A Framework towards a Participatory, Transparent, and Continuous Process." IDRC, Ottawa.

Ströbele-Gregor, Juliane. 1996. "Culture and Political Practice of the Aymara and Quechua in Bolivia: Autonomous Forms of Modernity in the Andes." *Latin American Perspectives* 23 (2): 72.

Tacchi, Jo, Don Slater, and Gregory Hearn. 2003. *Ethnographic Action Research: A User's Handbook Developed to Innovate and Research ICT Applications for Poverty Eradication.* New Delhi: UNESCO.

Tacchi, Jo, Dan Slater, and Peter Lewis. 2003. "Evaluating Community-Based Media Initiatives: An Ethnographic Action Research Approach." Paper presented at the OUR Media III, Barranquilla, Colombia, May 19–21.

Tettey, Wisdom J. 2002. "ICT, Local Government Capacity Building, and Civic Engagement: An Evaluation of the Sample Initiative in Ghana." *Perspectives on Global Development and Technology* 1 (2): 165–92.

Thompson, Marc. 2004. "Discourse, 'Development,' and the 'Digital Divide': ICT and the World Bank." *Review of African Political Economy* 31 (99): 103–23.

UNCTAD (United Nations Conference on Trade and Development). 2013. *The Information Economy Report.* New York: United Nations.

UNDP (United Nations Development Programme). 2001. *Human Development Report 2001: Making New Technologies Work for Human Development.* Oxford: UNDP.

VAIPO (Vice Ministerio de Asuntos Indígenas y Pueblos Originarios). 1999. *Proyecto iniciativa plan de desarrollo indígena originario: Diagnostico nacional.* Informe Final. La Paz: VAIPO, January.

Wade, Robert. 2002. "Bridging the Digital Divide: New Route to Development or New Form of Dependency?" *Global Governance* 8 (4, October-December): 443–66.

Walsham, Geoffrey. 1993a. "The Emergence of Interpretivism in IS Research." *Information Systems Research* 6 (4): 376–94.

———. 1993b. *Interpreting Information Systems in Organizations.* Chichester: John Wiley and Sons.

———. 1995. "Interpretative Case Studies in IS Research: Nature and Method." *European Journal of Information Systems* 4 (2): 74–81.

———. 1998. "IT and Changing Professional Identity: Micro Studies and Macro Theory." *Journal of American Society for Information Science* 49 (12): 1081–89.

Wang, Georgette, and Wimal Dissanayake, eds. 1994. *Continuity and Change in Communication Systems: A Cross-Cultural Perspective.* New York: East-West Communication Institute.

Warren, Dennis M. 1991. "Using Indigenous Knowledge in Agricultural Development." Discussion Paper 127, World Bank, Washington, DC.

Warschauer, Mark. 2002. "Reconceptualizing the Digital Divide." *First Monday* 7 (7): n.p.

———. 2003. "Demystifying the Digital Divide." *Scientific American* 289 (2): 42–47.

———. 2004. *Technology and Social Inclusion: Rethinking the Digital Divide.* Cambridge, MA: MIT Press.

Whyte, Anne. 2000. *Assessing Community Telecenters.* Ottawa: IDRC, Acacia Initiative.

Wilson, Ernest. 1999. "Development of National Information and Communications Services: A Comparison of Malaysia and South Africa." *Journal of Developing Societies* 15 (April): 47–60.

Wilson, Ernest, John Daly, and J. M. Griffiths. 1998. "Internet Counts, Measuring the Impact of the Internet." National Research Council, Office of International Affairs, Washington, DC.

Wilson, Gordon, and Richard Heeks. 2000. "Technology, Poverty, and Development." In *Poverty and Development: Into the 21st Century,* edited by Tim Allen and Alan Thomas. New York: Open University; Oxford: Oxford University Press.

World Bank. 1995. *Harnessing Information for Development: Vision and Strategy.* Washington, DC: World Bank.

———. 1998. *World Development Report 1998–99: Knowledge for Development.* New York: Oxford University Press.

———. 2004. *World Development Report 2004: Making Services Work for Poor People.* New York: Oxford University Press.

————. 2006. *Information and Communications for Development 2006: Global Trends and Policies.* Washington, DC: World Bank.

Wresch, William. 1998. "Information Access in Africa: Problems with Every Channel." *Information Society* 14 (4): 295–300.

Yigitcanlar, Tan. 2003. "Bridging the Gap between Citizens and Local Authorities Via e-Government." In *Symposium on E-government, 10–12 May 2003.* Muscat, Oman.

Yin, Robert K. 1994. *Case Study Research: Design and Methods,* 2d ed. Thousand Oaks, CA: Sage.

Young, Judy, Gail Ridley, and Jeff Ridley. 2001. "A Preliminary Evaluation of Online Access Centres: Promoting Micro E-Business Activity in Small, Isolated Communities." *Electronic Journal of Information Systems in Developing Countries* 4 (1): 1–16.

Zheng, Yingqin. 2009. "Different Spaces for e-Development: What Can We Learn from the Capability Approach?" *Information Technology for Development* 15 (2): 66–82.

Zurkowski, Paul. 1974. "The Information Service Environment, Relationship, and Priorities." National Commission on Libraries and Information Science, Washington, DC.

CHAPTER 2

The Methodology

This chapter describes the methodology and methods used in the study. It begins by describing the case study approach, which uses triangulation—a combination of quantitative and qualitative methods of data collection—to cross-validate the findings. It then explains the definition of information and communication technologies (ICTs) adopted and describes the logistics regression analysis. This is followed by a description of the data collection, including information about the quantitative methods (non-user and user surveys) and the qualitative methods (focus groups, in-depth interviews, and participatory observations). It ends with a brief discussion of the methodological challenges encountered in conducting field research with indigenous communities in rural Bolivia.

Methodological Approach

In this study, I adopt an interpretivist research approach (Walsham 1995; Galliers and Land 1987) that considers the application of ICTs as a process intertwined with social change—that is, a sociotechnical process as understood by social construction theory (Callon and Law 1989). The research focuses on the sociopolitical, economic, and cultural context within which ICTs are shaped and interact with human actors (Orlikowski and Icono 2001; Walsham 1995).

This approach is influenced by naturalistic methods, which recognize reality as constructed, multiple, and holistic, rejecting assumptions of causality and value neutrality (Glymour et al. 1987). Thus the research focuses on interpretation rather than deduction (Rubin and Rubin 1995). It uses quantitative methods to study the effects of ICTs on indigenous peoples. The survey data are based on people's perceptions of reality and thus reflect their interpretation of reality. The results from the quantitative research are then triangulated with qualitative research to contextualize the findings socially and politically.

The epistemology of the research stems from social institutional constructivism, contending that reality is socially conditioned and constructed. It analyzes the use of ICTs and the effects on indigenous peoples' well-being through the lens of Bolivia's broader sociopolitical framework, environment, and institutions.

Within the contextual view of the social embeddedness of technology, I use a case study approach that allows me to study in more depth the specific factors that influence whether or not ICTs can enhance indigenous peoples' human and social capabilities. Defined by Yin (1994), the case study approach investigates a "contemporary phenomenon within its real-life context, addresses a situation in which the boundaries between phenomenon and context are not clearly evident, and uses multiple sources of evidence" (Yin 1994, 23).

A principal hypothesis of the study is that the intermediation process of "how" ICTs are introduced to local communities is a critical factor for ICT programs to enhance indigenous peoples' well-being. Thus it is significant "who" or "which organization" is supporting the ICT intervention. Both of these questions lend themselves to carrying out case studies, since the process itself is at the core of the investigation, as are the conditions under which a community or organization is successful in making use of ICTs (Yin 1994). Furthermore, the theoretical framework highlights that the extent to which ICTs can enhance indigenous peoples' well-being is closely related to the broader socioeconomic, political, and cultural context in which the ICT program is embedded.

The case study approach has been criticized for its overt specificity, intense focus, and—by definition—inability to say anything useful beyond that focus. However, according to Yin—a great advocate of case studies—"case studies, like experiments, are generalizable to theoretical propositions and not to populations or universes," such that "the investigator's goal is to expand and generalize theories and not to enumerate frequencies" (Yin 1994, 10). I use the case study approach for the following reasons: (a) its emphasis on studying the broader context is consistent with my theoretical framework, (b) it enables me to make analytical generalizations about the effects of ICT use on indigenous peoples' well-being, and (c) it allows me to investigate in more depth the process of "how" ICTs have an impact on individuals and groups at the community level, emphasizing the linkages between local processes and the broader sociopolitical and economic context (Pettigrew 1990; Rossman and Wilson 1985). I therefore broaden the focus from the immediate local context to include the wider social, political, and cultural conditions (Avgerou 2002; Orlikowski 2000; Walsham and Sahay 2006; Ciborra and Lanzara 1994; Madon 1993).

Based on the theoretical framework, the methodological approach of the study combines conventional approaches to program evaluation (surveys) with participatory impact evaluation methodologies, using both quantitative and qualitative research methods (Bamberger 2000; Marsland et al. 1998). The principal advantage of using a mixed-methodological approach is that it improves the trustworthiness (internal validity), transferability (external validity), and reliability (dependability) of the results (Carvalho and White 1997; Rossman and Wilson 1985). Described by Denzin (1978), this methodological approach employs multiple methods to reveal various aspects of a single reality. The variety of research methods and research levels enables each to complement and verify each other and thus paints a fuller picture (Silverman 2000). Therefore, using a combination of quantitative and qualitative methods offers complementary ways of approaching the same research problem and helps to reveal multiple aspects of a single reality (Silverman 2000; Denzin 1978). Box 2.1 briefly describes the key methodological aspects of logistic regression analysis.

Combining different methods enables me to address my principal research questions better in several ways. First, the quantitative research methods enable me to understand better the conditions under which indigenous peoples

---

BOX 2.1 **Some notes on logistic regression analysis**

In chapters 6 and 7, I use logistic regression analysis to analyze the various factors that influence indigenous peoples' uses of ICTs and their ICT capabilities. This method allows me to develop a series of analytical models in order to discern the effect of each independent variable—such as users' demographic factors (age, gender, profession, ethnicity) and socioeconomic attributes (income)—on the dependent variables of indigenous peoples' ICT uses and ICT capabilities.

An important strength of regression analysis is that it allows the researcher to predict the probability of arriving at a particular dependent variable on the basis of continuous or categorical independent variables (Cohen et al. 2003; Menard 2002; Pampel 2000). It also enables the researcher to measure the magnitude of the interrelationship between the independent and dependent variables (Cohen et al. 2003). In general, a logistic regression model describes for a dichotomous Y variable (dependent variable) with values of 1 and 0 the log odds that $Y = 1$ is a function of the values of the predictors (independent variables). Providing the values of the beta coefficients and odds ratio enables the researcher to assess the relative importance of the independent variables in determining the model's dependent variable. The beta coefficients in logistic regression indicate the amount of change in the standardized scores of Y (dependent variable) for a one-unit change in the standardized scores of each predictor (independent variable), while controlling for the effects of all other independent variables. The odds ratio signifies the amount by which the odds of being in the case group are *multiplied* when the predictor is increased by a value of one unit (Cohen et al. 2003). An odds ratio of 1.00 corresponds to a regression coefficient of 0, indicating the absence of a relationship with Y. An odds ratio of 1.50, for example, indicates that the odds of a relationship are multiplied by 1.50 each time the independent variable X increases by one unit. Odds ratios below 1.00 relate to negative coefficients and signify that the odds of a relationship decrease as the independent variable X increases. According to Cohen et al. (2003, 493), "Each of the regression coefficients in multiple logistic regression models is only a *partial regression coefficient*; each needs to be interpreted adjusting for other effects in the model."

In logistic regression model analysis, it is critical to evaluate the model's adequacy by using alternative models to determine which one best explains the relationship between the independent and dependent variables. The analysis uses the statistical measure of the log likelihood R^2 in order to determine the extent to which a specific variable enhances the model fit of a particular logistic regression. According to DeMaris (1995, 963), the log likelihood R^2 "indicates the relative improvement in the likelihood of observing the sample data under the hypothesized model by including additional variables as predictors, compared with a model that contains only one intercept." The log likelihood R^2 ranges between the values of 0 and 1, where 0 means that the independent variables are useless in predicting the dependent variable and 1 indicates a model that predicts the dependent variable with perfect accuracy. R^2 is a useful measure for comparing across different models and between different sets of independent variables in a single hierarchical model.

can access and make meaningful use of ICTs. As such, the survey results reveal detailed information about the different factors such as socioeconomic status, gender, and literacy that influence people's uses of and capabilities to use ICTs. Second, the qualitative methods allow me to analyze in more depth indigenous peoples' expectations and perceptions about the impact of ICTs on their well-being and to understand whether or not the process of introducing ICTs into indigenous communities has a positive impact on their human and social capabilities. In particular, to understand the social and cultural impact of ICTs on indigenous communities, it is critical to analyze the process of how ICTs are being introduced, rather than to study the situation in one specific moment of time, as is done through a survey. The qualitative methods also allow me to investigate how ICTs affect existing power relationships within the community and whether or not the potential benefits resulting from the use of ICTs are being shared equally within communities. The qualitative methods strengthen the validity and reliability of the findings from the quantitative data by enabling me to interpret some of the associations found through the quantitative results and to understand some of the patterns identified in the surveys. Furthermore, the qualitative methods allow me to identify gaps and inconsistencies in the quantitative data and to explore more complex issues related to indigenous peoples' well-being, such as cultural identity, trust, and power relationships, which cannot be analyzed with quantitative research methods.

Due to the dynamic interaction between technology and people, the methodology for assessing impact has to incorporate strong elements of participatory evaluation. The intangible character of information and ICTs as well as the complex relationship between ICTs and well-being makes it impossible to measure direct impact. Instead, the effects of introducing ICTs into communities are much more complex and generate an indirect, if lasting, impact (Williams and Edge 1996).

I therefore draw on the participatory evaluation literature and apply its principal concepts to the study of ICTs (Oakley, Pratt, and Clayton 1998; Roche 1999; Pretty et al. 1995; Mayoux and Chambers 2005; Hulme 2000). I use the following definition from Oakley, Pratt, and Clayton (1998, 36): impact assessment refers to "an evaluation of how and to what extent development interventions cause sustainable changes in living conditions and behaviors of beneficiaries and influence the socioeconomic and political situation in society. In addition it refers to understanding, analyzing, and explaining processes and problems involved in bringing about change."

This approach is consistent with the alternative evaluation framework (AEF), which incorporates the informational dimension of the sustainable livelihoods approach and places at the center of the analysis the institutional processes of the changes that can result from the introduction of ICTs. People, technology, and the existing local institutional and socioeconomic context are closely intertwined with each other. Oakley's definition of impact assessment focuses the evaluation of development interventions on the process of change at both the individual and the society-wide level.

In particular, I draw on the participatory impact assessment methodology developed by Michel Menou (1993, 1999) for evaluating ICT projects funded by Canada's International Development Research Centre, a research and donor agency with an important program of ICT for development. Menou defines an ICT program's impact as "the change in the ability of people to improve their living conditions brought by the outcomes of the use of the Internet or other means of ICTs" (Menou 1999, 48). This approach stresses that a program's beneficiaries have to play a central role in the evaluation process itself. The main reason for choosing such a participatory approach has been that it allows me to place indigenous peoples' views and perceptions about the various effects of ICTs at the center of the evaluation (Oakley, Pratt, and Clayton 1998). Furthermore, it is consistent with the AEF, which places the expansion of people's human capabilities and development outcomes—not immediately quantifiable outputs—at the center of the evaluation of ICT programs.

Moreover, it stresses the need to analyze the effects of ICTs through a more holistic approach, emphasizing the importance of socioeconomic conditions and the cultural and local context of communities. For the data analysis, emphasizing development outcomes instead of outputs means that changes in quantitative indicators are not assessed in their own right, but through the contribution they make to the capabilities and well-being of indigenous communities. This approach allows me to investigate the lasting transformation of people's capabilities and livelihood opportunities, rather than focus on the informational and technological issues involved with ICT use.

According to Menou (1993, 1999), this methodology has the following features:

- The importance of including participatory mechanisms in order to involve representatives of all stakeholder groups in the assessment at different stages
- The necessity to gather detailed baseline data, in order to have a thorough understanding of the community's needs and development goals at the outset of the ICT project
- The emphasis on the ambivalence of the impact phenomena, thus tying together both positive and negative aspects of ICTs
- The use of a variety of methods and perspectives
- The emphasis on the assessment as a collective, participatory, and continuing learning process among all stakeholders, rather than exclusively on the internal management requirements of funding agencies.

In sum, the methodology used in this study combines features of more conventional evaluations with participatory approaches. From the conventional approach, it includes quantitative methods in the analysis and uses a conceptual framework that defines at the outset the expected relationships between independent (program outputs) and dependent (well-being outcomes) variables. At the same time, it is people centered, underscoring the need to base ICT interventions on real needs and on the communities' own development goals and priorities. This view stresses the importance of understanding the local and cultural context at the outset of the ICT intervention and the need to carry out detailed baseline studies before introducing ICTs. The consultative process is critical, allowing community members to express their own views on development and to develop community plans for how to reach these desired outcomes. During the fieldwork, it also facilitated an environment of mutual learning and trust both for community members and for me.

I use the results from the national indigenous peoples profile report (VAIPO 1999) as a baseline. As mentioned in chapter 1, the VAIPO report provides an in-depth description of indigenous peoples' own definitions and perceptions of human development and well-being based on the results of 46 focus groups in indigenous communities from all nine departments of Bolivia. The focus groups were facilitated by indigenous professionals who were fluent in the specific indigenous language of the consulted group and were trained and experienced in applying participatory rural appraisal methodologies in indigenous communities in rural Bolivia (Chambers 1997; Cornwall and Pratt 2002). Furthermore, the facilitators were thoroughly familiar with the specific cultural, social, and sociopolitical circumstances of the participating communities. The participants included indigenous men (60 percent), women (40 percent), and youth. I also participated in the study as the consultant responsible for coordinating the local consultation workshops (1997–99). The project's reach was national, representing the diversity of indigenous peoples in Bolivia, and I use the results as my baseline. In particular, I use the report's description of indigenous peoples' perceptions of and views on human development and well-being as the foundation for applying the AEF at the community level (see appendix B).

Definition of ICTs

Although the study uses Hamelink's broad definition of ICTs as "encompassing all technologies that enable the handling of information and facilitate different forms of communication among human actors, between human beings and electronic systems, and among electronic systems" (Hamelink 1997, 3), the survey instruments (user and non-user) specifically studied indigenous peoples' views on the impact of the Internet on their well-being. This decision was made in order to focus the questionnaire on the specific issues related to indigenous peoples' Internet uses and to include clear and concrete questions in the survey. Moreover, different forms of ICTs—radio, personal digital assistants, mobile phones, and the Internet—are bound to have different impacts on people's well-being, and I wanted to gain an in-depth understanding of one technology's impact on well-being and to unpack the process as such. Finally, due to time and financial constraints, it was not possible to carry out a survey on all forms of ICTs. This approach enabled me to concentrate my quantitative data analysis on issues related to the impact of the Internet on indigenous peoples' well-being.

Nevertheless, the non-user survey included questions related to traditional forms of ICTs (that is, community radio, amateur radio), in order to obtain a comprehensive overview of the existing information and communication ecologies within indigenous communities. The qualitative research gathered indigenous peoples' perceptions of ICTs more generally. The primary focus, however, continued to be the impact of the Internet on well-being, taking into account the important role that traditional forms of ICTs play in most indigenous communities. It was possible to take this broader approach since the qualitative methods, in contrast to the quantitative methods, provided me with the flexibility to discuss with indigenous peoples what is meant by ICTs *before* exploring their perceptions of the impact of different technologies on their well-being.

Methods of Data Collection

One of the key challenges I encountered in operationalizing my research stemmed from the limited empirical evidence available on factors that influence the impact of ICTs on the well-being of indigenous peoples. Figure 2.1 provides an overview of the methods used to gather primary data to validate the propositions made in the AEF. During my 20-month period of fieldwork in Bolivia, I used four methods to gather primary data: surveys, focus groups, participatory observation, and in-depth interviews.

Surveys with ICT Users and Non-users

From December 2004 to August 2006, I conducted a series of surveys with ICT users and non-users. The non-user survey was conducted to assess the existing information ecology and penetration of ICT use within indigenous communities prior to an ICT intervention. Specifically, it was used (a) to explore the local information needs and informational capital of indigenous peoples and communities and (b) to investigate the existing information and communication patterns as well as uses of ICTs by indigenous peoples. The user survey was conducted to study indigenous peoples' perceptions about the impact of the uses of ICTs on their well-being. These objectives are derived from the theoretical framework, which has placed the study of people's individual and collective capabilities at the center of the study. Specifically, the user survey was conducted (a) to evaluate the extent to which an ICT program has enhanced indigenous peoples' ICT capabilities, (b) to investigate the extent to which use of the Internet has improved the participants' informational capabilities, and

FIGURE 2.1   Research methods used to gather data to validate alternative evaluation framework at different stages of ICT projects

a. Research methods

LOCAL CONTEXT	LIVELIHOOD RESOURCES	INSTITUTIONAL PROCESSES	CAPABILITIES	LIVELIHOOD OUTCOMES
Non-user surveys Participatory observation	Non-user surveys Participatory observation Focus groups with non-users	ICT user surveys Focus groups with ICT users	ICT impact survey Focus groups with ICT users In-depth interviews	ICT impact survey In-depth interviews Participatory observation

b. Stages of ICT project

INFORMATION ENVIRONMENT	INFORMATION NEEDS ASSESSMENT	ICT USES ICT CAPABILITIES	INFORMATIONAL CAPABILITIES	HUMAN AND SOCIAL CAPABILITIES
Existing information Systems and environments	Assess Information needs Informational capital	Community ICT access Local and relevant content Capacity-building	Local appropriation and use of ICTs	Ownership Sustainability

(c) to investigate indigenous peoples' perceptions of how these acquired informational capabilities have expanded their human and social capabilities and enhanced their individual and collective well-being.

The research thus uses surveys as a means to gain insight into the factors that introduce indigenous peoples' non-use and use of ICTs, as well as to gain a basic understanding of their perceptions and views about the impact that ICTs have had on their well-being. This is the area where I felt that surveys could add the greatest value for my research.

At the same time the use of surveys as a data collection method has several challenges. First, to obtain good results the questionnaire needs to be as clear and simple as possible for the participants to be able to understand the questions easily. Second, the validity of survey results depends on a high response rate. Furthermore, surveys generally offer informants few opportunities to reflect on their responses and to check the accuracy of their answers (Hoggart, Lees, and Davies 2002, 199).

I addressed these challenges in the following ways. First, I used personal surveys to collect the quantitative data. This method allowed me to interact directly with the respondents and yielded a very high response rate. The respondents were all participants in ICT programs who attended either a training workshop or a focus group organized by one of the intermediary organizations with which I worked closely together during my research.

Second, I conducted a pilot survey, which enabled me to test this instrument and to make important adjustments to my questionnaire before applying it to a much broader range of participants. I carried out the pilot user survey in December 2004 with 60 indigenous leaders of Latin America who were attending the general assembly of the Fondo Indígena, the principal indigenous-led international organization representing indigenous peoples in Latin America.

The survey contained questions related to (a) demographics (gender and age), (b) information and communications needs, (c) access to ICTs, (d) experiences in ICT uses, (e) general attitudes, and (f) main obstacles to ICT uses. The preliminary results allowed me to gain a general overview of how and for what purposes indigenous leaders were using ICTs and to identify key issues surrounding their use. This pilot also provided me with the opportunity to test my preliminary questionnaire, and the data analysis enabled me to improve the final survey. Finally, it provided an opportunity to discuss with indigenous leaders the main objectives of my research and to consult them on the design of my questionnaire. I carried out a small focus group during the general assembly with 10 indigenous leaders to discuss in detail the objectives and design of the survey. As a result of this focus group, I added several questions suggested by indigenous leaders.

I also used triangulation as a means to validate my findings. The research combined the quantitative research methods with several qualitative research methods in order to offset the bias in any one method by using the other method. The key here is the systematic application of qualitative methods. "The accuracy of a method hereby comes from its systematic application, but rarely does the inaccuracy of one approach to the data complement the accuracy of another" (Fielding and Fielding 1986). The overall purpose of such a mixed-methods approach is to recognize the need for a well-structured methodology, the importance of not being confined to a structured set of tools and question areas, and the need to be open to exploring both positive and negative impacts of ICTs on people's well-being (Oakley, Pratt, and Clayton 1998).

The total sample size of the surveys is 513, and almost three-fourths of the sample ($n = 365$, or 71 percent) are *non-ICT users*, and only one-fourth ($n = 148$, or 29 percent) are *ICT users*. The analysis of the user survey is divided into two parts because I used two different user questionnaires during my fieldwork. After applying a first series of user surveys ($n = 57$), I expanded the questionnaire to include more specific questions on the impact of ICTs on people's human and social capabilities ($n = 91$). During the data analysis it was possible to aggregate both questionnaires' data for the first set of questions, which had remained unchanged and were related to demographic and socioeconomic characteristics as well as level of ICT use. Thus the total sample for the first sections of the survey amounted to 148. The second set of questions, which asked about the impact of ICTs on various aspects of people's well-being, were too different and thus were analyzed separately. This subset of the sample totaled 91, so a meaningful data analysis was still possible.

I consulted a broad range of intermediary organizations to obtain lists of Bolivia's indigenous leaders and representatives and lists of participants in different ICT programs. These contact lists served as the sampling frame for defining a sample for both the user and non-user surveys. The sampling frame is the list of all subjects in the sample, wherein, ideally, every person should be listed only once. The accuracy of a sample depends to a large extent on the quality of the sampling frame from which it is obtained (Bailey 1994). The research used a stratified cluster sampling to identify the sample. For collecting the survey data, the intermediary organizations facilitated lists of respondents from the different ICT programs. These lists were representative of Bolivia's overall indigenous population.

Indigenous organizations such as the Fondo Indígena, Confederación de Pueblos Indígenas de Bolivia (CIDOB), and Consejo Nacional de Ayllus y Markas del Qullasuyu (CONAMAQ) base their work on support from different national, regional, and local indigenous organizations. These are largely membership-driven organizations; their leadership is elected by a general assembly held every two years. Their main purpose is to "represent" the interests of various indigenous groups. They face significant pressures from the grassroots to engage in policy dialogue with the government and to implement programs beneficial to their communities. Thus these organizations are in permanent contact with the different levels of indigenous leadership and are experienced in balancing the diverse interests of indigenous communities and in organizing workshops that include a broad range of actors and indigenous representatives from all regions and ethnic groups in Bolivia. I was quite aware during my fieldwork that this method might introduce a certain bias in the sample, since each intermediary organization might have a loyal base and tend to work with certain groups of indigenous representatives. For instance, the Fondo Indígena—an international indigenous organization working across Latin America—is frequently perceived as being too far removed from the grassroots and as representing the new "elite indigenous leadership." However, it has perhaps the most comprehensive list of indigenous organizations in Bolivia and is careful not to insult its constituency; it sent workshop invitations to all of them. This still indicates a bias toward indigenous leaders and the Fondo Indígena's definition of Bolivia's most important indigenous organizations. To address these possible gaps, I collaborated with a broad range of intermediary organizations that work at different levels, including in selected communities at the grassroots level (*Organización de Mujeres Aymarás del Kollasuyo* [OMAK], *Coordinadora de Integración de Organizaciones Económicas Campesinas de Bolivia* [CIOEC]), in a specific subregion of Bolivia (*Instituto de Capacitación del Oriente* [ICO]), at the national level (CONAMAQ), and at the international level (Fondo Indígena). Finally, I cross-validated the demographics and socioeconomic statistics of my survey sample against the overall indigenous population of Bolivia and found that it was fairly representative, except in terms of gender and youth.

There is a significant gender gap in the survey sample: 70 percent of the non-user sample and 74 percent of the user sample are male. Unfortunately, this gender gap is not the result of a sampling problem; instead, it speaks to the unequal access of indigenous women to spheres of power and leadership positions in Latin America, similar to the plight of non-indigenous women in the region. To address this issue, I carried out several focus groups specifically targeting indigenous women and have included a case study of an OMAK project that promotes the use of ICTs for women's empowerment.

A second potential bias in my sample could be attributed to low levels of youth participation; this is because youth are traditionally excluded from positions of power and leadership within the indigenous movement. I therefore made special efforts to include youth by carrying out two targeted surveys, one with OMAK, the grassroots organization in the highlands, and one with CIDOB, the indigenous organization in the lowlands. As a consequence, 19 percent of the full sample of non-users is under 18 years of age and 24 percent of the sample of users is under 25 years of age. The relatively high percentage of youth in the user sample is due to the relative ease with which younger people use new technologies such as the Internet (Norris 2001).

Focus Groups

The surveys were complemented by 16 focus groups targeted to specific clusters—indigenous leaders, indigenous youth, and indigenous women—in order to understand their specific perspectives. Appendix J provides a detailed list of the 16 focus groups, as well as the location, target group, date, and number of participants for each group. During the research, it became clear that the views of indigenous men and women vary significantly and that the focus groups would play an important role in enriching the analysis and survey results. The focus groups enabled me to deepen my understanding of the expectations, attitudes, and perceptions about ICTs of different groups within indigenous communities. The group discussions were well suited to capture the opinions and perspectives of certain groups regarding the process of introducing ICTs into their communities. Furthermore, they were an effective method of soliciting the responses of a group and not just individuals. Focus groups frequently show that the perceptions of a group are more than the sum of individual views; during group conversations, one set of ideas can be set off by other thoughts and exchanges (Hoggart, Lees, and Davies 2002, 213). The research benefited greatly from these features of focus groups, since the views of a certain subgroup (for example, the women's group) often have very strong value for the individuals in it.

The focus groups were carried out with both non-users and users of ICTs in order to understand the informational ecology in indigenous communities prior to the introduction of ICTs and to analyze the effects of ICT use on individuals and communities. In the focus groups with non-users, the main topics discussed were related to the existing patterns of information and communication, channels of information, and principal information needs in their communities. Furthermore, all of the focus groups discussed expectations about the role that ICTs—in particular, the Internet—could play in providing indigenous peoples with new opportunities for their development and whether ICTs could help to reduce their social exclusion within Bolivian society.

The main topics discussed in the focus groups with ICT users were related to people's perceptions about how their participation in an ICT program has enhanced their ICT and informational capabilities and to what extent these newly acquired capabilities have improved their human capabilities in the multiple areas of their lives and thus enhanced their well-being. The focus groups also discussed the main barriers and obstacles that indigenous people have faced in the use of ICTs and the role that intermediary organizations have played in supporting them to overcome these obstacles.

As shown in appendix J, the focus groups covered all five types of intermediary organizations, in order to obtain a broad range of opinions. The groups were made up of 9 to 15 people who were asked to discuss in greater depth the issues associated with ICT intervention.

A key challenge of focus groups is the need to facilitate the interaction among group members carefully in order to draw out those who are not participating and to manage those who might otherwise dominate the discussions. Another issue is that the dynamic of the group discussion is frequently shaped by the intervention of the researcher (Hoggart, Lee, and Davies 2002, 217).

To address these challenges, each focus group was supported by a local indigenous facilitator who intervened as little as possible; my role as the researcher was limited to being available to clarify questions. The focus groups were organized with a subsample of the same individuals who had participated in the surveys. The sample was much smaller than the survey sample, was spread across fewer communities, and included representatives of national indigenous organizations (indigenous leaders from CIOEC) and more grassroots representation from four communities in which the research carried out its case studies (for example, Tiwanaku, Batallas, Comarapa and Saipina).

To enhance the validity of the data, I cross-checked the main findings from the focus groups with the data gathered from the in-depth interviews and participatory observations. Furthermore, the focus groups allowed me to capture the opinion and perspective of indigenous women, who were significantly underrepresented in the surveys due to their unequal access to spheres of power in the indigenous leadership.

In-Depth Interviews

I also conducted 30 open-ended in-depth interviews with a broad range of stakeholders, including government policy makers, national indigenous leaders, local mayors, representatives of nongovernmental organizations (NGOs), representatives of donor agencies, and community leaders. Appendix I provides a detailed list of the interviewees, including their names, title, affiliation, and date of the interview. All of the interviewees have extensive experience in the area of ICTs for indigenous development in Bolivia, have worked directly in one of the ICT programs studied, represent the position of a policy maker who has influenced the outcome of an ICT program, or form part of the newly emerging indigenous leadership in Bolivia.

A key objective of the interviews was to test the propositions and principal assumptions formulated in the AEF before beginning the fieldwork, as well as to validate and place in a much broader context the findings from the quantitative data gathered through the non-user and user surveys. Based on the theoretical framework, the specific objectives of the interviews were to analyze (a) institutional processes of introducing ICTs into indigenous communities, (b) major barriers in providing ICT services to rural and indigenous communities, (c) the effects of a specific ICT program on indigenous communities, and (d) the expectations and perspectives of different stakeholders regarding the role of ICTs for economic and social development.

In-depth interviews allowed me to gain a much richer and in-depth understanding of people's perceptions and views than surveys and helped me to overcome the abstraction inherent in quantitative studies (Firestone 1987). According to McCracken (1988, 21), in-depth interviews enable the researcher to establish a close personal relationship with the interviewee and are "most useful and powerful when they are used to discover how the respondent sees the world." He also emphasizes the need to avoid any interference on the part of interviewers in this process: "This objective of the method makes it essential that testimony be elicited in as unobtrusive, nondirective manner as possible … . An error can prevent the capture of the categories and the logic used by the respondent. It can mean that the project ends up 'capturing' nothing more that the investigator's own logic and categories" (McCracken 1988, 21).

To address this challenge, throughout the book I employ direct quotations from the people interviewed to minimize the influence of my personal view. Although I chose from a variety of quotes, as far as possible, I inform the reader whether the chosen statement represents a typical or an anomalous position. To enhance the validity of the data, I cross-checked the main findings from the

in-depth interviews with the data gathered from the focus groups and participatory observations.

Participatory Observation

I have been professionally involved in programs related to ICT for indigenous
peoples' development in Bolivia since 1999. At the time of the fieldwork, my
continuous 10-year engagement had allowed me to build relationships with a
broad range of contacts from government, nongovernmental, and indigenous
organizations. During my fieldwork, I collaborated closely with a broad range
of intermediary organizations and spent a considerable amount of time with
project staff and beneficiaries.

The techniques used in the participant observations included informal
interviews, direct observations, and informal conversations. I kept in frequent
contact with project managers and indigenous leaders, participated in planning meetings and seminars, and had many informal conversations with them
about my research—especially for gathering data for the case studies. On
five occasions, I spent three to four weeks in indigenous communities. This
time improved my understanding of the socioeconomic and political context
of these communities and allowed me to discover the everyday practices of
community members with regard to the use of new forms of ICTs. My frequent visits to computer centers (telecenters) allowed me to learn more about
"how" and "for what purposes" people were using the Internet and computers. Finally, the insights gained through participant observations helped me
to develop the specific questions asked in the survey, interviews, and survey
groups. A key challenge of participatory observation is that it requires the
researcher to establish a relationship of mutual trust with the participants and
that the presence of the researcher frequently has an important influence on
the subject being studied.

Collaboration with Intermediary Organizations

Given critical constraints of budget, time, and practical logistics due to the significant geographic distances within Bolivia, I collaborated with several organizations that were implementing ICT programs. This collaboration allowed
me an "inside look" into the different strategies, processes, and approaches
used by intermediary organizations in implementing ICT programs. My data
collection strategy was linked directly to the hypothesis that intermediary
organizations are crucial in determining whether or not an ICT program has a
positive impact on the well-being of indigenous peoples.

Furthermore, an important reason for taking this approach is that, due to
the long history of social and political exclusion of indigenous peoples as well
as their very high levels of poverty, many have a deep-seeded mistrust of outsiders (*mestizos* and foreigners) and are often reluctant to open up to people
outside their communities. Moreover, many indigenous communities are linked
to regional and national indigenous organizations and will only engage with
outsiders who have been "endorsed" by them. In addition, an important reason for collaborating with different organizations was to communicate better

with local indigenous leaders and community members. Although the majority of indigenous people with whom I interacted during my fieldwork were fluent in Spanish and their indigenous language, it was helpful to be accompanied by someone who spoke the local indigenous language. Frequently, addressing community members in their own language—even when this is done by a third party—is important in "breaking the ice" and establishing a level of trust with community members.

This approach also entailed several challenges. First, it required me to spend a significant amount of time developing a relationship with each intermediary organization before starting the fieldwork. Second, during the fieldwork, indigenous peoples frequently associated me with a specific program, even though I explicitly explained that I was an independent researcher and only collaborated with the organization on certain tasks (trainings and capacity-building workshops) unrelated to the interviews and survey work. This misconception may have skewed the results of the surveys: respondents may have viewed the surveys and interviews as possibly leading to future projects and financing for their communities. In addition, Bolivia's extremely high dependence on aid has led to the perception that foreigners are potential donors, a factor that may have led respondents to present problems more negatively or to describe the impact of programs in a more positive light in hopes of making a case for additional support.

Typology of Intermediary Organizations

I collaborated with eight intermediary organizations during my fieldwork (table 2.1). The organizations are grouped according to the following five types: (a) grassroots indigenous organizations, (b) national indigenous organizations, (c) international indigenous organizations, (d) NGOs, and

TABLE 2.1   Type of organization, by region and level of intermediation

Type of organization	Highlands	Lowlands
Grassroots organizations (high-level intermediation)	CIOEC: Oruro CIOEC: Potosí OMAK: Tiwanaku[a] OMAK youth	CIDOB APCOB:Concepción
NGOs (intermediate-level intermediation)		ICO Comarapa
National indigenous organization (intermediate-level intermediation)	CONAMAQ	CONAMAQ
International indigenous organization (low-level intermediation)	Fondo Indígena	Fondo Indígena
Government (low-level intermediation)	EnlaRed Municipal: Batallas, Machacamaraca[a]	EnlaRed Municipal: Comarapa[a]

a. These are the case studies presented in chapters 8 (OMAK) and 9 (EnlaRed).

(e) government organizations. The typology refers to the process of intermediation used within the organization's ICT program and how the organization designs and implements projects.

The factors considered in evaluating the level of intermediation include (a) the manner in which ICTs are introduced, (b) the type of technical support (training, content development) provided to the community, and (c) the extent to which the ICT program is embedded within existing social and organizational structures (the relationship between existing informal information systems and the ICT intervention).

The investigation distinguishes between low, intermediate, and high levels of intermediation. Low-level intermediation is characterized as a centrally managed, top-down approach: project staff has limited interaction with the local community, and local needs and realities do not frame either the design or implementation of the project, which instead respond to external factors such as tied financing and donor priorities. Medium-level intermediation includes locals in the planning and implementation process, but is not particularly responsive to or flexible regarding local realities. High-level intermediation is characterized by direct involvement at the local community level: local community members are included in the planning and implementation process, and a great degree of flexibility is incorporated into the program to respond to local needs, not just at the outset, but as needs arise.

Indigenous Grassroots Organizations

I worked with two indigenous grassroots organizations from the highlands.

The *Organización de Mujeres Aymarás del Kollasuyo* (OMAK),[1] located in the city of El Alto, works predominantly with indigenous women in eight Aymara communities on issues related to empowerment, sustainable development, and enhanced access to information. The main objective of its ICT program is to strengthen the network of women's organizations and to train indigenous women in the use of ICTs (interview with Andrea Flores, OMAK director, June 2006).

The *Coordinadora de Integración de Organizaciones Económicas Campesinas de Bolivia* (CIOEC),[2] including two of its programs in Potosí and Oruro, is a small-scale producer association that aims to strengthen the position of indigenous farmers by (a) providing institutional strengthening and technical support in production and marketing to farmer organizations and small cooperatives in various sectors, (b) strengthening the network of producer organizations, and (c) lobbying the government to enhance the legal framework and overall conditions in support of small producers. CIOEC has integrated into its overall capacity-building program a module providing training to community leaders in the use of information and ICTs. The main aims of the training module are (a) to promote local economic development by providing small-scale producers with improved access to information about government programs in the productive sectors and (b) to enhance the capabilities of local leaders to use ICTs as a tool for improved product marketing and increased political participation in local government decision-making processes (interview with Renato García, CIOEC director, July 2005).

In the Bolivian lowlands, I collaborated with the following two indigenous grassroots organizations.

Apoyo para el Campesino Indígena del Oriente Boliviano (APCOB)[3] is a technical support organization for indigenous peoples of the Bolivian lowlands. While APCOB is formally registered as an NGO, I have categorized it as a grassroots program because of the way it carries out its intermediation, including (a) a bottom-up approach at the grassroots level, where most programs are implemented by local indigenous people with the support of indigenous professionals, (b) a number of decentralized offices in rural areas that are staffed by local actors, and (c) a high degree of respect among indigenous communities due to its instrumental role in uniting and empowering indigenous peoples of the lowlands under the formation of one common umbrella organization—the Confederation of Indigenous Peoples of Eastern Bolivia. In spite of its emphasis on providing technical assistance, APCOB continues to represent the interests of many indigenous organizations and grassroots groups. Indeed, indigenous representatives continue to have a voice in its decision-making processes. APCOB supports a program entitled ICT for Agriculture in the Chiquitania region, which aims to (a) strengthen the capacity of small indigenous farmers to participate in the local economy through improved access to information and ICT use, (b) enhance the participation of indigenous leaders in local development planning, and (c) strengthen knowledge exchanges and networking among the different indigenous groups of the Chiquitania region (interview with Juan Burgos, ICT program coordinator, APCOB, April 2005).

The *Confederación de Pueblos Indígenas de Bolivia* (CIDOB)[4] is an indigenous organization representing all of the indigenous groups from the lowlands of Bolivia. CIDOB was founded in October 1982 in Santa Cruz de la Sierra with the participation of representatives from the four main indigenous groups in eastern Bolivia—Guarani-Izoceños, Chiquitanos, Ayoreos, and Guarayos. Currently, CIDOB represents all 34 indigenous groups of the Bolivian lowlands from seven of the nine departments, including Santa Cruz, Beni, Pando, Tarija, Chuquisaca, Cochabamba, and La Paz. Since 2001, CIDOB has promoted an ICT program based on an information system that tracks land titling of indigenous lands in Bolivia. Furthermore, the program supports an ICT capacity-building component for indigenous leaders aiming to improve information sharing and exchanges among its network of regional organizations and central offices (interview with Eliana Rojas, ICT program coordinator, CIDOB, April 2005). While CIDOB has recently extended its reach to the national level, it is categorized as an indigenous organization from the lowlands, since it remains the central organization representing all indigenous groups of the Bolivian lowlands and has relatively weak representation in the highlands.

National Indigenous Organizations

I also worked with one national indigenous organization. The *Consejo Nacional de Ayllus y Markas del Qullasuyu* (CONAMAQ)[5] is a national indigenous organization that represents the interests of indigenous communities across Bolivia.

One of the most important political organizations in the indigenous movement, CONAMAQ gives indigenous peoples a "voice" in national policy dialogues and defends their rights vis-à-vis the state and the private sector. Its political strength emanates from its mandate to form a National Council including all of the traditional indigenous local governments—*ayllus*[6]—of the Bolivian highlands and lowlands. Within this context, CONAMAQ leaders view access to information and ICTs as an important citizenship right and consider inequitable access a form of political and social exclusion of indigenous peoples by the state (focus group with CONAMAQ leaders, March 2005). In March 2005, CONAMAQ lobbied the Bolivian government for improved access for indigenous communities to ICT services, signing an agreement with the Superintendent for Telecommunications and the United Nations Development Programme (UNDP) to establish six community telecenters in indigenous communities (focus group with CONAMAQ leaders, March 2005). As the first step of the agreement, UNDP and the superintendent jointly organized a training workshop for indigenous leaders on the role of ICTs in indigenous community development. Unfortunately, changes in the leadership of the government and CONAMAQ prevented both implementation of the agreement and construction of the telecenters.

International Indigenous Organizations

I also worked with one international indigenous organization. *Fondo Indígena*[7] is an indigenous-led international organization representing indigenous peoples from 18 countries in Latin America. It was founded during the Second Ibero-America Summit of the Head of States held in Spain in 1992. Its main role is to raise funds for sustainable development projects for indigenous peoples in the region from multilateral and bilateral aid agencies and to facilitate policy dialogue between indigenous organizations and governments. To meet one of its key objectives—strengthening indigenous organizations and their networks in Latin America—it has incorporated an ICT project within its ongoing work program since 2001. The project's goals are to (a) develop a portal or clearinghouse on issues affecting Latin America's indigenous peoples, (b) improve access to information by indigenous peoples, in particular, information about government and international donor aid programs, (c) enhance the flow of communications and information between the Fondo Indígena and indigenous organizations, and (d) train indigenous leaders on issues related to the role of information for development and use of the Internet as a networking and advocacy tool for indigenous organizations (interview with Mateo Martínez, general secretary of the Fondo Indígena, January 2005).

Nongovernmental Organizations

I worked with one NGO. The *Instituto de Capacitación del Oriente* (ICO)[8] has more than 25 years of experience in providing technical assistance, agricultural extension, and support in local economic development to small agricultural producers in the Santa Cruz Valley. Since October 2001, ICO has been implementing an ICT program that has established three community information centers in the region. The program provides small farmers across the region with daily information on market prices of the principal agricultural products and their selling prices at the Santa Cruz market in the largest urban center of the region. The program uses a mixed-technology approach, integrating the Internet and community radio: ICO's central office in Santa Cruz uses e-mail to send the information on market prices to the regional offices in Vallegrande, Comarapa, and Mairana, and these offices give the information to community radio stations, which broadcast the prices via radio to the local population across the entire Santa Cruz Valley. This mixed-technology approach allows ICO to reach even the most remote communities in the region and is a highly effective means of disseminating market price information (interview with Edwin Rocha, ICO director, June 2005).

Government Organizations

Finally, I collaborated with several central government agencies during my field-work, including the Vice Ministry for Telecommunications, the Superintendent for Telecommunications, and the Ministry of Agriculture. However, none of the ambitious plans of the central government agencies came to fruition during my 20-month fieldwork in Bolivia. This led me to collaborate with the *Federación de Asociaciones Municipales de Bolivia* (FAM),[9] the national umbrella association of small and medium-size Bolivian municipalities that was responsible for EnlaRed Municipal, a program aiming to strengthen the provision of e-government services across all Bolivian municipalities. Its objectives included (a) increasing ICT use by local governments, (b) improving service provision by local governments through ICTs, and (c) improving communication flows between citizens, communities, and local government. The project's components included capacity building, institutional strengthening, and increased access to infrastructure. Some of the activities carried out under the project included (a) training workshops for municipal government staff on the use of ICTs and development of local e-government strategies and service provision and (b) development of sustainable local e-government portals through mixed public-private partnerships. Municipalities were encouraged to join the program in exchange for joint access to an all-member database with legal information and a rich library of documents relevant to municipal management.

Methodological Challenges

Due to the intangible nature of information and communication, as well as its indirect long-term effects, it is impossible to establish a direct causal relationship between ICTs and poor people's well-being. The very nature of information and ICTs makes it very difficult to quantify or measure their impact on people's well-being. The relationship between technology and people is characterized by dynamic interaction. Consequently, the users of technologies are able to shape ICTs to fit their local needs, while the embedded value of technology solutions affects people's ability to appropriate them (Williams and Edge 1996; Menou 2001). In this sense, the evaluation methodology has to be very flexible to account for unexpected outcomes caused by the introduction of technologies into a specific local sociopolitical context.

For these reasons, I acknowledged at the outset of the research that complicated processes can occur once people and technology start to interact with each other. During the investigation, I was fully aware of the challenge of differentiating between external influences and the impact factors that are related to the ICT interventions. Therefore, I took an adaptive systems approach, which considers that social systems are constantly changing and that conceptual frameworks about expected outcomes inevitably represent a gross simplification of reality and are subject to uncertainties.

The AEF adopted in this book does not describe linear relationships; instead, it forms part of an adaptive learning process. For instance, making information available through ICTs does not guarantee that people will make use of them or that, when they access information, it is possible to observe the utility they derive from use of that information. Due to the innovative nature of the research conducted, I have attempted to approach the subject matter with an open mind and to present both the expected and unexpected findings of the research.

Notes

1. The Aymara Women's Organization of the Kollasuyo.
2. The Coordinating Agency for the Producer Organizations of Indigenous Farmers of Bolivia.
3. Assistance for the Indigenous Peoples of the Eastern Lowlands of Bolivia.
4. Confederation of Indigenous Peoples of Bolivia.
5. National Federation of the Ayullus and Indigenous Communities.
6. *Ayllu* is an Aymara word referring to a network of families in a given area. It is the traditional local government model of indigenous communities in the Andean region of Bolivia. The head of the *ayllu* is the *mallku*, the traditional representative of the community, who frequently is the political and religious leader of the community.
7. Indigenous Peoples Fund.
8. Institute for Capacity-Building of the Eastern Lowlands.
9. National Federation of Municipal Associations of Bolivia.

References

Avgerou, Chrisanthi. 2002. *Information Systems and Global Diversity.* Oxford: Oxford University Press.

Bailey, Kenneth D. 1994. *Methods of Social Research*, 4th ed. New York: Free Press.

Bamberger, Michael. 2000. *Integrating Quantitative and Qualitative Research in Development Projects.* Washington, DC: World Bank.

Callon, Michel, and John Law. 1989. "On the Construction of Sociotechnical Networks: Content and Context Revisited." *Knowledge and Society* 8: 57–83.

Carvalho, Soniya, and Howard White. 1997. "Combining the Quantitative and Qualitative Approaches to Poverty Measurement and Analysis: The Practice and the Potential." Technical Paper 366, World Bank, Washington, DC.

Chambers, Robert. 1997. *Whose Reality Counts: Putting the Last First.* London: Intermediate Technology Publications.

Ciborra, Claudio, and Giovan F. Lanzara. 1994. "Formative Contexts and Information Technology: Understanding the Dynamics of Innovation in Organizations." *Accounting, Management, and Information Technology* 4 (2): 61–86.

Cohen, Jacob, Patricia Cohen, Stephen G. West, and Leona S. Aiken. 2003. *Applied Multiple Regression/Correlation Analysis for the Behavioral Sciences.* London: Lawrence Erlbaum Associates.

Cornwall, Andrea, and Garett Pratt. 2002. "Pathways to Participation: Critical Reflections on PRA." Summary Report, IDS, Brighton.

DeMaris, Alfred. 1995. "A Tutorial in Logistic Regression." *Journal of Marriage and Family* 57 (4): 956–68.

Denzin, N. K. 1978. *Sociological Methods: A Sourcebook.* New York: McGraw-Hill.

Fielding, Nigel, and Jane Fielding. 1986. *Linking Data: The Articulation of Qualitative and Quantitative Methods in Social Research.* London: Sage.

Firestone, W. A. 1987. "Meaning in Method: The Rhetoric of Quantitative and Qualitative Research." *Educational Researcher* 16 (7): 16–21.

Galliers, R. D., and F. L. Land. 1987. "Choosing Appropriate Information System Research Methodologies." *Communication of the ACM* 30 (11): 900–20.

Glymour, Clark, Richard Scheines, Peter Spirtes, and Kevin Kelly. 1987. *Discovering Causal Structure: Artificial Intelligence, Philosophy of Science, and Statistical Modeling.* Orlando, FL: Academic Press.

Hamelink, Cees J. 1997. "New Information and Communication Technologies, Social Development, and Cultural Change." Discussion Paper 86, UNRISD, Geneva.

Hoggart, Keith, Loretta Lees, and Anna Davies. 2002. *Researching Human Geography.* London: Arnold.

Hulme, David. 2000. "Impact Assessment Methodologies for Microfinance: Theory, Experience, and Better Practice." *World Development* 28 (1): 79–98.

Madon, Shirin. 1993. "Introducing Administrative Reform through the Application of Computer-Based Information Systems: A Case Study in India." *Public Administration and Development* 13 (1): 37–48.

Marsland, Neil, I. Wilson, S. Abeyasekera, and U. Kleih. 1998. "A Methodological Framework for Combining Quantitative and Qualitative Survey Methods: Background Paper; Types of Combinations." Report written for DFID Research Project R7033.

Mayoux, Linda, and Robert Chambers. 2005. "Reversing the Paradigm: Quantification, Participatory Methods, and Pro-poor Growth." *Journal of International Development* 17 (2): 271–98.

McCracken, Grant. 1988. *The Long Interview.* Newbury Park, CA: Sage.

Menard, Scott. 2002. *Applied Logistic Regression Analysis.* Thousand Oaks, CA: Sage Publications.

Menou, Michel. 1993. *Measuring the Impact of Information on Development.* Ottawa: International Development Research Centre.

———. 1999. "Impact of the Internet: Some Conceptual and Methodological Issues, or How to Hit a Moving Target behind the Smoke Screen." In *The Internet: Its Impact and Evaluation,* edited by David Nicholas and Ian Rowlands. Proceedings of an international forum held at Cumberland Lodge, Windsor Park, London, July.

———. 2001. "Digital and Social Equity? Opportunities and Threats on the Road to Empowerment." Paper presented at the Libraries in the Digital Age (LIDA) 2001 Annual Course and Conference, Dubrovnik, Croatia.

Norris, Pipa. 2001. *Digital Divide: Civic Engagement, Information Poverty, and the Internet Worldwide.* Cambridge, MA: Harvard University, John F. Kennedy School of Government.

Oakley, Peter, Brian Pratt, and Andrew Clayton. 1998. *Outcomes and Impact: Evaluating Change in Social Development.* NGO Management and Policy Series 6. Oxford: INTRAC.

Orlikowski, Wanda J. 2000. "Using Technology and Constituting Structures: A Practice Lens for Studying Technology in Organizations." *Organizational Science* 11 (4): 404–28.

Orlikowski, Wanda J., and Suzanne C. Icono. 2001. "Research Commentary: Desperately Seeking the 'IT' Artifact in IT Research; a Call to Theorizing the IT Artifact." *Information Systems Research* 12 (2): 121–34.

Pampel, Fred C. 2000. "Logistic Regression: A Primer." Quantitative Applications in Social Sciences. Thousand Oaks, CA: Sage Publications.

Pettigrew, Andrew M. 1990. "Longitudinal Field Research on Change: Theory and Practice." *Organizational Science* 1 (3): 267–92.

Pretty, Jules N., Irene Guijit, John Thompson, and Ian Scoones. 1995. "Participatory Learning and Action: A Trainer's Guide." METH PRA Series, International Institute for Environment and Development, London.

Roche, Chris. 1999. *Impact Assessment for Development Agencies: Learning to Value Change.* Oxford: Oxfam.

Rossman, G. B., and B. L. Wilson. 1985. "Numbers and Words: Combining Quantitative and Qualitative Methods in a Single Large-Scale Evaluation Study." *Evaluation Review* 9 (5): 627–43.

Rubin, Herbert, and Irene Rubin. 1995. *Qualitative Interviewing: The Art of Hearing Data.* Thousand Oaks, CA: Sage.

Silverman, David. 2000. *Doing Qualitative Research: A Practical Handbook.* Thousand Oaks, CA: Sage.

VAIPO (Vice Ministerio de Asuntos Indígenas y Pueblos Originarios). 1999. *Proyecto iniciativa plan de desarrollo indígena originario: Diagnostico nacional.* Informe Final. La Paz: VAIPO, January.

Walsham, Geoffrey. 1995. "Interpretative Case Studies in IS Research: Nature and Method." *European Journal of Information Systems* 4 (2): 74–81.

Walsham, Geoffrey, and Sundeep Sahay. 2006. "Research on Information Systems in Developing Countries: Current Landscape and Future Prospects." *Information Technology for Development* 12 (1): 7–24.

Williams, Robin, and David Edge. 1996. "The Social Shaping of Technology." *Research Policy* 25 (6): 865–99.

Yin, R. K. 1994. *Case Study Research: Design and Methods,* 2d ed. Thousand Oaks, CA: Sage.

CHAPTER 3

Poverty, Inequality, and Human Development of Indigenous Peoples in Bolivia

Bolivia, with gross domestic product (GDP) per capita of US$2,867 in 2013, is the poorest country in South America (World Bank 2014a). Since its independence in 1825, this land-locked country has had high levels of extreme poverty, high economic and social inequality, and political instability (Klein 1982). With a population of about 10.67 million people and 35 distinct ethnic groups, Bolivia is one of the most culturally diverse countries in Latin America (World Bank 2014a). In 2001 approximately 4.6 million people, or 62 percent of the population, were indigenous (INE 2001).

Since the early 1990s, there has been a strong trend toward reaffirming indigenous identity and culture, and important advances have been made in the protection of indigenous rights and the participation of indigenous peoples in national policy making. These advances have focused mainly on political, social, and cultural aspects, culminating in a redrafting of the 1994 Constitution to recognize, for the first time, Bolivia's pluricultural and multiethnic character and the rights of indigenous peoples:

>> To the extent covered by the Law the social, economic, and cultural rights of indigenous peoples are respected and protected, especially those rights relating to their communal territories of origin, guaranteeing the sustainable use of natural resources, and their identities, values, languages, customs, and institutions. (Bolivian Constitution 1994, Article 171) <<

In spite of their important political advances, in particular since the 1990s, indigenous peoples remain marginalized in economic terms and continue to live in extreme poverty (Jiménez Pozo, Landa Casazola, and Yáñez Aguilar 2006; Psacharopoulos and Patrinos 1994). This chapter provides an overview of the demographic characteristics and socioeconomic situation of indigenous peoples in Bolivia, examines the persistent social and economic gaps between the indigenous and non-indigenous populations, and analyzes the key factors underlying the economic and social exclusion of Bolivia's indigenous peoples.

Identifying Indigenous Peoples

Indigenous peoples have historically been among the poorest and most excluded demographic groups throughout Latin America. They have faced serious discrimination not only in terms of their basic rights to ancestral property, languages, cultures, and forms of governance, but also in terms of their access to basic social services (education, health and nutrition, water and sanitation, and housing) and the essential material conditions for a satisfying life (Partridge, Uquillas, and Johns 1998; Plant 1998).

There is no universal definition of indigenous peoples and, because of the history of political repression, discrimination, and assimilation policies by states,[1] indigenous peoples often refuse to be defined by external agencies. In order to identify rather than "define" indigenous identities, I use the working definitions provided by the International Labour Organization Convention 169 (Tomei and Swepston 1996) and the draft United Nations Resolution on Indigenous Rights,[2] according to which self-identification by indigenous peoples should be the main criterion. These documents highlight the following commonalities of indigenous identities: (a) historical continuity with precolonial societies; (b) strong link to territories; (c) distinct social, economic, and political systems; (d) distinct language, culture, and beliefs; and (e) self-identification as distinct from the national society. From an indigenous perspective, the right to self-identification is essential to ensuring that they are respected as peoples with identities, cultures, languages, worldviews, and religions distinct from those of the rest of society (Stavenhagen 2002; Cárdenas 1997).

The Demographics of Indigenous Peoples

The number of indigenous peoples in Bolivia rose sharply from 1.7 million in 1950 to 4.6 million in 2001 due to a fertility rate higher than that of the non-indigenous population (INE 2001). The Aymara and Quechua people who live in the Andean highlands (*altiplano*) and central valleys (*valles*) form the majority of the indigenous population in Bolivia, representing 25 and 31 percent, respectively. Other indigenous groups inhabit the eastern lowlands and the Chaco region, accounting for only 6 percent of the population; however, they are extremely rich in cultural diversity and are distributed among 32 distinct ethnic groups. The indigenous peoples of the Amazon, such as the Chuiquitano—a group of 112,000 hunter-gatherers—are very different from the Aymara and Quechua populations of the highlands, not only culturally but also in their social organization, subsistence economy, and settlement patterns. The Guaraní are the largest indigenous group of the eastern lowlands (with a population estimated at 35,488); they inhabit the southern lowlands of the Chaco region, bordering Paraguay (Molina Barrios and Albó 2006).

The demographics of the indigenous population differ from those of the non-indigenous. The fertility rate per woman in 1994 was significantly higher for the indigenous than for the non-indigenous population. For indigenous peoples, the fertility rate was 6.4 in the Andean highlands and 7.1 in the eastern lowlands, while for non-indigenous peoples overall, it was 3.8, according to the Encuesta Nacional Demográfica y de Salud (INE 1994). While the majority of indigenous peoples are bilingual and speak Spanish as well as their own indigenous language, 40 percent are monolingual (speaking only their native language). In particular, many indigenous women are monolingual.

An important factor influencing the demographics of indigenous peoples is the increasing urbanization of Bolivia. In 1976 only 41 percent of the population lived in urban centers; by 2012, approximately 67 percent lived in major cities or towns (World Bank 2014b). This rapid urbanization is the result largely of the migration of indigenous peoples from the countryside to the cities. By 2002, a majority (52 percent) of Bolivia's indigenous peoples lived in urban centers, while 48 percent lived in rural communities (MAIPO 2003). Nevertheless, rural areas are still inhabited predominantly by indigenous peoples (78 percent), and urban centers are inhabited largely by the non-indigenous or *mestizo* population (66 percent).[3]

As shown on map 3.1, a majority of the population is indigenous in the highlands (89 percent in La Paz Department) and in the sub-Andean valleys (74 and 73 percent, respectively, in Cochabamba and Chuquisaca). In contrast, in the eastern lowlands, the majority of the population is non-indigenous, with an indigenous population in Santa Cruz and Tarija estimated, respectively, at 23 and 16 percent. The number of indigenous peoples in the lowlands, about 250,000, is relatively small (Valenzuela Fernández 2004, 13–14).

Because of the lack of census data on indigenous peoples in Bolivia, studies frequently use "rural" as a proxy for "indigenous" and "urban" as a proxy for "non-indigenous." This conceptualization has become outdated as a result of the rapid urbanization of indigenous peoples over the past decade, but most studies use this correlation, and the majority population in rural areas continues to be heavily indigenous. For these reasons, I also assume this correlation in my research. This means that the data may obfuscate the situation of indigenous peoples living in urban areas, where poorer, recent migrants live in informal settlements with limited access to basic infrastructure and social services.

National and Indigenous Poverty

Economic indicators reveal a high degree of poverty and income inequality in Bolivia. The poverty headcount ratio in 2002 indicates that more than 65 percent of the population was living below the national poverty line and 41 percent of the population was living on less than US$1 a day (World Bank 2005). From 1993 to 1999, Bolivia experienced a period of stable economic growth and relative political stability during which poverty rates fell from about 70 to 62 percent (Klasen et al. 2004). At the turn of the new millennium, however, the economy slowed down. Between 1998 and 2002, GDP per capita fell from about US$1,100 to approximately US$900, and poverty rates rose from 62 to 65 percent (figure 3.1). Moreover, extreme poverty rose slightly from 40 to 41 percent (World Bank 2014c).

MAP 3.1   **Regional distribution of indigenous peoples in Bolivia, 2001**

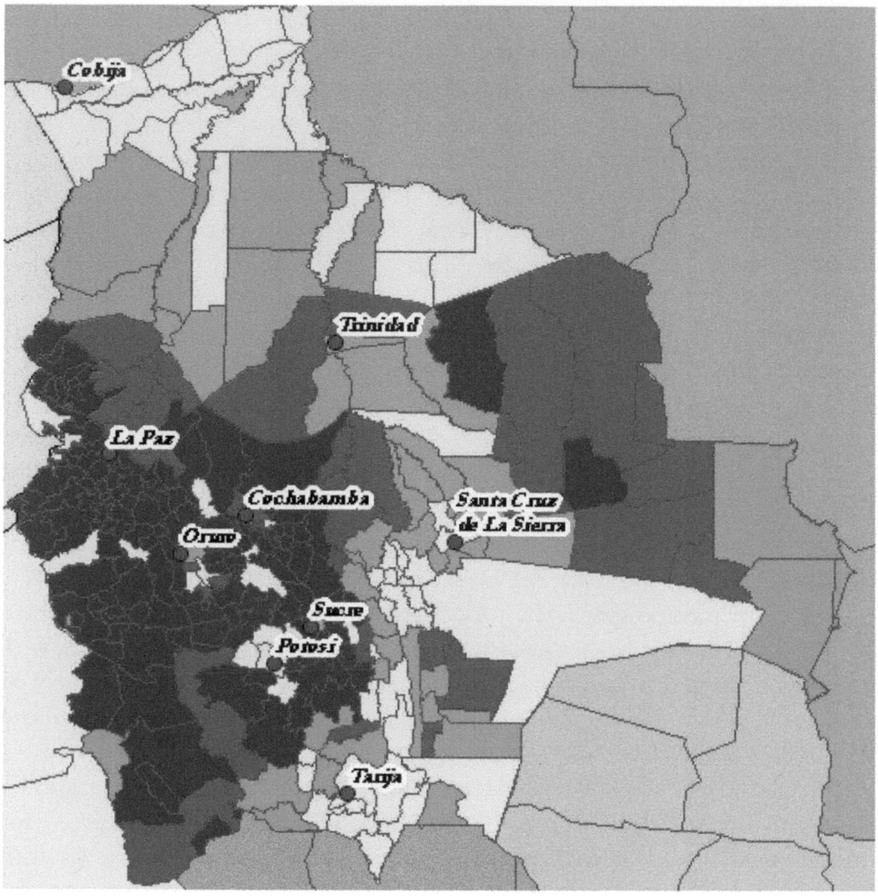

Population over age 15 in 2001 (%)

- ☐ 0–26.64
- ☐ 26.64–61.50
- ■ 61.50–85.47
- ■ 85.47–100

Source: Based on census data from INE 2001.

These values surpass by far the rates in 1999–2005, when moderate poverty fell 4.5 percent and extreme poverty fell 6.1 percent. These findings are buttressed by the thorough analysis of Uribe and Hernani-Limarino (2013), who have measured average rates of poverty reduction. They find that the average rates of poverty reduction for the period 2005–11 were double the average rates registered for the period 1999–2005.

There is a high and persistent correlation between being poor and being indigenous. Using urban household data, Psacharopoulos and Patrinos (1994) demonstrate that indigenous peoples are systematically impoverished.

FIGURE 3.1   Moderate and extreme poverty in Bolivia, 1997–2011

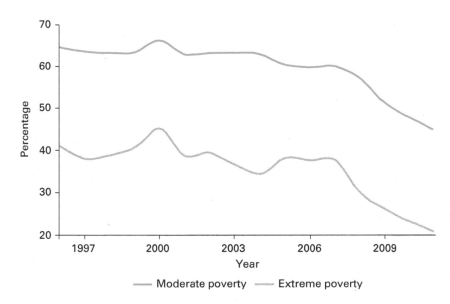

Source: Based on data from UDAPE 2014.

Among the urban population, approximately 65 percent of indigenous peoples were below the poverty line in 1989, compared with less than 50 percent of non-indigenous peoples. Thus the incidence of poverty was about 15 percentage points higher for indigenous than for non-indigenous people in urban areas. At the same time, the average per capita income of indigenous peoples was, on average, less than two-thirds the income of Spanish-speaking persons. In 1999 nationally, 75 percent of the indigenous population lived under the moderate poverty line, and 37 percent of these people were categorized as extremely poor. In comparison, only 45 percent of the non-indigenous population lived below the moderate poverty line (Jiménez Pozo, Landa Casazola, and Yáñez Aguilar 2006). The national poverty statistics from 2002 confirm this large gap (Klasen et al. 2004, 7). The predominantly indigenous rural areas still experienced very high poverty rates (83 percent were poor, and 67 percent were extremely poor), while the predominantly non-indigenous urban areas had lower poverty rates (54 percent were poor and 26 percent were extremely poor).

In terms of poverty trends, the data indicate, however, that the poverty gap between the indigenous and non-indigenous populations has been narrowing in recent years. The poverty rate of the indigenous population fell to 68 percent in 2005 and even further to 39 percent in 2011, compared with 49 and 32 percent, respectively, for the non-indigenous population (figure 3.2). In 1999 the indigenous–non-indigenous poverty gap was 18 points for both moderate and extreme poverty.[4] After a notable decline in 2000 and 2001, the gap widened slightly from 20 to 23 points between 2001 and 2005, before falling sharply after 2008 to 4.7 points for extreme poverty and to 7.3 points for moderate poverty in 2011.

FIGURE 3.2   Gap in the rates of extreme and moderate poverty between the non-indigenous and the indigenous populations in Bolivia, 1999–2011

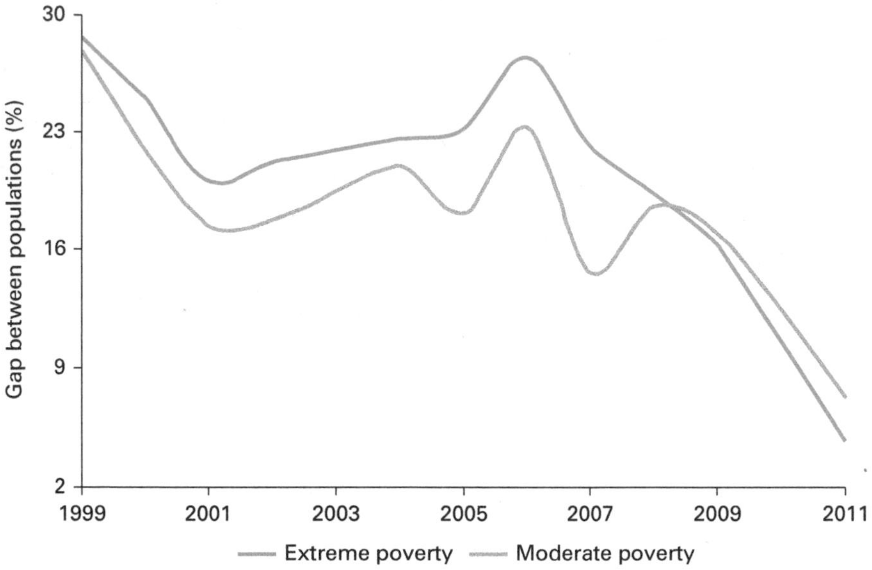

Source: Based on data from UDAPE 2014.

In sum, poverty has decreased substantially, the gap between the indigenous and non-indigenous poor has been closing, and these changes accelerated after 2005. Nevertheless, Bolivia continues to face important challenges with regard to poverty alleviation. Almost half of its population (45 percent) still lived below the national poverty line in 2011 (data from the UDAPE). Moreover, the income poverty gap of 13 percent in 2008 was higher than the average for all Latin American developing countries (World Bank 2014c).[5]

Regional Distribution of Poverty

Bolivia has three main geographic zones: highlands (*altiplano*), valleys (*valles*), and plains (*llanos*), also known as the Andean region, the sub-Andean region, and the lowlands, respectively. Today, approximately 42 percent of the population lives in the highlands, 29 percent in the valleys, and 29 percent in the plains.

In 2001, the Unidad Económica de Políticas Sociales y Económicas (UDAPE) and the Instituto Nacional de Estadísticas (INE) developed a consumption-based poverty map (map 3.2). While poverty is widespread, the regional distribution of poverty has one constant feature: the poorest regions are the ones with the most indigenous peoples. The highlands have a poverty incidence of 70 percent, followed by the valleys (69 percent) and the plains (54 percent; Kay and Urioste 2007). In addition, the departments of Potosí (66 percent indigenous) in the highlands and Chuquisaca (74 percent indigenous) in the

MAP 3.2   Extreme poverty (less than US$1 a day) in Bolivia, 2001

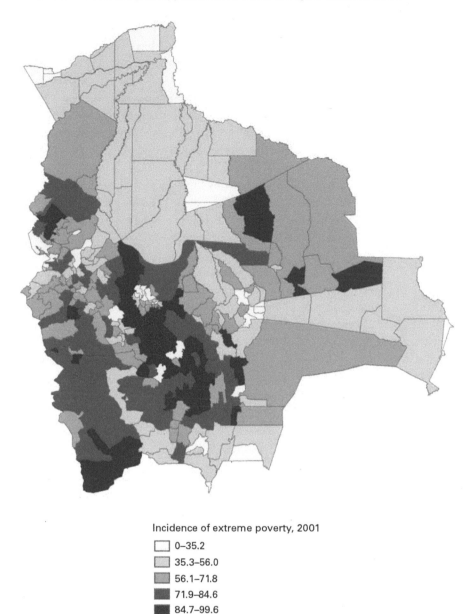

Incidence of extreme poverty, 2001

☐ 0–35.2

▢ 35.3–56.0

▢ 56.1–71.8

▢ 71.9–84.6

■ 84.7–99.6

Source: Based on data from INE.

sub-Andean region have the highest incidence of poverty, followed by Beni, La Paz, and Oruro, while areas such as Santa Cruz (32 percent indigenous), Tarija (16 percent indigenous), Pando, and Cochabamba are the least poor (World Bank 2005; Jiménez Pozo, Landa Casazola, and Yáñez Aguilar 2006). Rates of extreme poverty are also very high in many indigenous municipalities (above 72 percent). Comparing the map of indigenous peoples with the poverty map demonstrates a strong and direct correlation between extreme poverty and indigenous municipalities. The map further highlights that the large majority of extreme poverty is concentrated in the western highlands, in particular, in the departments of Chuquisaca, Oruro, La Paz, and Potosí. At the same time, the majority of municipalities in the eastern lowlands have lower extreme poverty, with either 0–35 percent or 35–56 percent of the population living in extreme poverty.

Poverty levels in rural areas have declined significantly in recent years (figure 3.3), but the gap between urban and rural areas is still wide. In 2011, 61 percent of the population in rural areas was moderately poor, compared with 37 percent of the population in urban areas, while extreme poverty was 41 and 11 percent, respectively (UDAPE 2014).

Although poverty in rural areas remains high, the speed of poverty reduction has been faster in rural than in urban areas. During the period 1999–2005, poverty rates fell an average of 6 percent in the cities, compared with 16 percent in the countryside (Uribe and Hernani-Limarino 2013). Likewise, between 2005 and 2011 poverty also fell faster in rural than in urban areas (18 and 11 percent, respectively), with extreme poverty in rural areas declining at an average rate nearly double that in urban areas.[6] What do these figures mean for indigenous poverty? First, poverty is declining in a geographic area that is, overall, predominantly indigenous. At the same time, the progress toward reducing poverty has not been geographically homogeneous. While certain areas, such as Tarija or Santa Cruz, have made significant progress in

FIGURE 3.3 Trends in extreme and moderate poverty in rural and urban areas of Bolivia, 1996–2010

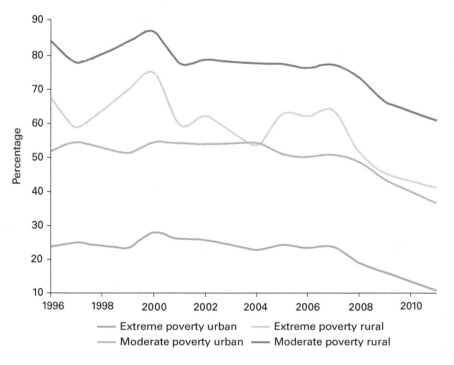

Source: Based on data from UDAPE 2014.

FIGURE 3.4   Poverty in Bolivia, by department, 1999, 2005, and 2012
Regional distribution of poverty ($2.50 line)

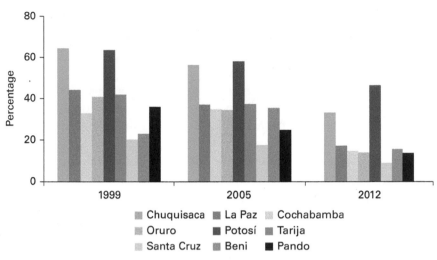

Source: Based on data from SEDLAC (World Bank 2014d).
Note: Poverty is defined as living on less than US$2.50 a day.

reducing poverty, others, such as Chuquisaca or Potosí, have made only modest progress in the past decade. Figure 3.4 illustrates regional disparities in poverty incidence by department.

The department of Tarija has made surprising progress by reducing its poverty incidence from 42 percent in 1999 to 0.6 percent in 2012. However, departments such as Potosí and Chuquisaca have not: 46 and 33 percent, respectively, of their populations were poor in 2012. While most regions have reduced their poverty indexes approximately 50 percent, the poorest regions have experienced the smallest declines: Potosí (20 percent) and Chuquisaca (40 percent).[7] Thus pockets of extreme poverty remain and are lagging behind the national and regional trends in poverty reduction. These areas—Potosí and highland Chuquisaca—also have a strong indigenous presence.

The Unsatisfied Basic Needs Approach

Recognizing that there are many more dimensions to poverty than income or consumption, this section focuses on the nonincome, nonconsumption dimensions of poverty. The analysis uses the unsatisfied basic needs approach, which defines poverty in terms of a household's living conditions through six social indicators: (a) quality of housing material, (b) number of household members, (c) access to water and sanitation services, (d) access to energy services, (e) access to education, and (f) access to health care. These indicators yield the unsatisfied basic needs (UBN) index.

Bolivia has made considerable progress in improving the living conditions of the poor. In terms of unsatisfied basic needs, poverty at the national level declined steadily from 71 percent in 1992 to 59 percent in 2001 and to 52 percent in 2008. In 1992 about 76 percent of the population did not have access to water and sanitation services, compared with only 58 percent in 2001. Major improvements have also been made in access to health care, which was available to 63 percent of the population in 2001 (from 47 percent in 1992). However, these improvements have been achieved primarily in urban areas and the eastern lowlands; the quality of the living conditions in rural indigenous areas is still extremely low.

Regional inequalities in terms of living conditions worsened significantly in Bolivia during the 1990s. Map 3.3 and table 3.1 illustrate the opposing trends in the UBN index between the municipalities of the western highlands and the eastern lowlands, based on a comparison of census data from 1992 and 2001.

The data analysis reveals that while Bolivia has made important progress in enhancing people's living conditions at the national level, the gains have been divided unequally within the country. In municipalities located in the eastern departments of Santa Cruz, Beni, Pando, and Tarija, where approximately 65–75 percent of the population is non-indigenous, the UBN index rose between 10 and 36 percent (light and dark green areas on the map). In contrast, in the municipalities in the western departments of La Paz, Oruro, and Potosí and in the central Andean department of Cochabamba, living conditions have either stagnated or worsened by up to 24 percent (orange or red

MAP 3.3   Change in unsatisfied basic needs in Bolivia, 1992–2001

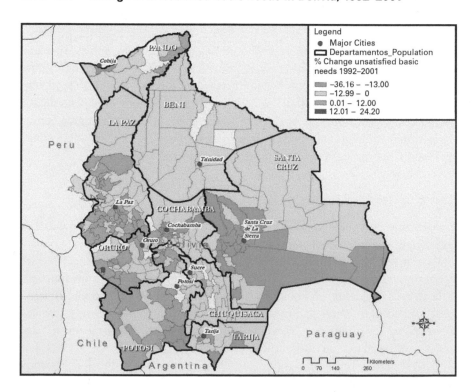

Source: Based on data from INE.

TABLE 3.1   Poverty according to unsatisfied basic needs in Bolivia, 1992 and 2001

Indicator	1992			2001		
	National	Urban	Rural	National	Urban	Rural
Overall index	70.9	53.1	95.3	58.6	39.0	90.8
Housing materials	48.2	22.5	83.6	39.6	15.6	75.7
Housing crowding	80.0	76.3	85.1	70.8	68.9	76.3
Sanitary services	75.9	60.0	97.6	58.0	44.3	78.9
Energy services	51.8	21.2	93.8	43.7	14.1	91.2
Education	69.1	53.9	90.1	52.5	36.5	70.9
Health care	53.6	44.2	66.6	37.9	31.0	54.5

Source: World Bank 2005.

areas on the map). All of the 39 municipalities in which the UBN index either worsened or stagnated (a) are located in the western highlands and (b) have a majority indigenous population of at least 79 percent. Thus, in spite of the significant improvements made in Bolivia at the national level, the living conditions of rural indigenous communities either stagnated or worsened, while those of urban areas improved.

Although the UBN index at the national level fell from 53 to 39 percent between 1992 and 2001, it improved only slightly in rural areas in this period. Indeed, 91 percent of the rural population continued to suffer from unsatisfied basic needs in 2001. Most of the improvement took place in urban areas, with rural areas lagging behind (World Bank 2005; INE 2001). For instance, 44 percent of the urban population lacked access to water and sanitation services in 2001, compared with 79 percent of the rural population.

Regional disparity can also be assessed by looking at the UBN index across departments. The indicator with the most unequal distribution in 2001 was health care. While Santa Cruz and Tarija had low levels of unsatisfied basic needs in this area (6 and 15 percent, respectively), La Paz, Oruro, and Potosí had very high levels: 65, 59, and 59 percent, respectively (INE 2014).

Income Inequality

Bolivia is among the most unequal countries in the world. The bottom 20 percent of households have only 4 percent of national income, while the top 20 percent have 62 percent (World Bank 2003). This inequality affects indigenous peoples disproportionately, since almost two-thirds of them are in the poorest five deciles—the poorest 50 percent of the population. Between 1990 and 2000, the gap between the rich and the poor rose sharply, particularly in rural areas, suggesting that indigenous peoples' living conditions worsened during the 1990s. This is reflected in the substantial increase in the Gini coefficient, which measures inequality. The Gini coefficient rose approximately 10 percent from 0.53 in 1989 to 0.62 in 2000, placing Bolivia second only to Brazil—the Latin American country with the highest income inequality—even though Bolivia's GDP per capita is only about one-third of Brazil's (Glassman and Handa 2005; Gray 2004). In 2000, income inequality was also more severe in rural than in urban areas, at 0.632 and 0.554, respectively (ECLAC 2004, 301–05).

This trend has been reversed since the early 2000s. According to UDAPE data, the level of income inequality spiked in 2000 and has fallen steadily from its highest level in 2000, with a Gini coefficient of 0.62, to a value of 0.60 in 2005, reaching its lowest level of 0.46 in 2011. Most of the improvement occurred after 2005, with the Gini coefficient dropping 14 points.

Eid and Aguirre (2013) also measure consumption inequality in Bolivia, finding that this indicator dropped from 0.47 to 0.37 between 1999 and 2011, which represents a 20 percent decrease. It is notable that 17.85 percent of this drop occurred between 2005 and 2011. Nevertheless, the average measure of inequality hides the divergent trends in rural and urban areas (figure 3.5). In 1999, urban inequality was at 0.49, while rural inequality was at 0.64. As inequality fell overall, the gap between rural and urban areas widened. In 2011, the Gini coefficient was 0.41 in urban areas, compared with 0.54 in rural areas. Indigenous peoples, particularly those living in rural areas, continue to have difficulty improving their living conditions.

FIGURE 3.5   Income inequality in urban and rural areas of Bolivia, 1996–2011

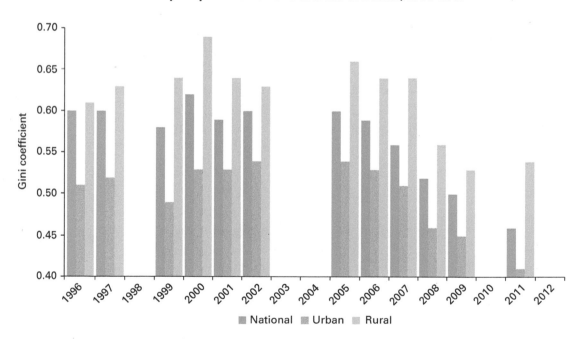

Source: Based on data from UDAPE 2014.

The Human Development of Indigenous Peoples

In this section, I use the human development index (HDI) and the Millennium Development Goals (MDGs) to broaden the analysis of indigenous peoples' socioeconomic conditions and to examine various aspects of poverty. According to the human development approach, development consists of enlarging people's choices. When assessing deprivation, it focuses on aspects other than income and monetary indicators, such as health, education, political freedom, and safety, among others. As Mahbub ul Haq puts it, "The objective of development is to create an enabling environment for people to enjoy long, healthy, and creative lives."[8] Specifically, the HDI, developed by the United Nations Development Programme (UNDP), measures four basic indicators: life expectancy at birth; adult literacy (over 15 years); enrollment in primary, secondary, and postsecondary education; and GDP per capita.

Bolivia has improved its indicators of human development, moving from 114 in the UNDP ranking in 2004 to 108 in 2012 (UNDP 2004, 2014). For instance, the national child mortality rate (deaths before the age of five) was cut almost in half, from 120 deaths per 1,000 live births in 1990 to 66 per 1,000 in 2003. This positive trend continued throughout the decade, declining to about 41 per 1,000 in 2011. At the same time, however, Bolivia has the highest maternal and infant mortality rates in Latin America and suffers from high rates of malnutrition, malaria, chagas, and tuberculosis. Most important, the progress achieved in improving human development has been extremely uneven (map 3.4). Within Bolivia, 62 municipalities continue to have low levels of human development (below 0.5) and 18 municipalities have extremely low levels (below 0.4). This compares to the levels in several Sub-Saharan African countries, such as Mali (0.391) and Ethiopia (0.389). All of the municipalities with either low or extremely low levels of human development are located in the western highlands and have predominantly indigenous populations.

Furthermore, Bolivia's progress in reaching the MDGs has been uneven, with stark differences within the country. The critical reasons that Bolivia is facing difficulties in meeting the MDGs until 2015 are (a) the tremendous differences between geographic areas and (b) the tremendous gaps in access to basic social services between the indigenous and non-indigenous populations (Valenzuela Fernández 2004). For instance, Bolivia's national infant mortality rate was cut almost in half from 96 deaths per 1,000 live births in 1989 to 54 in 2003 and to 33 in 2012 (World Bank 2014c). However, huge gaps in this indicator persist within Bolivia, with infant mortality rates declining dramatically in the departments of the eastern lowlands but remaining unacceptably high in the western highlands (figure 3.6). For instance, infant mortality rates were three times as high in the department of Oruro as in the department of Tarija in the eastern lowlands. Important differences in infant mortality also remain between income groups, with infant mortality falling 27 percent for the richest households and only 5 percent for the poorest (World Bank 2003).

MAP 3.4   Human development in Bolivia, 2002

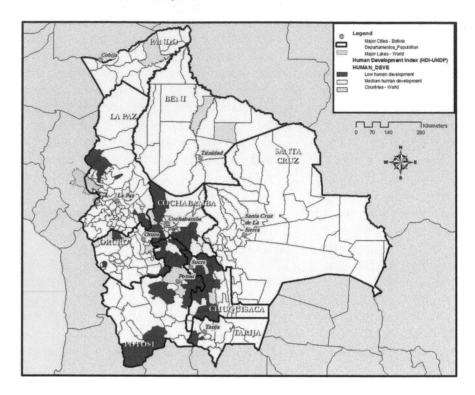

Source: Based on data from UNDP 2002.

FIGURE 3.6   Infant mortality in Bolivia, by department, 2003

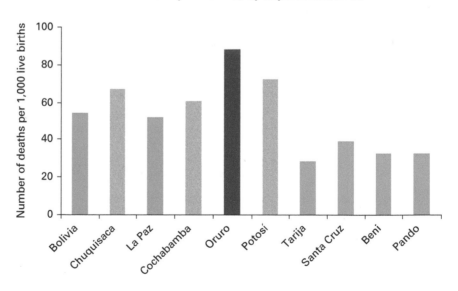

Source: Based on data from INE 2003.

Under-five mortality rates rose in 26 municipalities during the 1990s (map 3.5). An in-depth data analysis revealed that this indicator worsened 11–30 percent in 19 municipalities and increased 31–52 percent in 7 municipalities. Thus tremendous gaps in living conditions persist between indigenous and non-indigenous communities. Basic living conditions as measured by infant and under-five mortality rates stagnated or even worsened in many indigenous communities in the 1990s.

Education is another key aspect of human development. Literacy rates have risen since 1999, with 93 percent of young people (ages 15 to 24) literate in 1999, compared with 99 percent in 2012. The improvements were

MAP 3.5 Change in under-five mortality rates at the municipal level, 1992–2003

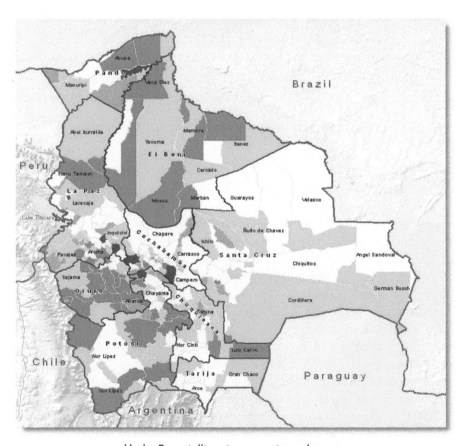

Under-5 mortality rate, percentage change

- −116 – −30
- −29 – −10
- −9 – 10
- 11 – 30
- 31 – 52

Source: Based on data from INE.

FIGURE 3.7 **Literacy rates in rural and urban areas of Bolivia, by age group, 1997–2012**

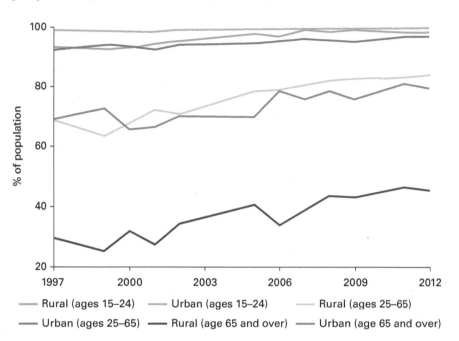

Source: Based on data from SEDLAC (World Bank 2014d).

proportionally even greater in other age groups. Among adults 25 to 65 years of age, literacy rose from 83 to 93 percent in the same period, with literacy among females improving 14 percentage points (from 75 to 89 percent). Among the elderly (65 years of age and older), literacy rose from 50 percent in 1997 to 64 percent in 2012. Among elderly females, literacy improved 40 percent between 2005 and 2012, with most of the change occurring after 2005 (World Bank 2014d).

Once again, rural and urban areas are quite different in their levels of literacy and education (figure 3.7). In 1999, 92 percent of adults 24 to 65 years of age living in cities were literate, compared with 68 percent of those living in rural areas. Also in 1999, 69 percent of the elderly living in urban areas were literate, compared with 29 percent in rural areas. Thus rural areas, which are predominantly indigenous, have relatively low levels of education.

Much of this gap has narrowed since 1999. In rural areas, literacy among adults (24–65) rose from 68 percent in 1999 to 73 percent in 2005 and to 83 percent in 2012. Among the elderly in rural areas (65 and older), it rose to 40 percent in 2005 and 45 percent in 2012. Moreover, the gap between literacy rates in urban and rural areas has been closing since the late 1990s.

Data regarding years of schooling also illustrate the inequalities among the indigenous and non-indigenous populations (figure 3.8). On average, Bolivians had 7.6 years of schooling in 1999, but indigenous peoples had only 6.1 years, while non-indigenous people had 10 years. These conditions have improved

FIGURE 3.8 **Years of education in Bolivia, by ethnicity and geographic location, 1998–2008**

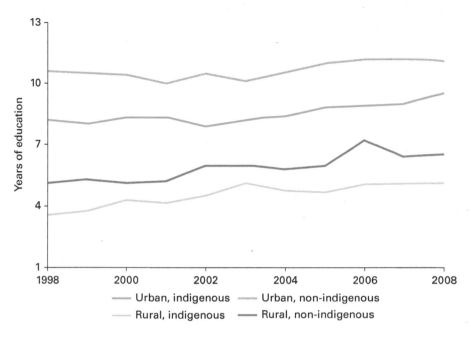

Source: Based on data from UDAPE 2014.

significantly in the past decade: the indigenous population had 6.8 years of schooling in 2005 and 7.6 years in 2009, compared with 9.8 and 10.3 years, respectively, for the non-indigenous population (UDAPE 2014).

There are, however, also wide disparities in the level of education between urban and rural areas. In urban areas, the average number of years of schooling was 10.6 for the non-indigenous population and 8.2 for the indigenous population in 1999 (figure 3.8); indigenous peoples living in urban areas increased their average years of schooling to 8.4 in 2005 and to 9.5 in 2009. In rural areas, the non-indigenous population had 5.1 years of schooling in 1999, while the indigenous population had only 3.6. Although both the indigenous and the non-indigenous populations improved their indicators after 1999, the gap remains wide. In 2009 in rural areas, indigenous peoples had just 5.1 years of schooling, on average, compared with 6.5 years for the non-indigenous population (UDAPE 2014).

Underlying Causes of Indigenous Poverty

Given the stark differences between the indigenous and non-indigenous populations in terms of poverty, human development, and access to basic services, the following section analyzes some of the main causes of indigenous poverty. As the data presented in this chapter have shown, Bolivia is a highly fragmented society, with sharp distinctions along a formal-informal, rural-urban, and indigenous–non-indigenous divide. Based on the historic context, the Bolivian economy has developed a high degree of dualism. The economy is divided into (a) a formal sector dominated by *mestizos* and closely linked to global markets through the export of natural resources and the inflow of international investments and (b) a rural and informal sector consisting primarily of indigenous peoples reliant on a combination of subsistence agriculture, informal commerce, and temporary or permanent labor.

In this section, I argue that the long-term lack of access to social services (education and health) and productive assets (land) are critical factors that have impeded indigenous peoples from significantly improving their living conditions. The main social indicators (illiteracy, mother and child mortality, and primary and secondary enrollment rates) indicate that indigenous peoples have significantly less access to social services than non-indigenous peoples.

Limited Access to Education

The human capital approach has demonstrated an important and positive relationship between education, earnings, and poverty (Schultz 1961; Mincer 1974). A key reason for the persistently high levels of poverty among indigenous peoples in Bolivia is that their level of schooling is low compared to that of the non-indigenous population. As pointed out above, there exists a large gap in years of schooling between the indigenous and non-indigenous peoples in Bolivia. In 2009 school enrollment was, on average, 3.0 years less for indigenous peoples than for the non-indigenous population. A study by Psachoropoulos and Patrinos (1994) finds that this difference was even greater for indigenous women, who only completed, on average, 5.5 years of schooling—suggesting that indigenous women are the most disadvantaged group within Bolivian society. Furthermore, a study covering four isolated indigenous communities in the eastern lowlands finds that this relationship holds even for remote communities, where the positive impact of education on income increases the closer an individual is to a town with a market (Godoy et al. 2005). Furthermore, illiteracy remains widespread in the indigenous population. While the overall illiteracy rate in Bolivia has decreased to about 15 percent for the population as a whole, it remains at well over 40 percent for indigenous people and ranges as high as 58–64 percent for indigenous women (Jiménez, Candia, and Mercado 2005, 59).

In spite of the significant progress made in enhancing access to education in terms of school enrollment rates in the last two decades,[9] several factors are responsible for the persistent gap in schooling levels. First, indigenous children often leave school earlier than their wealthier classmates in order to contribute to household income (McEwan 2004; Valenzuela Fernández 2004, 23–24). The household data from 2002 show that the incidence of child labor was almost four times higher among indigenous children than among non-indigenous children. While only 8.4 percent of non-indigenous children between the ages of 9 and 11 worked, 31 percent of indigenous children worked (Jiménez, Candia, and Mercado 2005, 54). Second, in many rural areas, access to schools—particularly bilingual and intercultural education—remains limited. Many indigenous children still cannot take full advantage of schooling resources because of language and cultural barriers (McEwan 2004). Third, rural areas often suffer from low quality of schooling, exacerbated by teacher shortages and a lack of teaching materials and classroom equipment. While recent test scores for language and mathematics place Bolivian students overall well below the Latin American average, indigenous students score even lower than non-indigenous students on math and Spanish exams (McEwan 2004). Finally, in urban areas many indigenous households do not have the resources to cover the direct costs of schooling. Although public schooling is free, related expenses, including uniforms, school materials, and transportation, can be as high as US$120 a year in urban areas, which is not affordable for many indigenous families, whose annual income might be no more than US$500 (Inchauste 2000).

Lack of Access to Health Care

Major inequalities in access to basic health services persist between indigenous and non-indigenous peoples (table 3.2). In particular, indigenous women and children continue to have limited access to basic health services.

Bolivia's public health expenditure amounted to less than 1 percent of GDP in 1998, compared with 2.6 percent for all lower-middle-income countries (Thiele 2003, 316). Despite this neglect, most basic health indicators have improved significantly at the national level since the mid-1980s. Infant mortality, for instance, declined almost 40 percent between 1985 and 1998. However, health conditions for poor, indigenous peoples are still dismal. In the mid-1990s, for example, infant and child mortality rates were more than twice as high for the poorest quintile, which consists largely of indigenous peoples (World Bank 2000). A key reason for this is their lack of access to health care. While in 2002, indigenous peoples were more likely to have suffered an illness or injury in the past 30 days than non-indigenous people (21 and 13 percent, respectively), they were much less likely to have received medical attention (37 and 50 percent, respectively). More than 50 percent of the sick or injured non-indigenous population received medical care, compared with only 37 percent of the sick or injured indigenous population. Moreover, 55 percent of non-indigenous women have their babies in hospitals, compared with only 30 percent of indigenous women (Jiménez Pozo, Landa Casazola, and Yáñez Aguilar 2006, 62–64). Finally, indigenous children are twice as likely to be malnourished as non-indigenous children—40 and 20 percent, respectively, in 1998 (ECLAC 2004).

TABLE 3.2   Selected health indicators in Bolivia, 1985–2000

Indicator	1985	1989	1994	1998	2000
Infant mortality rate (per 1,000 live births)	108	96	75	67	59
Under-five mortality rate (per 1,000 live births)	148	130	116	92	69
Maternal mortality rate (per 100,000 births)	—	416	390	—	—
% of births in medical facilities	—	37.6	42.3	52.9	—
Child malnutrition	—	13.3	15.7	7.6	—

Source: World Bank 2000.
Note: — = not available.

Lack of Access to Land

One factor that clearly influences poverty rates is access to land (Plant and Hvalkof 2001). In Bolivia's geographic extremes, arable land is scarce and equivalent to a mere 2.67 percent of the national territory. In spite of the 1953 agrarian reform, which abolished the old *hacienda* system and established a new class of communal peasant owners, distribution of land remains extremely unequal. According to official data from the National Council for Agrarian Reform, 7 percent of landowners own 87 percent of usable land (28 million hectares), while the remaining 93 percent own only 4 million hectares. Moreover, only 5 percent of the land belonging to big landowners is being put to productive use (Urioste 1994, 184–85). As Kevin Healy has shown, the 1953 agrarian reform had opposite effects in the western and eastern regions of Bolivia. While the *latifundia* disappeared in the west, they reappeared in the east as a result of the giveaway of huge tracts of "public" land (Healy 2004). As a result, the Gini coefficient for land inequality was 0.768 in 1989, indicating a very high concentration of landownership (Deininger and Squire 1998).

The government's land policies, particularly between 1953 and 1996, clearly benefited the elites and the *mestizos* and significantly hurt the indigenous peoples from both the highlands and the lowlands (Ballivián and Zeballos 2003; Vargas 2003). In the Andean highlands, the scarcity of arable land and demographic pressures led to the subdivision and fragmentation of land. This trend has increasingly forced indigenous peoples into landlessness. In the eastern lowlands, where the state used colonization and infrastructure programs to promote the expansion of the agricultural frontier toward the east, many indigenous communities lost their lands to the commercial agroindustry and colonizers (Gill 1987).

Since the livelihoods of indigenous peoples continue to depend largely on either subsistence farming (mostly in the highlands) or hunting and gathering (in the eastern lowlands), secure access and the right to collective land management are essential to them. Thus indigenous peoples struggle to protect their ancestral land and territories, resulting in frequent confrontations with large landowners, logging companies, cattle ranchers, and the state.

Neglect of the Rural Sector

In addition to the lack of access to agricultural land by *minifundistas*, there is a detrimental bias in public investment against the small-scale agriculture characteristic of the highlands and in favor of the large-scale commercial agriculture that is common in Santa Cruz and the eastern lowlands (Healy 2001). The policies of structural adjustment during the mid-1980s favored an economic model based on export-led growth. Government promoted the rapid growth of commercial agriculture and the extraction of natural resources, especially natural gas from the eastern lowlands (Kay and Urioste 2007). According to Demeure (1999), this policy led to a dramatic increase in Santa Cruz's land cultivation—from 60,000 hectares in 1950 to more than 1 million hectares at the end of the 1980s. Agribusinesses grew new cash crops including soya, wheat, cotton, and sunflowers. In contrast, small-scale subsistence agriculture stagnated in the highlands.

A study carried out by the Vice Ministry of Popular Participation in 40 communities of the Bolivian highlands and valleys found that rural productivity declined sharply in the 1980s (World Bank 1998). Many Aymara and Quechua small farmers were immersed in a vicious cycle of severe land fragmentation, declining soil fertility, and environmental degradation. Indigenous farmers were constrained by the lack of public investment in rural infrastructure (for example, rural roads) and productive agricultural activities (irrigation systems and extension services).

Over the past two decades, the government has sought to reduce poverty by investing in the social sectors—particularly education—and has put relatively few resources into the productive sector. Bolivia's expenditures on education more than doubled, from 2.5 percent of GDP in 1985 to 5.7 percent in 1998. Public investments remained constant during the 1990s. The numbers demonstrate that the government has neglected the agriculture sector; public investments in the agriculture sector accounted for a mere 0.48 percent of GDP in 1985, falling to 0.27 percent in 1998. The consequence of this neglect in the rural sector is the severe economic crisis in the Andean highlands and valleys and the stagnation of indigenous poverty in rural areas at very high levels (Thiele 2003). Most small-scale indigenous farmers face the immense challenges of low rural productivity, growing population, division and subdivision of land, migration and emigration, and degradation of natural resources.

Labor Market Discrimination

Indigenous peoples' labor force participation, at 80 percent, is significantly higher than that of the non-indigenous population, at 64 percent (Valenzuela 2004). The reason for this is the noticeably higher participation of indigenous women in the labor market (70 and 50 percent, respectively). However, 67 percent of unskilled workers are indigenous, and only 4 percent of indigenous workers occupy skilled positions (Zamora-Jiménez 1999). As a consequence, there are marked earnings differentials between the indigenous and the *mestizo* populations. Table 3.3 shows the average

TABLE 3.3   Average income in Bolivia, by geographic area and language spoken at home
Bolivianos

Language spoken (at home)	Urban	Rural	Country as a whole
Quechua	577	213	302
Aymara	584	168	306
Guaraní	419	262	289
Other indigenous	236	243	242
Spanish	931	373	783

Source: Valenzuela Fernández 2004, 22.

earnings for urban and rural areas and for the indigenous and non-indigenous populations.

Table 3.3 demonstrates the tremendous income differentials between urban and rural areas in all categories and large earnings differentials between the Spanish-speaking and indigenous populations. A stark fact, illustrating this point, estimates that the Aymara earn only about 45 percent of what Spanish-speakers earn, reflecting the unequal participation of indigenous peoples in the labor market. A World Bank study analyzing earnings differentials with regard to education and labor market discrimination concludes that 73 percent of the earnings differential is due to observable factors (differences in schooling), while the remaining 27 percent is due to discrimination against indigenous peoples in the labor market (Patrinos and Grootaert 2002).

The labor earnings that indigenous peoples derive from each year of schooling are lower than those of non-indigenous people. In Bolivia, the returns to schooling are significant, whereby each additional year of education results in a 9 percent increase in earnings for non-indigenous peoples, but only a 6 percent increase for indigenous peoples. Patrinos and Grootaert, using 1989 and 1993 data, estimate that 23 to 29 percent of earnings differentials in Bolivian cities between indigenous and non-indigenous men are due to discrimination (Patrinos and Grootaert 2002). Among both men and women, controlling for all observable characteristics, monolingual Spanish speakers earn about 25 percent more than bilingual speakers (Chiswick, Patrinos, and Hurst 2000). Open discrimination in Bolivian society is an important underlying cause of the urban poverty of indigenous peoples.

Indigenous Peoples and Natural Resources

The Bolivian economy has been based for centuries on the exploitation of natural resources, particularly in the mining sector and more recently in the hydrocarbons and the agribusiness sectors. Bolivia's high dependency on natural resources is an important factor in the impoverishment of indigenous peoples.

First, the Spanish conquest forced indigenous peoples to provide free labor (*mita*) for the rapidly growing mining industry in the western highlands. This systematic exploitation of indigenous labor had severe consequences for many indigenous communities, which had to endure the very difficult living conditions in the highlands and the extremely dangerous and labor-intensive work in the mines. Even after the mines were nationalized following the 1952 revolution, the highland communities continued to suffer from extremely adverse living conditions, with very low life expectancy rates, high levels of child labor, and generally high levels of poverty (Nash 1979). Furthermore, miners were almost completely dependent on the state for their livelihoods (Sanabria 2000).

Second, dependency on natural resources made Bolivia, and in particular the indigenous population, extremely vulnerable to external shocks. Consequently, indigenous peoples directly suffered from the volatility of a "boom and bust" economy, in which the prices of minerals and other natural resources fluctuate on the global markets. For instance, the world economic crisis in the early nineteenth century led to a sharp decline in global mineral prices and forced many mines to close. During the structural adjustment program of the mid-1980s, the government privatized the state-owned mining company. As a consequence, more than 35,000 miners and their families lost their livelihood and were forced to migrate to the urban centers of La Paz and El Alto to find new employment opportunities, mainly in the informal sector.

In addition, these extractive industries frequently negatively affect the environment of indigenous communities and undermine their ability to secure their livelihoods based on subsistence farming and hunting and gathering activities. For instance, the rubber boom in the early nineteenth century led to the invasion and destruction of indigenous lands in the Amazon. While local communities benefited from this boom for a while, they suffered the long-term consequences of environmental degradation and loss of territory to *mestizo* farmers, who continued to settle on indigenous lands even after the boom ended. The agribusiness boom of the 1990s had similar consequences, as both soybean production and cattle ranching displaced indigenous communities from their traditional lands. The boom of the 1990s was spurred by the discovery of large natural gas deposits in the eastern lowlands, which are believed to be the second largest in South America and are expected to be worth US$70 billion. Hydrocarbons, which accounted for only 5.6 percent of exports in 1999, accounted for 21 percent in 2001. Based on their experience with the mining sector, most indigenous peoples are extremely critical of the government's policies promoting the exploitation of natural gas.

Notes

1. Assimilation policies, applied by all nation-states of Latin America, were justified by the need for a unique "national identity" after countries gained independence from the colonial powers. They attempted to "assimilate" indigenous peoples into the dominant sectors of society, leading to the systematic discrimination against indigenous peoples, with the objective of undermining their cultures, identities, and traditional subsistence economies.
2. United Nations Document no. E/CN.4/Sub.2/1986/87.
3. *Mestizo* designates the people of mixed European and indigenous non-European ancestry.
4. The indigenous–non-indigenous poverty gap is measured as the difference between the non-indigenous and the indigenous poverty rate.
5. The World Bank's Development Research Group (World Bank 2014c) defines the poverty gap as "the mean shortfall from the poverty line (counting the nonpoor as having zero shortfall), expressed as a percentage of the poverty line. This measure reflects the depth of poverty as well as its incidence."

6. During the first period (1999–2005), extreme poverty declined 12 percent in rural areas and 6 percent in urban areas, compared with 18 and 11 percent, respectively, during the second period (2005–11).
7. These numbers are according to a US$2.50 poverty line as measured by World Bank (2014d).
8. See http://hdr.undp.org/en/humandev.
9. The primary enrollment rates rose significantly from 87 percent in 1985 to 91 percent in 1990 and to 95 percent in 2003 (World Bank 2000).

References

Ballivián, Danielo P., and Huntardo Zeballos. 2003. *Diagnóstico de la reforma agraria boliviana 50 años después de la promulgación de la ley.* La Paz: Instituto Nacional de Reforma Agraria.

Bebbington, Anthony. 1996. "Organizations and Intensifications: Campesino Federations, Rural Livelihoods, and Agricultural Technology in the Andes and Amazonia." *World Development* 24 (7): 1161–78.

Butz, David, Steven Lonergan, and Barry Smit. 1991. "Why International Development Neglects Indigenous Social Reality." *Canadian Journal of Development Studies/Revue canadienne d'études du développement* 12 (1): 143–57.

Cárdenas, Victor Hugo. 1997. "Indigenous Peoples, Development, and Democracy in Latin America." Inter-American Development Bank, SDS/IND, Washington, DC.

Chiswick, Barry R., Harry A. Patrinos, and M. E. Hurst. 2000. "Indigenous Language Skills and the Labor Market in a Developing Economy: Bolivia." *Economic Development and Cultural Change* 48 (2): 349–67.

Deininger, Klaus, and Lyn Squire. 1998. "New Ways of Looking at Old Issues: Inequality and Growth." *Journal of Development Economics* 57 (2): 259–87.

Demeure, Juan. 1999. "Agricultura, de la subsistencia a la competencia internacional." In *Bolivia en el siglo XX: La formación de la Bolivia contemporánea,* edited by Fernando Campero, 269–90. La Paz: Harvard Club de Bolivia.

ECLAC (Economic Commission for Latin America and the Caribbean). 2004. *Social Panorama of Latin America, 2002–2003.* Santiago: United Nations, ECLAC.

Eid, Ahmed, and Rodrigo Aguirre. 2013. "Trends in Income and Consumption Inequality in Bolivia: A Fairy Tale of Growing Dwarfs and Shrinking Giants." *Revista Latinoamericana de Desarrollo Económico* 20 (November): 75–110.

Gill, Lesley. 1987. *Peasants, Entrepreneurs, and Social Change: Frontier Development in Lowland Bolivia.* Boulder, CO: Westview Press.

Glassman, Amanda, and Sudhanshu Handa. 2005. "Poverty, Inequality, and Public Policy in the Andean Region: A Comparative Perspective." In *Political Crises, Social Conflict, and Economic Development: The Political Economy of the Andean Region,* edited by Andrés Solimano. Northampton, MA: Edward Elgar.

Godoy, Ricardo, Victoria Reyes-García, Elizabeth Byron, William R. Leonardo, and Vincent Vades. 2005. "The Effect of Market Economies on the Well-Being of Indigenous Peoples and on Their Use of Renewable Natural Resources." *Annual Review of Anthropology* 34 (October): 121–38.

Gray, Andrew. 2004. *Indigenous Rights and Development: Self-Determination in an Amazonian Community.* Oxfordshire: Marston Lindsay Ross International.

Healy, Kevin. 2001. *Llamas, Weavings, and Organic Chocolate: Multicultural Grassroots Development in the Andes and Amazon of Bolivia.* South Bend, IN: Notre Dame University Press.

———. 2004. "Towards an Andean Rural Development Paradigm?" *NACLA Report on the Americas* 38 (3): 28–33.

Inchauste, Gabriela. 2000. "Educational Choices and Educational Constraints: Evidence from Bolivia." IMF Working Paper 00/42, International Monetary Fund, Washington, DC.

INE (Instituto Nacional de Estadística). 1994. *Encuesta nacional demográfica y de salud (ENDSA) 1994.* Bolivia: INE.

———. 2001. *National Census of Housing and Population 2001.* La Paz: Ministry of Planning and Coordination.

———. 2014. *Censo de población y vivienda.* La Paz: Ministry of Planning and Coordination. http://www.inc.gob.bo/.

Jiménez, Elizabeth, Gaby Candia, and Marcelo Mercado. 2005. "Economic Growth, Poverty, and Institutions: A Case Study of Bolivia." Unidad Económica de Políticas Sociales y Económicas, UDAPE, La Paz.

Jiménez Pozo, Fernando, Wilson Landa Casazola, and Ernesto Yáñez Aguilar. 2006. "Bolivia." In *Indigenous Peoples, Poverty, and Human Development in Latin America*, edited by Gillette Hall and Harry Patrinos. London: Palgrave Macmillan.

Kay, Cristobal, and Miguel Urioste. 2007. "Bolivia's Unfinished Reform: Rural Poverty and Development Policies." In *Land, Poverty, and Livelihoods in an Era of Neoliberal Globalization: Perspective from Developing and Transition Nations,* edited by A. Haroon Akram-Lodhi, Saturnino Borras, and Cristóbal Kay. London: Routledge.

Klasen, Stephan, Melanie Grosse, Rainer Thiele, Jann Lay, Julius Spatz, and Manfred Wiebelt. 2004. "Operationalising Pro-Poor Growth: A Country Case Study of Bolivia." Georg-August-Universität, Ibero-Amerika Institut für Wirtschaftsforschung, Göttingen.

Klein, Herbert S. 1982. *Bolivia: The Evolution of a Multi-Ethnic Society.* New York: Oxford University Press.

MAIPO (Ministerio de Asuntos Indígenas y Pueblos Originarios de Bolivia). 2003. *Plan nacional del desarollo indigena.* La Paz: MAIPO.

McEwan, Patrick J. 2004. "The Indigenous Test Score Gap in Bolivia and Chile." *Economic Development and Cultural Change* 53 (1): 157–90.

Mincer, Jacob. 1974. *Schooling, Experience, and Earnings*. New York: National Bureau of Economic Research.

Molina Barrios, Ramiro, and Xavier Albó. 2006. *Gama étnica y lingüística de la población boliviana*. La Paz: Sistema de las Naciones Unidas en Bolivia.

Nash, John. 1979. *We Eat the Mines and the Mines Eat Us: Dependency and Exploitation in Bolivian Tin Mines*. New York: Columbia University Press.

Partridge, William, Jorge Uquillas, and Katherine Johns. 1998. "Including the Excluded: Ethnodevelopment in Latin America." In *Poverty and Inequality: Proceedings of the Annual World Bank Conference on Development in Latin America and the Caribbean*, 229–50. Washington, DC: World Bank.

Patrinos, Harry Anthony, and Christiaan Grootaert. 2002. "Child Labor in Bolivia and Colombia." World Bank, Washington, DC.

Plant, Roger. 1998. "Issues in Indigenous Poverty and Development." Technical Study, Inter-American Development Bank, Sustainable Development Department, Indigenous Peoples and Community Development Unit, Washington, DC.

Plant, Roger, and Soren Hvalkof. 2001. "Land Titling and Indigenous Peoples." Sustainable Development Department Technical Paper Series, Inter-American Development Bank, Washington, DC.

Psacharopoulos, George, and Harry Patrinos. 1994. *Indigenous Peoples and Poverty in Latin America: An Empirical Analysis*. Washington, DC: World Bank.

Sanabria, Harry. 2000. "Resistance and the Arts of Domination: Miners and the Bolivian State." *Latin American Perspectives* 26 (1): 56–81.

Schultz, Theodore W. 1961. "Investment in Human Capital." *American Economic Review* 51 (1): 1–17.

Stavenhagen, Rodolfo. 2002. "Indigenous Peoples and the State in Latin America: An Ongoing Debate." In *Multiculturalism in Latin America: Indigenous Rights, Diversity, and Democracy*, edited by Rachel Sieder, 24–44. New York: Palgrave.

Thiele, Rainer. 2003. "The Social Impact of Structural Adjustment in Bolivia." *Journal of International Development* 15 (1): 1–21.

Tomei, Manuela, and Lee Swepston. 1996. "Indigenous and Tribal Peoples: A Guide to ILO Convention No. 169." International Labour Organization, Office and the Center for Human Rights and Democratic Development, Geneva.

UDAPE (Unidad de Análisis de Políticas Económicas y Sociales). 2014. "Dossier de estadísticas sociales y económicas." Dossier 23, UDAPE, La Paz. http://www.udape.gob.bo/index.php?option=com_wrapper&view=wrapper&Itemid=38.

UNDP (United Nations Development Programme). 2002. "Desarrollo humano en Bolivia." UNDP, La Paz.

———. 2004. *Human Development Report 2004: Cultural Liberty in Today's Diverse World.* New York: UNDP.

———. 2014. "International Human Development Indicators." UNDP, New York. http://hdr.undp.org/en/countries.

Uribe, Alejandra, and Werner L. Hernani-Limarino. 2013. "Pobreza monetaria: Crecimiento y redistribución." *Revista Latinoamericana de Desarrollo Económico* 20 (November): 149–230.

Urioste, Miguel. 1994. "The Lost Mines of Llallagua." *Los Angeles Times,* March 13.

Valenzuela, María Elena. 2004. "Desigualdades entrecruzadas: Pobreza, género, etnia y raza en América Latina." In *Desigualdades entre cruzadas: Pobreza, género, etnia y raza en América Latina,* edited by María Elena Valenzuela and Marta Rangel. Santiago: Oficina Internacional del Trabajo.

Valenzuela Fernández, Rodrigo. 2004. "Inequidad, ciudadania y pueblos indígenas en Bolivia." CEPAL, Santiago de Chile.

Vargas, John D. 2003. *Proceso agrario en Bolivia y América Latina: 50 años de la reforma agraria en Bolivia.* La Paz: Plural Editores.

World Bank. 1998. "Bolivia Rural Productivity Study." World Bank, Washington, DC.

———. 2000. "Bolivia Poverty Diagnostic." World Bank, Washington, DC, June 30.

———. 2003. "World Development Indicators 2003." World Bank, Washington, DC.

———. 2005. "Bolivia: Poverty Assessment; Establishing the Basis for More Pro-Poor Growth." World Bank, Washington, DC.

———. 2014a. "Bolivia." World Bank, Washington, DC. http://data.worldbank.org/country/bolivia#cp_wdi.

———. 2014b. "Data Bank." World Bank, Washington, DC. http://databank.worldbank.org/data/Popular_indicators/id/af3ce82b.

———. 2014c. "Data: Poverty Gap at $1.25 a Day (PPP)." World Bank, Development Research Group, Washington, DC. http://data.worldbank.org/indicator/SI.POV.GAPS.

———. 2014d. Socio-Economic Database for Latin America and the Caribbean (SEDLAC). World Bank, Washington, DC. http://sedlac.econo.unlp.edu.ar/eng/statistics.php.

Zamora-Jiménez, Elizabeth. 1999. *Labor Market Segmentation and Migrant Labor: A Case Study of Indigenous and Mestizo Migrants in Bolivia.* South Bend, IN: University of Notre Dame.

Part II

The Empirical Evidence

CHAPTER 4

Poor People's Information Needs, Perceptions, and Expectations about the Internet

Prior to evaluating the impact of information and communication technologies (ICTs) on people's well-being, it is necessary to assess the existing "information ecologies" in local communities. Applying the alternative evaluation framework presented in chapter 1, it is essential to carry out a baseline study of the existing patterns of information and communication in a community before introducing an ICT program. This chapter identifies the information needs of local communities in Bolivia based on the results of a non-ICT user survey conducted during my fieldwork in rural Bolivia before ICTs were introduced into local communities, complemented by data collected in focus groups held in 2005 and 2006. The survey (n = 365) collected quantitative data from 365 predominantly indigenous peoples (81 percent) on (a) local information needs, (b) existing communication patterns, (c) awareness of and access to different forms of ICTs, (d) availability of information and existing information gaps, and (e) perceptions and expectations of the impact of the Internet on their well-being. In analyzing the data, I am particularly interested in answering the following questions:

1. How aware of and exposed to ICTs are indigenous peoples?
2. What do the current "information ecologies" look like at the local level, and what are the existing communication patterns?
3. What are the information and communication needs and priorities of indigenous peoples?
4. What important gaps exist between the information currently available, and what information is needed?
5. To what extent do indigenous peoples have access to information from policy makers and the central government?
6. Does the indigenous leadership have a culture of information sharing (at both the vertical and horizontal levels)?

7. What expectations do indigenous peoples have regarding the role that ICTs may play in enhancing their livelihoods?

8. What perceptions do indigenous peoples have regarding the negative impact, if any, that ICTs could have on their communities?

9. What role, if any, can ICTs play in enhancing the two-way flow of information between the indigenous and non-indigenous populations?

Principal Findings

A principal finding of this chapter is that indigenous people's existing patterns of information and communication are thoroughly embedded in their community's traditional social structures. They extensively use community radio and traditional channels of communication, such as community meetings and personal visits, to receive information. Therefore, it is particularly important for ICT programs that work in indigenous communities (a) to be based on the local need for information and communication, (b) to strengthen—instead of replace—existing traditional mechanisms of information exchanges and communications, and (c) to be embedded in the traditional social structures of the community.

The survey results demonstrate that indigenous peoples' overall awareness of and access to the Internet and other forms of new technologies (that is, cell phones) are very limited. Specifically, low access to the Internet among indigenous peoples is primarily a consequence of socioeconomic factors such as high illiteracy and poverty rates in indigenous communities, as well as the lack of social opportunities as a group. Contrary to the existing "digital divide" literature, the chapter finds that indigenous peoples' lack of access to the Internet and other forms of ICTs in general is associated with ethnicity and belonging to an indigenous group, rather than with a person's individual characteristics such as age, gender, and education. Thus indigenous peoples' low awareness of and access to ICTs and the Internet are a reflection of their broader social exclusion and marginalization within Bolivian society.

Finally, the analysis finds that indigenous peoples have very high expectations about the Internet's potential to enhance their well-being. An unexpected finding is that the most vulnerable groups within indigenous communities—in particular, indigenous women—have the highest expectations about the positive effects of the Internet on their lives, but they also face the most significant structural barriers, such as high illiteracy rates, to accessing and using this technology.

Organization of the Chapter

The chapter is organized as follows. It begins by describing the data analysis, outlining the main demographic and socioeconomic characteristics of the survey respondents, and outlining the main patterns of using and communicating information. It then analyzes the overall awareness of and access to ICTs in indigenous communities, followed by a more detailed analysis of the information needs of indigenous peoples and their expectations about the value of the Internet for their well-being, paying particular attention to their expectations about the Internet's potential to improve their relationship with the central government.

Data Analysis

The data analysis presents the principal results of the non-user survey, which focuses on issues related to information sources, information needs, communication patterns, and awareness and expectations of indigenous peoples in terms of ICTs. While the survey focused on non-users, it included two questions concerning the participant's basic access to and use of the Internet— *Do you have access to the Internet?* and *Do you currently use or have you ever used the Internet before?* These two questions were included to gauge their general awareness of and exposure to the Internet. However, the core of the chapter focuses on assessing the indigenous respondents' "informational capital."

For analyzing the survey results, I created five indexes in order to gain a better understanding about the existing information ecologies in local communities and indigenous peoples' communication patterns and information needs. The information source index (ISI) addresses issues related to indigenous people's uses of different information sources. The communications channel index (CCI) yields a communications profile of indigenous respondents. The information needs index (INI) captures the respondents' perceptions about which types of information they value and demand the most. The information availability index (IAI) allows for a detailed data analysis of the degree to which indigenous peoples receive information for which they have expressed a need. Finally, the information gap index (IGI) investigates the areas in which respondents face the greatest barriers to receiving information they consider important to their well-being.

The first four indexes were calculated based on a population-size-weighted average using a three-point rating scale with values from 0 to 2, indicating different levels of use of information sources and communications channels and different levels of need for and availability of information. The fifth index, the IGI, was calculated by subtracting the value of the IAI from the value of the INI.

In several sections of the chapter, the information collected on individual characteristics, demographics (age, gender, level of education, and so forth), and socioeconomic context (belonging to a specific indigenous group, level of poverty, geographic location) was cross-tabulated against the data on information needs and communication patterns. Chi-square analysis was used to test statistically significant factors that affect people's uses of different information sources, communications media, and expectations about the value of ICTs. Chi-square, a test of statistical significance for relationships between variables measured in the nominal and ordinal levels, was used to assess whether the observed relationship in a cross-tabulation of a sample of data is sufficiently strong to infer that a relationship exists in the full population (Meier and Brudney 1997). Cramer's V was then used to determine which among the significant variables had the strongest effect on the data.

The data for the full sample were analyzed before breaking the information into clusters. Based on the first hypothesis outlined in chapter 1—that access to and use of ICTs have to be facilitated by an effective and local intermediary organization in order to improve poor people's well-being—the type of intermediation process used is crucial to explaining the difference in outcomes of ICT use and the resulting impact on the well-being of indigenous peoples. For this reason, I established clusters of different types of intermediary organizations. For the non-user survey, I grouped the organizations—Apoyo para el Campesino Indígena del Oriente de Bolivia (APCOB), Confederación de Pueblos Indígenas de Bolivia (CIDOB), Coordinadora de Integración de Organizaciones Económicas Campesinas Indígenas y Originarias (CIOEC) Oruro, CIOEC Potosí, Consejo Nacional de Ayllus y Markas del Qullasuyu (CONAMAQ), Instituto de Capacitación del Oriente (ICO), Fondo Indígena, and the Organización de Mujeres Aymaras del Kollasuyo (OMAK)—into six clusters according to their type of intermediation, as shown in table 4.1.

In the first category—grassroots organizations—I combined the results from the OMAK surveys with those from CIOEC Oruro and CIOEC Potosí. All of the sample data represent indigenous peoples from the highlands who participated in an ICT program coordinated by a grassroots organization. The same logic was used to form the other five clusters. A separate cluster was established for youth, an important demographic group whose voice may be lost among the largely homogeneous and predominantly older indigenous leadership.

Survey Results of Non-ICT Users

This section is divided into five subsections that summarize the main findings of the non–ICT user survey in terms of respondents' (a) demographics and socioeconomic characteristics; (b) patterns of use or information ecology; (c) communications profile; (d) level of awareness of, access to, and use of ICTs; and (e) information and communication needs.

TABLE 4.1   Clusters of intermediaries

Cluster	Name of intermediary	Number of respondents	% of respondents
Grassroots organizations in the highlands	CIOEC Oruro; CIOEC Potosí; OMAK indigenous leaders	67	18
Grassroots organizations in the lowlands	APCOB; CIDOB	40	11
Nongovernmental organization in the lowlands	ICO Comarapa	49	13
National indigenous organizations	CONAMAQ; CIOEC leaders	56	15
International indigenous organization	Fondo Indígena	74	20
Indigenous youth from the highlands	OMAK indigenous youth	79	22
Total		365	100

Demographic and Socioeconomic Profile of Respondents

The sample is quite representative of the overall socioeconomic situation of indigenous peoples in Bolivia. The vast majority of the survey respondents (81 percent) identified themselves as indigenous—meaning that the primary language spoken in their household is an indigenous one (table 4.2). While 60 percent of the respondents indicated that they speak both an indigenous language and Spanish, only 19 percent said that they primarily speak Spanish in the household. Thus the methodology used was effective in targeting indigenous peoples.

The more detailed clustered analysis found significant variations in the level of ethnicity, proxied by language, across the various intermediaries. The largest number of non-indigenous respondents, almost half (49 percent) of whom identified themselves as speaking only Spanish, were participants in the ICO program in Comarapa, located in the Santa Cruz Valley—an area populated by a relatively low percentage of indigenous peoples. With the exception of the participants from Comarapa, the large majority of respondents were fluent in an indigenous language and Spanish.

With regard to poverty, the sample is highly representative of Bolivia's indigenous population (INE 2001), with a high correlation between being indigenous and being poor. Since the concept of "well-being" is central to this investigation, the research uses the unsatisfied basic needs (UBN) index and the human development index (HDI)—two sets of indicators that stress non-income dimensions of poverty—to analyze the poverty level of the sample. As the survey did not include data on poverty or on living conditions more generally, data from the UBN index gathered during the 2001 census and tabulated by Bolivia's National Statistics Institute (INE) were correlated with data for municipalities included in the sample. In using this method, I am aware that the respondents may not necessarily have the same levels of poverty as the municipalities' overall population. However, since in most indigenous rural municipalities the population is rather homogeneous in terms of its exposure to poverty, the overall level of poverty is used as a proxy for what respondents are most likely to experience. The overwhelming majority (96 percent) of the survey respondents are from municipalities with unsatisfied basic needs, and only 4 percent of the sample are from municipalities with satisfied basic needs

TABLE 4.2 Principal language spoken in the household (used as proxy for ethnic identity)

Principal language	Number of respondents	% of respondents
Spanish	68	19
Indigenous language (primary) and Spanish	224	60
Aymara	50	14
Quechua	21	6
Other indigenous language	2	1
Total	365	100

TABLE 4.3   Poverty level among respondents, by municipal UBN index

Poverty category	Number of respondents	% of respondents	Cumulative % of respondents
Satisfied basic needs (I)	14	4	4
At the threshold of poverty (II)	103	28	32
Moderate poverty (III)	88	24	56
Marginal population (IV)	127	35	91
Indigent population (V)	33	9	100
Total	365	100	n.a.

Note: n.a. = not applicable.

(table 4.3). Furthermore, almost half (44 percent) are from municipalities with levels of extreme poverty, measured as levels IV and V of the UBN index.

The sample mirrors the overall poverty situation of indigenous peoples in Bolivia, as measured by the UBN index (table 4.4). The correlation is very strong between being indigenous and being poor, and this correlation is even stronger in my sample than in Bolivia overall. In the 2001 census, 71 percent of the population older than 15 years of age identified themselves as being indigenous, compared with 81 percent of my sample. The strong correlation between having unsatisfied basic needs and being indigenous becomes clear when analyzing the last two categories of the UBN index, which represent the population living under extreme poverty. In the 2001 census data, 77 and 87 percent of the population living in municipalities with a UBN index of IV and V, respectively, identified themselves as indigenous. In my sample, 91 and 85 percent of the population from municipalities with a UBN index of IV and V, respectively, are indigenous.

The HDI confirms the high level of poverty among survey respondents. At the national level, Bolivia had an HDI of 0.69 in 2004; however, the large majority of the sample population (74 percent) is from municipalities with a lower HDI index, ranging between 0.5 and 0.6 (UNDP 2005).

Only a small minority of the respondents (17 percent) come from municipalities with an HDI of 0.6 and higher (table 4.5). Moreover, the levels of human development in the sample are significantly lower in the highlands than in the lowlands, which is similar to the national trend. Only 7 percent of respondents from the highlands are from municipalities with an HDI of more than 0.6. The number is significantly higher in the lowlands, where 38 percent of the survey respondents live in municipalities with an HDI higher than 0.6.

Almost two-thirds (63 percent) of respondents are under the age of 35 (table 4.6). This small bias toward youth is largely due to special efforts undertaken during my fieldwork to include young people's voices. However, if the sample is analyzed without the youth cohort (86 participants), the majority (53 percent) of the respondents are still younger than 35 years of age.

The reason for this relatively young average age is twofold: (a) the new emerging indigenous leadership includes a growing number of young leaders,

TABLE 4.4 UBN index correlated with ethnicity (comparison between the 2001 census data and survey data sample)

% of sample

UBN category	Population 15 years and older who identify themselves as indigenous, 2001[a]	Survey respondents who speak Spanish and an indigenous language
I	53	57
II	60	59
III	70	91
IV	77	91
V	87	85
Total	71	81

a. Census data from INE 2001.

TABLE 4.5 Share of survey participants, by municipal HDI and geographic region

% of respondents

Geographic area	Less than 0.5	0.5–0.6	More than 0.6	Total %	Total number
Highlands	8	85	7	100	224
Valleys	41	38	21	100	29
Lowlands	0	63	38	100	112
Total	8	74	17	100	365

TABLE 4.6 Use of ICTs, by age

Age	Number of respondents	% of respondents	Cumulative % of respondents
Less than 18	68	19	19
19–35	161	44	63
36–44	70	19	82
More than 45	55	15	97
No response	11	3	100
Total	365	100	n.a.

Note: n.a. = not applicable.

overturning past traditional practices, and (b) younger people are much more attracted to participating in ICT programs than their elders and are less insecure about using new technologies (Norris 2001). However, 15 percent of the respondents older than 45 said they are interested in learning more about how to use ICTs.

There is, however, a gender gap in the survey, with results biased toward males (table 4.7). The ratio of male to female is 2.4:1, and 70 percent of respondents are male. The main reason for this is that men are overrepresented in

TABLE 4.7   Gender of survey respondents

Gender	Number of respondents	% of respondents
Male	254	70
Female	106	29
No response	5	1
Total	365	100

TABLE 4.8   Education of survey respondents

Level of education	Number of respondents	% of respondents	Cumulative % of respondents
Less than 5 years (primary school)	42	11.5	11.5
Primary school	86	23.6	35.1
Secondary school	172	47.1	82.2
Technical schooling (intermediate level)	34	9.3	91.5
Technical schooling (advanced)	17	4.7	96.2
University	5	1.4	97.5
Other	4	1.1	98.6
No schooling or no response	5	1.4	100.0
Total	365	100.0	n.a.

Note: n.a. = not applicable.

the indigenous leadership. Although no female-targeted surveys were administered, focus groups were held and a detailed case study was conducted, both targeting women.

The education level of respondents is far above the national average (table 4.8). At the national level, the average number of years of schooling completed is seven for males and five for females; however, average schooling is significantly lower for indigenous peoples than for the non-indigenous population (Psacharopoulos and Patrinos 1994). Nonetheless, more than 60 percent of respondents have a secondary education or more. Like the data on youth, I tested the extent to which the youth sample is responsible for these high numbers.

Even without the respondents from the youth sample, 51 percent of respondents have some secondary education or more, and almost a third (31 percent) have completed secondary education. At the national level, only 33 percent of indigenous peoples have a secondary education or higher. Also at the national level, 12 percent of the indigenous population either have no formal education at all or have not completed primary education (Hall and Patrinos 2006).

What explains the exceptionally high levels of educational attainment in the sample compared with national figures for the indigenous population in rural Bolivia? The primary reason is that 49 percent of the respondents are indigenous leaders; in recent years, the indigenous leadership has made tremendous progress in terms of their education and professional development. As such, the new indigenous leadership includes an increasing number of professionals who combine a strong educational background with practical experience working on community development issues at the grassroots level. This new leadership recently gained access to political power in 2006, when Evo Morales was elected the first indigenous president of Bolivia.

At the same time, there is an ever-growing educational divide within the indigenous leadership. As shown on table 4.9, 41.8 percent of indigenous leaders have a primary education or less, while 21.7 percent have either technical training or a university education. It is also notable that while 14.5 percent of indigenous leaders have less than a primary education in comparison to 11.5 percent of the overall survey, a portion of them are even more educated than the respondents from the overall sample, with 21.7 percent of indigenous leaders compared with only 15.3 percent of the overall respondents having technical training or a university education.

Educational attainment is a key individual factor determining a person's information and communication behaviors, information needs, expectations about the role of ICTs, and capabilities to use these technologies effectively.

Patterns of Information Use: The "Information Ecology" of Indigenous Communities

Before analyzing the specific information needs and expectations of indigenous peoples, I describe the "information ecology" of Bolivia's indigenous communities to serve as a baseline. This approach is particularly relevant to the study of indigenous communities in rural Bolivia, where communication and information flows are deeply embedded in traditional institutions. The *ayllu*, for example, is an indigenous institution of local government that has for centuries used traditional means of communication and decision making, like community meetings and *encargos* (use of messengers), to share important information among members and neighboring communities.

TABLE 4.9 **Educational attainment of indigenous leaders and all respondents**
% of respondents

Level of education	Indigenous leaders	Cumulative	Total sample	Cumulative
Less than 5 years (primary education)	14.5	14.5	11.5	11.5
Primary education	27.3	41.8	23.6	35.1
Secondary education	35.2	77.0	47.1	82.2
Technical training (intermediate level)	14.5	91.5	9.3	91.5
Technical training (advanced level)	5.5	97.0	4.7	96.2
University	1.7	98.7	1.3	97.5
Other	0.7	99.4	1.3	98.8
No response	0.6	100.0	1.2	100.0
Total (%)	100.0		100.0	

The following data analysis is based on the alternative evaluation framework described in chapter 1. Contrary to conventional ICT impact evaluations, a critical component of the framework is to evaluate the existing patterns of information uses and communications channels *before* assessing the impact of ICTs on people's well-being (Slater, Tacchi, and Lewis 2002; O'Farrell, Norrish, and Scott 1999). This section therefore analyzes the empirical data from the non-user survey regarding the detailed patterns of information and communication of indigenous communities. In order to gain a greater understanding of the main sources of information that indigenous respondents use, I developed an information source index, which is constructed based on the respondent's answers to the following question: How frequently do you use the following sources to receive information? The index was calculated based on a population-size-weighted average using the following three-point scale, with values from 0 to 2. Specifically, the different values were employed in the following way:

- 0: this information source is not used to receive information
- 1: this information source is used only somewhat to receive information
- 2: this information source is used a lot to receive information.

In order to interpret the resulting data in terms of the differences in the frequency of use among the various information sources, four categories were created:

- Minimal use: 0.0–0.49
- Low use: 0.5–0.99
- Intermediate use: 1.0–1.49
- High use: 1.5–2.0.

The *minimal-use* category (0.0–0.49) indicates that respondents use a specific information source only rarely. Such a low value shows that they either consider this information source to be unimportant or are unable to use it due to an array of potential barriers (such as lack of access to telecommunications infrastructure).

The *low-use* category (0.5–0.99) indicates that respondents use a specific information source modestly. Scores falling within this category indicate relatively infrequent use.

The *intermediate-use* category (1.0–1.49) indicates that respondents use an information source fairly often, with a large number of respondents indicating that they frequently use this particular source to receive information. In this category, there are no major barriers to use.

The *high-use* category (1.5–2.0) indicates that the large majority of respondents use this information source often. A high-use score demonstrates that the information source has become fully integrated into the respondents' information ecology.

The three information sources considered to be most important are community radio, community meetings, and information exchanges through family and friends (figure 4.1).

FIGURE 4.1   Most frequently used information sources

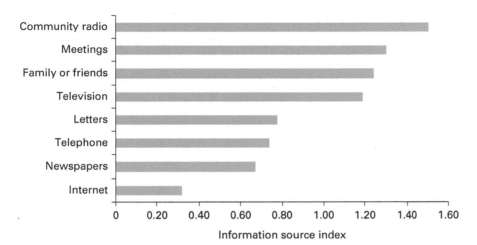

Information source index

Indigenous respondents most frequently use community radio; it is the only information source in the high-use category, with an ISI value of 1.51. A comparison with other ICTs indicates that respondents use community radio almost five times as often as the Internet and twice as much as the telephone. The importance of community radio in indigenous peoples' information and communication patterns is also confirmed by national ICT statistics, which show that Bolivia has 89 radios per 100 persons, one of the highest per capita rates in all of Latin America (ITU 2003). The survey results as well as the national statistics reflect the high levels of appropriation of community radios by indigenous peoples. Indigenous peoples have successfully incorporated community radios into their community structure. The overwhelming majority of community radio stations are owned, managed, and operated by local indigenous professionals; moreover, the majority of community radio programs are transmitted in indigenous languages, and their content is developed mostly by local correspondents, reflecting communities' local interests and information needs. Thus local radio stations have become an integral part of the information ecology in many indigenous communities across rural Bolivia (Girard 1992).

Furthermore, indigenous peoples continue to use community meetings and networks of family and friends, both considered to be important sources of information. These two information sources fall into the intermediate-use category, with community meetings (ISI = 1.30) being used only somewhat less than community radio and with networks of family and friends (ISI = 1.24) also being used relatively frequently. In contrast, the Internet plays a minor role: it is the only source of information in the minimal-use category (ISI = 0.32).

Finally, the no-response rate was very high for questions related to the Internet (with 67 percent of participants not answering the question), telephone (46 percent), and television (39 percent). These numbers confirm the low use of these technologies, possibly as a result of the very limited access to electricity and communications infrastructure services in rural indigenous communities.

In contrast, traditional means of communication continue to play an important role. The extent to which the respondents use meetings and networks of family members or friends (dependent variable) as their main source of information is positively associated with the type of intermediary organization (independent variable). The variable for type of intermediary organization was constructed based on the six clusters described in the introduction to this chapter. Figure 4.2 presents the results of a cluster analysis, which shows that national indigenous organizations (ISI = 1.67) and international indigenous organizations (ISI = 1.59) make the most use of community meetings.

Moreover, the nongovernmental organization (NGO) and OMAK youth make the least use of community meetings (ISI = 1.03 and 1.05, respectively). Both of these information sources still score relatively high and fall into the intermediate-use category. Furthermore, both grassroots and national indigenous organizations make effective use of kinship networks (family and friends) to receive information—in particular, the respondents from lowlands grassroots organizations use this information source the most (ISI = 1.59). Respondents from the national indigenous organization also use networks of family and friends relatively frequently to receive information: with an ISI of 1.40, this information source falls in the intermediate-use category. It is, however, noteworthy that while the respondents from the international indigenous

FIGURE 4.2   Most frequently used information sources, by type of intermediary organization

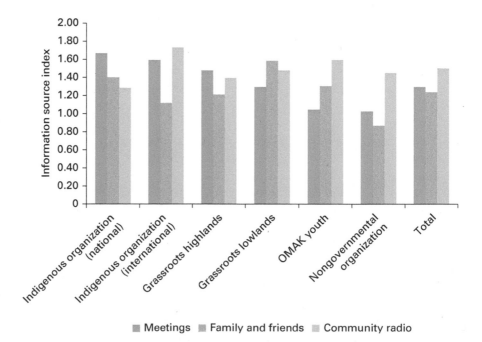

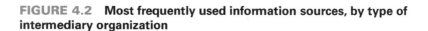

organization use their networks of family and friends less frequently, they nevertheless continue to use kinship networks at an intermediate level (ISI = 1.12). These findings are confirmed by the Kruskal-Wallis test,[1] which shows a significant difference in the use of meetings ($\chi^2 = 18.51$, $p = 0.00 < 0.05$) and the use of networks ($\chi^2 = 16.17$, $p = 0.01 < 0.05$), depending on the intermediary organization, as detailed in table E.1 in appendix E.

Therefore, indigenous peoples continue to use community meetings and face-to-face conversations as a key source of information at all levels of leadership, even at the national and international levels. The use of community radio also remains critical for all types of intermediary organizations, with respondents from the international indigenous organization (ISI = 1.73), the youth program (ISI = 1.60), and the NGO (ISI = 1.45) saying that they consider this medium to be their most important source of information. The Kruskal-Wallis test shows that the use of community radio is independent of the type of intermediary organization ($\chi^2 = 9.98$, $p = 0.08 > 0.05$).

Furthermore, an association exists between the use of meetings and networks and the type of ICT program (figure 4.3). In order to establish this association, the ICT program variable is broken down into the following nine categories: CIOEC, Fondo Indígena, OMAK leaders, CIOEC Potosí, CIOEC Oruro, APCOB, CIDOB, OMAK youth, and ICO.

The Kruskal-Wallis test reveals a significant difference in the use of meetings ($\chi^2 = 18.89$, $p = 0.02 < 0.05$) and the use of networks ($\chi^2 = 22.39$, $p = 0.00 < 0.05$) by the type of program, as shown in table E.2 in appendix E. The relationship between the use of community radio and ICT program is, however, not statistically significant ($\chi^2 = 13.02$, $p = 0.11 > 0.05$). The APCOB's program frequently uses kinship networks (ISI = 1.71), whereas the program run by ICO, which works primarily with non-indigenous small-scale farmers, uses both community meetings (ISI = 1.03) and kinship networks the least (ISI = 0.87). Thus indigenous respondents use traditional information sources to a larger extent than non-indigenous respondents.

FIGURE 4.3 Most frequently used source of information, by program

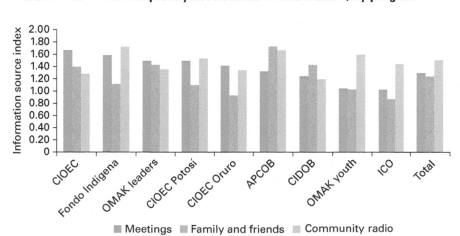

FIGURE 4.4   Use of television, telephone, and the Internet as information source, by type of organization

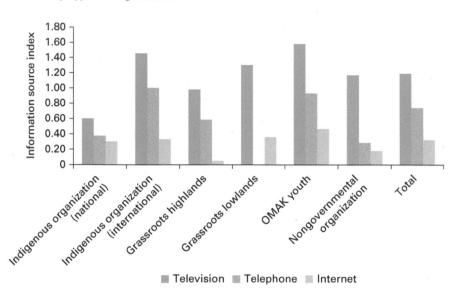

There is a significant association between both the use of telephone and television and the type of intermediary organization (figure 4.4). The Kruskal-Wallis test confirms a statistically significant association between both the medium and the intermediary organization, as indicated by the values for telephone use ($\chi^2 = 24.73$, $p = 0.00 < 0.05$) and television use ($\chi^2 = 34.99$, $p = 0.00 < 0.05$). Use of the Internet remains minimal for all intermediary organizations, and there is no statistically significant association between use of the Internet and type of intermediary organization ($\chi^2 = 8.04$, $p = 0.15 > 0.05$). The detailed results of the Kurskal-Wallis test are presented in table E.1 in appendix E.

The cluster analysis shows that the OMAK indigenous youth program differs significantly from the other types of programs. The data indicate that the youth cohort is making significantly more use of television (ISI = 1.58), telephone (ISI = 0.93), and the Internet (ISI = 0.46) than the overall sample, whereby respondents indicated a relatively low use of television (ISI = 1.19), telephone (ISI = 0.74), and the Internet (ISI = 0.32). Indigenous youth use community meetings as their source of information to a much lesser extent (ISI = 1.05) than respondents from the overall sample (ISI = 1.30).

With regard to ethnicity, there is no clear association between the use of various information sources and ethnicity (figure 4.5). The ethnicity variable was based on the following question: Which of the following languages do you primarily speak at home? The questionnaire differentiated between the following six categories: (1) Spanish, (2) Spanish and an indigenous language, (3) Aymara, (4) Quechua, (5) Guaraní, and (6) another indigenous language. In order to construct the ethnicity variable, a classification based on the following three categories was used: (1) Spanish, (2) Spanish and indigenous language,

FIGURE 4.5   Most frequently used source of information, by ethnicity

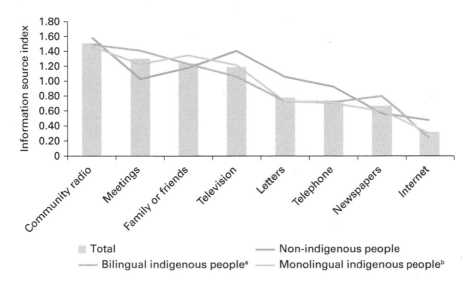

a. Respondents who speak Spanish and an indigenous language
b. Respondents who only speak an indigenous language at home

and (3) indigenous language. The questionnaire options three through six were aggregated into one category: "indigenous language." This was done to simplify the variable and because only a small number of respondents selected options three, four, five, or six.

The non-indigenous population uses television and ICTs such as the telephone and the Internet more frequently than the group of respondents who primarily speak an indigenous language and Spanish or the group who speak an indigenous language at home (figure 4.5). Moreover, respondents who speak Spanish and an indigenous language use broadcasting media such as television (ISI = 1.06) and ICTs such as telephone (ISI = 0.72) and the Internet (ISI = 0.24) to a lesser extent than the non-indigenous respondents, who rated the use of these information sources as follows: television (ISI = 1.41), telephone (ISI = 0.93), and the Internet (ISI = 0.48).

The Kruskal-Wallis test finds a statistically significant relationship only between the use of television and ethnicity (χ^2 = 7.79, p = 0.02 < 0.05), as shown in table E.3 in appendix E. The non-indigenous respondents use ICTs such as the telephone and the Internet more frequently than indigenous respondents, but the association between their use of these media and ethnicity is not statistically significant, as shown in table E.3 in appendix E.

In contrast, non-indigenous respondents use traditional ways of communication—community meetings and networks among family and friends—slightly less frequently than indigenous respondents. Respondents who primarily speak an indigenous language use community meetings as their main source of information more frequently than non-indigenous respondents (ISI = 1.23 and 1.03, respectively), and they use networks of family members and friends more extensively than non-indigenous respondents (ISI = 1.35 and 1.18, respectively). Only the association between the use of meetings and ethnicity is statistically significant ($\chi^2 = 7.62$, $p = 0.02 < 0.05$), as shown in table E.3 in appendix E. Furthermore, community radio is the most frequently used source of information by all respondents independent of their ethnicity. This finding is confirmed by the Kruskal-Wallis test, which shows no statistically significant association between the use of community radio and ethnicity ($\chi^2 = 1.04$, $p = 0.60 > 0.05$).

A Communications Profile

This section analyzes the survey-based results related to the main patterns and channels of communication of indigenous peoples in Bolivia in order to gain greater insight into their current communications profile. It considers the communications channel index, which was constructed based on respondents' answers to the following question: How frequently do you use the following communications channels to communicate with family and friends? Specifically, the index was constructed based on a population-size-weighted average using a three-point rating scale with the following values:

- 0: this type of medium is not used to communicate with family and friends
- 1: this type of medium is used only somewhat to communicate with family and friends
- 2: this type of medium is used a lot to communicate with family and friends.

In order to facilitate analysis of the resulting data, four categories were created:

- Minimal use: 0.0–0.49
- Low use: 0.5–0.99
- Intermediate use: 1.0–1.49
- High use: 1.5–2.0.

The first important finding is that indigenous peoples continue to use traditional means of communication frequently (figure 4.6). A majority of the respondents said that they use personal visits (CCI = 1.39) or *encargo* (CCI = 1.33)—a traditional practice of relaying messages through a trusted third-party, normally a friend or family member—to communicate with family members and friends.

On the one hand, indigenous respondents frequently use amateur radio (CCI = 1.24) to communicate with other communities and indigenous organizations at the local, regional, and national levels.[2] On the other hand, they rarely use the Internet (CCI = 0.49) and cell phones (CCI = 0.83) to communicate with family members and friends. The no-response rates for the Internet and cell phones are very high (63 and 52 percent, respectively), which may be attributed to the fact that the overall access of indigenous peoples to these media remains very limited—including widespread lack of Internet connectivity and cell phone coverage in rural areas.

For a more in-depth analysis of the data, I now assess the extent to which the use of communications channels is associated with other external variables.

FIGURE 4.6   Most frequently used channel of communication with family and friends

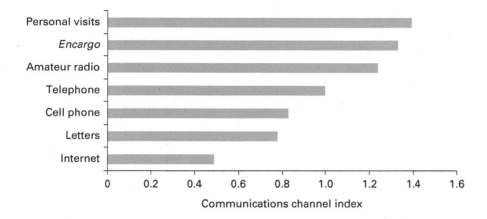

FIGURE 4.7   Use of personal visits as a means of communication, by program

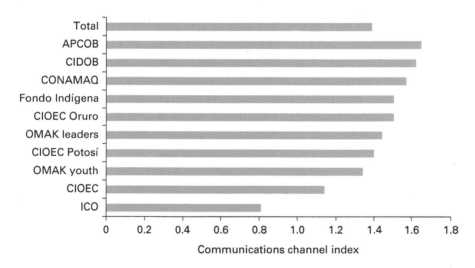

The respondents' preference for, and frequency of, using personal visits as a way to communicate with family and friends are significantly associated with both ICT program and profession. The Kruskal-Wallis test confirms a statistically significant association between visits and the two variables (appendix E). Respondents from APCOB (CCI = 1.65), CIDOB (CCI = 1.62), and CONAMAQ (CCI = 1.57) expressed the strongest preference for this type of communication (figure 4.7).

While it is not surprising that respondents who are associated with both of the grassroots organizations included in the sample (APCOB and CIDOB) indicated a preference for personal visits, it is unexpected that the leaders associated with CONAMAQ (CCI = 1.57) and Fondo Indígena

(CCI = 1.50)—who are based in the capital city, represent indigenous peoples at the national and international levels, and travel frequently—also continue to prefer personal visits. In contrast, the primarily non-indigenous respondents from ICO (CCI = 0.81) rarely use personal visits to communicate with family members and friends. In fact, respondents from the indigenous grassroots programs APCOB and CIDOB use visits twice as frequently as the respondents from ICO.

The finding that indigenous peoples use personal visits more frequently than non-indigenous peoples is also confirmed by the finding that use is associated with profession (summarized in table E.5 in appendix E). Indigenous leaders indicated that personal visits are a critical channel of communication. Indeed, this is the only group where the communications channel index reaches a high level of use (CCI = 1.50). For the respondents who are council members and students, this index reflects an intermediate level of use (CCI = 1.40 and 1.35, respectively). However, respondents who are farmers point to a low use of personal visits (CCI = 0.84). The Kruskal-Wallis test (table E.5 in appendix E) confirms the statistically significant association between the use of personal visits and profession ($\chi^2 = 17.14$, $p = 0.00 < 0.05$).

The use of amateur radio is widespread across the ICT programs studied (figure 4.8). Respondents from one program (APCOB) indicated a high level of use of amateur radio (CCI = 1.67), and respondents from three programs (OMAK youth, CIOEC, CIDOB) indicated an intermediate level of use. Respondents in only two programs (CIOEC Oruro and ICO) indicated a low level of use of amateur radio. The widespread use of amateur radio across programs is also confirmed by the Kruskal-Wallis test, which indicates that the association between the use of amateur radio and program is not statistically significant ($\chi^2 = 15.21$, $p = 0.09 > 0.05$), as shown in table E.4 in appendix E.

FIGURE 4.8 Use of amateur radio as a channel of communication, by program

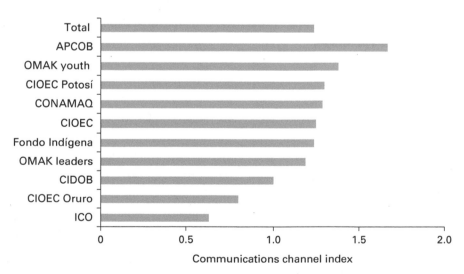

Communications channel index

A principal finding from this section is that indigenous communities have incorporated both community radio and amateur radio into their "information ecology" and increasingly use these technologies to receive information and to communicate with other indigenous communities or with family and friends.

How have these technologies been integrated so successfully into the everyday lives of indigenous communities? A key factor is that both of these technologies are based on the "oral" tradition of indigenous communities and are seen as culturally "appropriate" mechanisms for exchanging information in indigenous communities. Second, radio technology is relatively simple and inexpensive to maintain, and indigenous peoples have been able both to shape the technologies and to adapt them to their specific needs. Third, both of these channels have responded to specific local information and communication needs and are able to (partially) address them. Finally, intermediary organizations, such as NGOs, and support organizations, such as APCOB, have played an important role in providing technical assistance, financing, and training to indigenous communities in the use of these technologies. Today, the majority of communities are able to secure the financial and social sustainability of these services by themselves.

Finally, indigenous peoples continue to use traditional information sources and communications channels extensively, such as community meetings and kinship-based networks of family and friends. However, their use of the Internet is very limited; these technologies have not been integrated into the "information ecology" of indigenous communities.

Awareness, Access, and Use

What awareness do indigenous peoples have of ICTs? And what level of access to the Internet is available to indigenous peoples in rural areas?

The awareness of the Internet is relatively high in indigenous communities. Results from focus groups and the survey show, however, that indigenous peoples have only a vague idea about what the Internet can and cannot do. One important factor is the low level of access to this technology in indigenous communities. There is a strong association between having "no access" to the Internet and being indigenous (figure 4.9). Cross-tabulating the survey results and the ethnicity index developed by Xavier Albó and Ramiro Molina (Molina and Albó 2006; Albó and Quispe 2004)—described in appendix E— shows that more than 75 percent of the municipalities with a high degree of ethnicity have no access to Internet services, compared with only 22 percent of the municipalities with a low degree of ethnicity.

However, having access to the Internet in a municipality does not mean that people are actually using it. For this reason, having access to the Internet is a necessary, but not a sufficient, condition for its use.

Why don't more indigenous Bolivians use the Internet? There are several barriers to Internet use. One reason is that access to educational services is limited. The majority of indigenous women and elders attain only very low levels of education, as measured by years of formal schooling. The same municipalities that have a high degree of ethnicity and a very low rate of access to the

FIGURE 4.9   **Municipalities with and without access to the Internet, by degree of ethnicity**

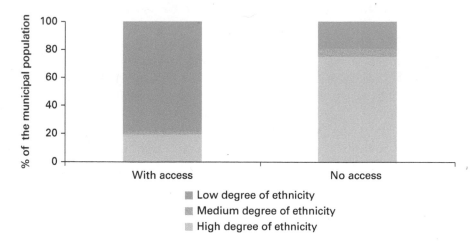

FIGURE 4.10   **Illiteracy rates in municipalities, by degree of ethnicity**

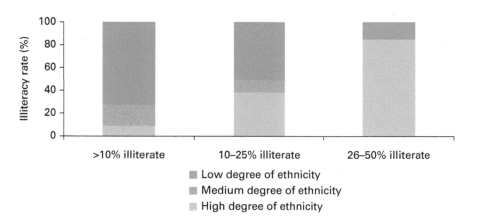

Internet also have a very high rate of illiteracy (figure 4.10). About 85 percent of the municipalities with a high degree of ethnicity have a high rate of illiteracy of between 26 and 50 percent. In contrast, the large majority of municipalities with a low degree of ethnicity (72 percent) have a low rate of illiteracy of less than 10 percent. Because of the high rate of illiteracy and the low level of Internet access, a large percentage of indigenous peoples have never used the Internet.

This finding is confirmed by the survey results. The large majority (71 percent) of the respondents said that they have never used the Internet, while only 24 percent said they have used it (figure 4.11). Out of the total respondents, 84 people (24 percent) said they have used the Internet before,

FIGURE 4.11   Prior use of the Internet

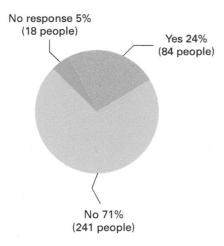

No response 5%
(18 people)

Yes 24%
(84 people)

No 71%
(241 people)

FIGURE 4.12   Principal reasons for not using the Internet

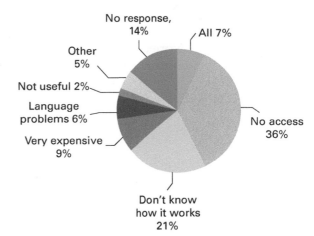

No response,
14%

All 7%

Other
5%

Not useful 2%

Language
problems 6%

Very expensive
9%

No access
36%

Don't know
how it works
21%

but 49 (13 percent) said they used it for the first time during the training workshop they were attending. This means that only 35 people (9 percent of all respondents) had used the Internet before attending the training workshops.

Respondents frequently cited lack of access as a reason why they had not used the Internet before (36 percent; figure 4.12). Another reason given was the lack of human capacity or training (21 percent); only a small number said that the Internet is not useful (2 percent). These numbers suggest that lack of access is not the only reason why the large majority of indigenous peoples have not used the Internet and that the lack of training, language challenges, and the high costs of Internet are equally important. The data also show that the large majority of indigenous peoples are generally open to this new technology.

Information and Communication Needs

This section analyzes in more detail the information needs of indigenous peoples. Analyzing the specific information and communication needs is essential for understanding the effects of ICT programs on poor people's human and social capabilities. Therefore, this section analyzes in more detail the information needs of indigenous peoples and presents the results of an information needs index that was constructed based on the respondents' answers to the following question: Which type of information do you need? The index is based on a population-seize-weighted average using a three-point rating scale with values from 0 to 2, as follows:

- 0: the respondents do not express any demand for this type of information

- 1: the respondents only somewhat need this type of information
- 2: the respondents have a very strong demand for this type of information.

The index thus captures the respondents' perceptions about which types of information they value the most. For the data analysis, four categories were created, reflecting differences in the respondents' need for information:

- Minimal need: 0.0–0.49
- Low need: 0.5–0.99
- Intermediate need: 1.0–1.49
- High need: 1.5–2.0.

Figure 4.13 presents the types of information that indigenous peoples would like to receive. They have the greatest need for information about educational materials (INI = 1.57), agricultural practices (INI = 1.53), and local news (INI = 1.43). They have limited need for information on festivities and religious events (INI = 0.92) and job opportunities (INI = 1.15) and relatively modest need for information on government policies (INI = 1.22).

Furthermore, an information availability index (IAI) was developed in order to measure the extent to which different types of information are available to indigenous peoples. Similar to the information needs index, it was constructed on the basis of the question: How much information is available to you about the following? The index was based on a weighted average using a three-point rating scale ranging from 0 to 2.

- 0: this type of information is not available
- 1: this type of information is only somewhat available
- 2: this type of information is abundantly available.

FIGURE 4.13 **Information needs**

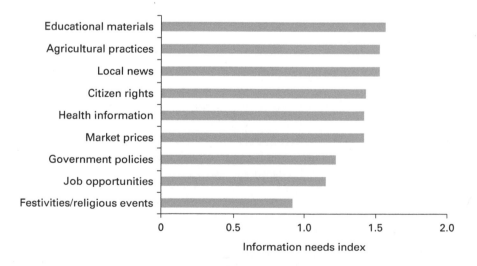

Information needs index

In order to facilitate analysis of the resulting data, four categories were created:

- Minimal availability: 0.0–0.49
- Low availability: 0.5–0.99
- Intermediate availability: 1.0–1.49
- High availability: 1.5–2.0.

Figure 4.14 shows the type of information that the respondents are currently receiving in their communities. There is a significant information gap for indigenous peoples—a mismatch between their information needs and the availability of a particular type of information. For instance, respondents expressed a very high demand for information on education (INI = 1.57), but this type of information is not readily available (IAI = 0.98). However, indigenous respondents indicated that market prices are their sixth most important information need (INI = 1.42), but the second most available type of information (IAI = 1.03).

These findings were also confirmed in the focus groups, where indigenous community members expressed frustration about being "isolated" from the rest of society and strongly criticized the national government for ignoring their interests and concerns (focus group with indigenous leaders in Tiwanaku in April 2005 and in Concepción in 2005).

As a last step, the section describes the results of an "information gap" analysis, in order to investigate the areas in which indigenous peoples face the greatest barriers to access the information they consider important for their well-being. It develops an information gap index (IGI) that was constructed based on the results for the IAI and the INI just described. It is calculated by subtracting the value of the information availability index from the value of the information needs index.

FIGURE 4.14 Information received

Information availability index

Education, agriculture, citizenship rights, and health are the four areas with the greatest information gaps (figure 4.15). The gap for local news (IGI = 0.32), government policies (IGI = 0.27), and information on festivities and religious events (IGI = 0.16) is rather marginal.

What accounts for the large variations in information gap by type of information? The top line of figure 4.16 indicates the information needs of indigenous peoples, and the bottom line indicates the extent to which they currently have access to different types of information. In certain areas

FIGURE 4.15   Information gap, by type of information (full sample)

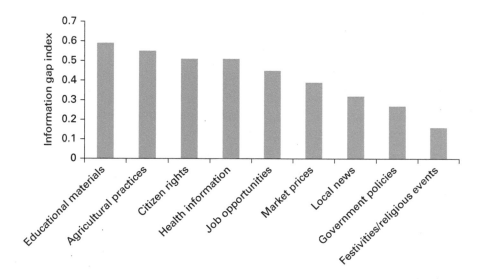

FIGURE 4.16   Detailed information gap, by type of information

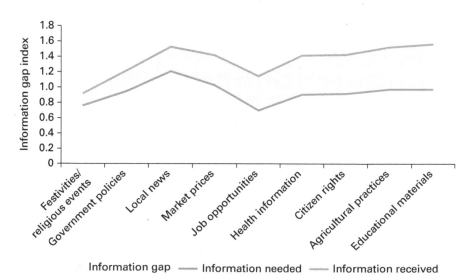

(education and agriculture), the information gap can be attributed primarily to the large demand for this type of information, while in other areas (job opportunities and market prices), the gap can be attributed mainly to the limited capacity to access this type of information.

The differences in information gap arise for various reasons. For instance, indigenous peoples have a strong demand for local news (INI = 1.57), and that demand is almost met by their significantly higher capacity to receive information (IAI = 1.21) in comparison to other topics, resulting in a small information gap (IGI = 0.32). In the case of festivities and religious events (INI = 0.92) and government policies (INI = 1.22), the information gap is small (0.16 and 0.27, respectively), primarily due to relatively weak demand for information on these topics.

The size of the information gap is statistically significant for all of the eight topics listed in figure 4.17. The Wilcoxon test finds a statistically significant information gap for all types of information. Even the small information gap for the topic of festivities and religious events (IGI = 0.16) is statistically significant (χ^2 = –2.283, p = 0.02 < 0.05). Details are presented in table E.6 in appendix E.

A noteworthy variation with regard to the perceived ability to satisfy their information needs is distinct to the OMAK youth program. The youth cohort has a significantly lower demand for information than the rest of the sample. In fact, in six out of nine topics, respondents indicated that they receive more information than they would like, resulting in a negative information gap. For instance, in relation to job opportunities and government policies, the information gap is equivalent to –0.22 and –0.25, respectively.

FIGURE 4.17 Information gap for OMAK youth, by type of information

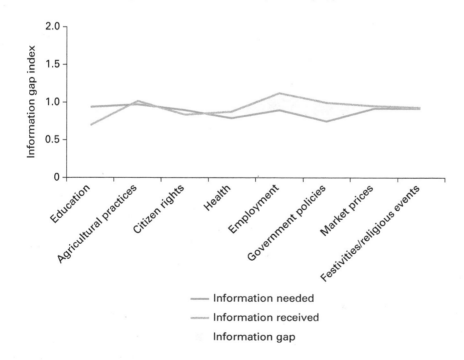

Information gap index

— Information needed

– Information received

Information gap

The Wilcoxon test also reveals the existence of a positive and statistically significant information gap in the area of educational material (χ^2 = 2.507, p = 0.01 < 0.05), but a negative and statistically significant information gap (χ^2 = –2.011, p = 0.04 < 0.05) in the area of government information. This negative information gap (IGI = –0.25) indicates that indigenous youth respondents believe that they receive an abundance of government information. For all of the other topics, the Wilcoxon test finds no statistically significant information gap (table E.7 in appendix E).

The two focus groups carried out with OMAK youth representatives provide a possible explanation for this surprising finding. Youth said that they receive enough information from their schools, whereas the rest of the indigenous respondents expressed concerns about being unable to access needed information (focus groups with OMAK indigenous youth and other OMAK participants). At the time of the research, the youth participants were enrolled in either their junior or senior year of secondary school—a factor contributing to their greater access to information.

Because of the distinct results from the youth program, I subsequently conducted an information gap analysis without the youth cohort. The results are summarized in figure 4.18. After removing the youth cohort, the information gap for all topics except for local news is significantly greater than in the sample with youth. For instance, the information gap in agricultural practices and educational materials rose from 0.55 to 0.75 and from 0.59 to 0.72, respectively. The greatest increase in the information gap is found for information about job opportunities, with the gap rising from 0.45 to 0.74.

In addition, when comparing the information gaps across all topics, job opportunities goes from being the sixth largest gap to being the second. In fact,

FIGURE 4.18 Information gap, without the youth cohort, by type of information

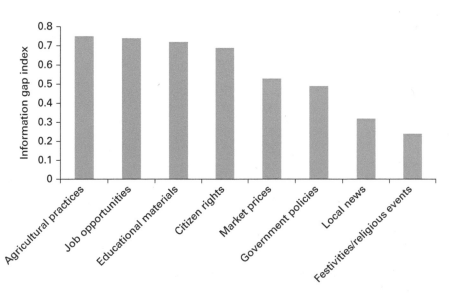

FIGURE 4.19   Detailed information gap, by type of information (without youth cohort)

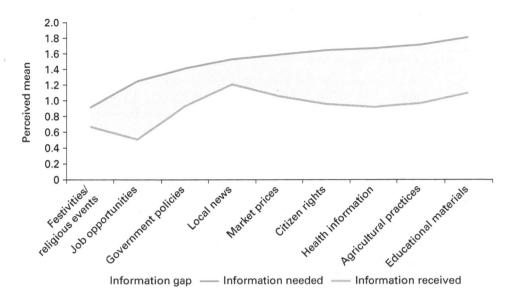

the detailed data analysis shows that four of the topics in the aggregate sample where the information gaps are the largest change slightly once the youth cohort is excluded. Here, the following topics exhibit the largest information gaps: agricultural practices, health information, job opportunities, and educational materials (figure 4.19).

The data illustrate that access to educational information is continuously given the highest priority, reflected by high demand (INI = 1.82). Subsequently, in order of importance and as measured by demand for information, agricultural information (INI = 1.72) is second, followed by health information (INI = 1.67) and citizenship rights (INI = 1.65). These results are similar to the findings from the aggregate sample, where these four topics also exhibited the greatest demand. Thus the differences between the overall sample and the sample excluding the youth cohort are due to differences in the perceived availability of information, while the perceived need for information between the two samples is very similar. Although youth respondents indicated that they have ample access to information in several areas, such as government information, the information gap related to job opportunities is high (IGI = 0.74) due to their minimal access to (IAI = 0.51, lowest value) and significant demand for (INI = 1.25) this type of information.

Information Gap by Program

This section analyzes the extent to which there is an association between variations in the information gap for a particular topic and ICT program. The relationship between these two variables is significant in the case of the OMAK youth program, which raises a similar question for the other programs. The analysis focuses on five topics: education, agriculture, citizenship rights, market prices, and government policies.

Educational Information

In education, there is a strong association between variations in the information gap and ICT program (figure 4.20). The Kruskal-Wallis significance test demonstrates a statistically significant association between the respondents' need for information on educational materials and the ICT program (χ^2 = 77.37, p = 0.00 < 0.05) and a statistically significant relationship between the extent to which the respondents receive information and the program variable (χ^2 = 26.34, p = 0.00 < 0.05), as shown in table E.9 in appendix E.

Both the CIOEC and OMAK grassroots programs exhibit a very small information gap in education, with an IGI value of 0.38 and 0.39, respectively. The CIOEC Oruro, CIOEC Potosí, and ICO programs have a relatively large information gap, with an IGI value of 1.08, 0.87, and 0.85, respectively. Moreover, with the exception of OMAK youth (INI = 0.98), the overall demand for educational information is very strong across all other programs, with the need for information being approximately 1.80 for all programs. Variations in the information gap can be attributed to disparities in the respondents' capacity to receive—in this case, educational—information. For instance, in the CIOEC Oruro program, respondents expressed a very high demand for educational information (INI = 1.87), similar to that of respondents in the national CIOEC program (INI = 1.88). However, in the remote region of Oruro, indigenous peoples have a much lower capacity to receive educational information, as evidenced by the very low information availability index of 0.79, compared with respondents from the national

FIGURE 4.20   Information gap regarding educational material, by program

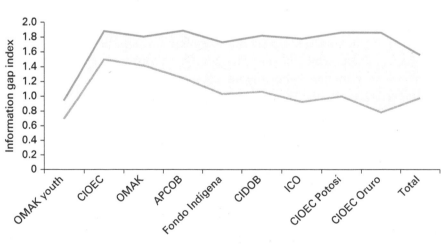

Information gap —— Information needed —— Information received

CIOEC program, who have the highest capacity of all the indigenous respondents to attain educational information (IAI = 1.50).

Moreover, the data analysis points to an association between demand for educational information and ethnicity, whereby respondents who speak both an indigenous language and Spanish in their household have the highest demand and those who primarily speak an indigenous language have the lowest demand. Figure 4.21 illustrates this finding, with the overall share of respondents expressing a substantial demand for informational education rising from 68.9 percent of the total sample to 81.5 percent of persons from bilingual (Spanish and indigenous language) households (table E.10 in appendix E).

The figure also shows that 34.4 percent of indigenous respondents who primarily speak an indigenous language at home demand educational information, compared with 12.4 percent of the total sample. The association between demand for educational information and ethnicity is statistically significant, as confirmed by the chi-square test ($p = 0.00 < 0.05$). However, the Cramer V test—with a value of 0.32—indicates that while these two variables are associated, the relationship is relatively weak (table E.10, panel b, in appendix E). The value of the Cramer V test varies between 0 and 1, with a value close to 0 showing little association between two variables and a value close to 1 indicating a very strong association.

Figure 4.22 confirms this finding, showing an educational information gap that is three times larger for respondents who speak both an indigenous language and Spanish at home (IGI = 0.71) than for those who primarily speak an indigenous language (IGI = 0.24). The Kruskal-Wallis test confirms a statistically significant association between ethnicity and both (a) the respondents' need for educational information ($\chi^2 = 50.26$, $p = 0.00 < 0.05$) and (b) the extent to which they receive educational material ($\chi^2 = 6.97$, $p = 0.04 < 0.05$). For details, see table E.11 in appendix E.

FIGURE 4.21 **Information need regarding educational material, by ethnicity**

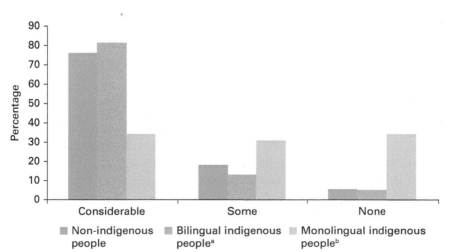

a. Respondents who speak Spanish and an indigenous language.
b. Respondents who only speak an indigenous language at home.

FIGURE 4.22   Information gap regarding educational material, by ethnicity

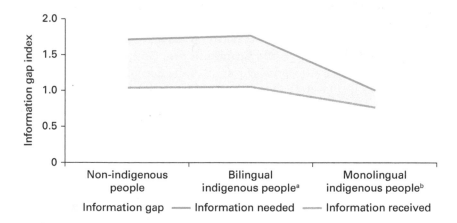

a. Respondents who speak Spanish and an indigenous language.
b. Respondents who only speak an indigenous language at home.

FIGURE 4.23   Information needs for education by ethnicity (without youth cohort)

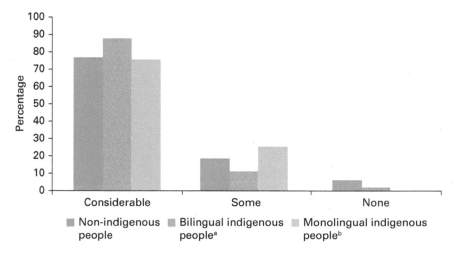

a. Respondents who speak Spanish and an indigenous language.
b. Respondents who only speak an indigenous language at home.

The small information gap is due to the low demand for educational information among respondents from indigenous-language households (INI = 1.00). This low demand is attributed to the inclusion of respondents from the OMAK youth program, who have a disproportionately low demand for information across all topics and represent 87 percent of the respondents who primarily speak an indigenous language at home. For this reason, the youth cohort is excluded from the subsequent analysis.

The results, presented in figure 4.23, indicate that the demand for educational information rises significantly for respondents from indigenous-language households when excluding the youth cohort. All of the remaining respondents from this group said that they want more educational information, with 75 percent indicating that they have a strong demand for

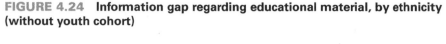

FIGURE 4.24   Information gap regarding educational material, by ethnicity (without youth cohort)

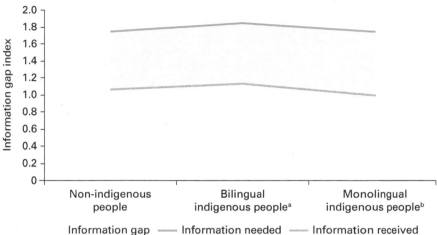

Information gap ——— Information needed ——— Information received

a. Respondents who speak Spanish and an indigenous language.
b. Respondents who only speak an indigenous language at home.

this type of information. Furthermore, the demand from respondents who speak an indigenous language and Spanish at home also rises—from 81.5 to 87.2 percent. Therefore, when excluding the youth cohort, (a) the overall demand for educational material increases, and (b) the demand for educational material is as high for respondents from indigenous-language households as for those from bilingual (Spanish and an indigenous language) and Spanish-speaking households. The chi-square test—shown in table E.12, panel b, in appendix E—confirms this finding, indicating that the relationship between the demand for educational information and ethnicity is statistically insignificant when the youth cohort is excluded ($\chi^2 = 4.345$, $p = 0.361 > 0.05$).

Both bilingual and indigenous-language-only respondents face a slightly larger educational information gap than the non-indigenous respondents (figure 4.24). In fact, persons who speak an indigenous language in the household have a larger gap than either of the other two groups (IGI = 0.75). While they have the lowest capacity to access educational information (IAI = 1.00), they also have a slightly lower demand for this type of information than the respondents who speak Spanish and an indigenous language (INI = 1.75). Persons in this group have the second largest information gap in this area (IGI = 0.72). They have the most access to educational material (IAI = 1.13) and the greatest demand for this type of information (INI = 1.85). The Spanish-speaking respondents have the smallest information gap (IGI = 0.65). These differences are not statistically significant. Both the results of the Kruskal-Wallis test for the association between educational information needs and ethnicity ($\chi^2=1.91$, $p = 0.38 > 0.05$) as well as the relationship between the respondents' ability to receive information and ethnicity are statistically insignificant ($\chi^2 = 0.62$, $p = 0.73 > 0.05$), as shown in table E.13 in appendix E.

Furthermore, women have significantly less access to educational information than men (figure 4.25). Female respondents said that they have a low level of information availability. Their information availability index of 0.78 is

FIGURE 4.25   Information gap regarding educational material, by gender

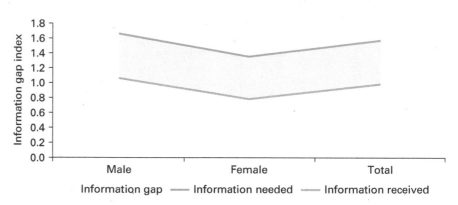

significantly lower than that of male respondents (IAI = 1.06). However, their demand for educational information is also relatively low (INI = 1.35), so the information gap in this area is even smaller than for male respondents. This finding is attributable to both their very restricted ability to access educational information and their low demand for this type of information. The Kruskal-Wallis test confirms a statistically significant association between both educational information needs and gender (χ^2 = 6.66, p = 0.01 > 0.05) and between the degree to which information is available and gender (χ^2 = 6.33, p = 0.01 > 0.05). For details, see table E.14 in appendix E.

The Aymara are also significantly disadvantaged regarding access to educational material. They have the lowest information availability index (IAI = 0.64) but also low demand (INI = 0.96), resulting in a small educational information gap (IGI = 0.31). As shown in table E.11, panel b, in appendix E, they are most disadvantaged in their ability to access educational materials (IAI = 0.64). For instance, the bilingual respondents (Spanish and indigenous language) said that the level of availability of information on educational materials is intermediate (IAI = 1.05).

The association between the need for educational information and the more detailed ethnicity variable—which disaggregates the indigenous language variable into two groups (the Aymara and Quechua)—is also statistically significant (χ^2 = 50.27, p = 0.00 > 0.05). The Kruskal-Wallis test confirms a statistically significant association between the degree of availability of educational material and the more detailed ethnicity variable (χ^2 = 9.97, p = 0.02 > 0.05), as shown in table E.11, panel b, in appendix E.

Information on Agriculture

Information on agriculture is the second most demanded type of information (INI = 1.57), but the need is inadequately met (IAI = 0.98). The association between the level of demand for agricultural information and the ICT program is significant (figure 4.26; for details, see table E.15 in appendix E). The CIOEC is salient in this regard, as all of its participants said that agricultural information is particularly important, with the INI reaching its highest possible value (2.00). At the same time, the large majority of CIOEC

FIGURE 4.26   Information gap regarding agricultural practices, by program

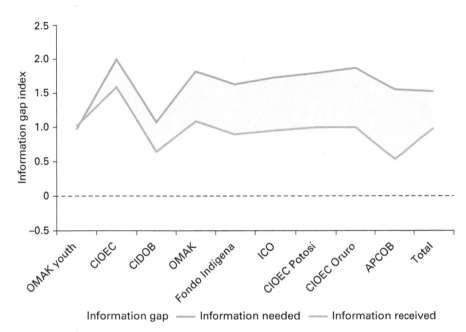

Information gap ——— Information needed ——— Information received

participants said that they have relatively high levels of access to agricultural information (IAI = 1.59). Consequently, of all the programs, CIOEC has the lowest information gap for agricultural information (IGI = 0.41).

The CIDOB results demonstrate a small information gap (IGI = 0.43), which can be attributed to the low demand for agricultural information (INI = 1.08), accompanied by limited capacity to access this type of information (IAI = 0.64).

The ICO program produced an unexpected finding: the agricultural information gap of ICO respondents is almost twice as large as the one for the CIOEC program (IGI = 0.78). While the mostly non-indigenous ICO participants have a relatively high demand for agricultural information (INI = 1.73), they lack adequate access, as evidenced by the relatively low IAI of 0.95. This is of particular relevance, as the ICO's principal objective is to provide participants with information on agricultural practices and market prices.

Citizenship Rights Information

The fourth most important priority for respondents is to improve their access to information on citizenship rights. Under this category, the information gap becomes markedly visible, as shown in table E.16 in appendix E. As shown in figure 4.27, respondents expressed a relatively strong interest in information on citizenship rights (INI = 1.43), with the majority indicating that they receive only very little information on this topic (IAI = 0.92). In other words, while they ranked improved access to citizenship rights as the fourth most important priority (figure 4.14), they only ranked it sixth in terms of availability, resulting in the third largest information gap overall (IGI = 0.51; figure 4.15).

FIGURE 4.27 **Information gap regarding citizen rights, by program**

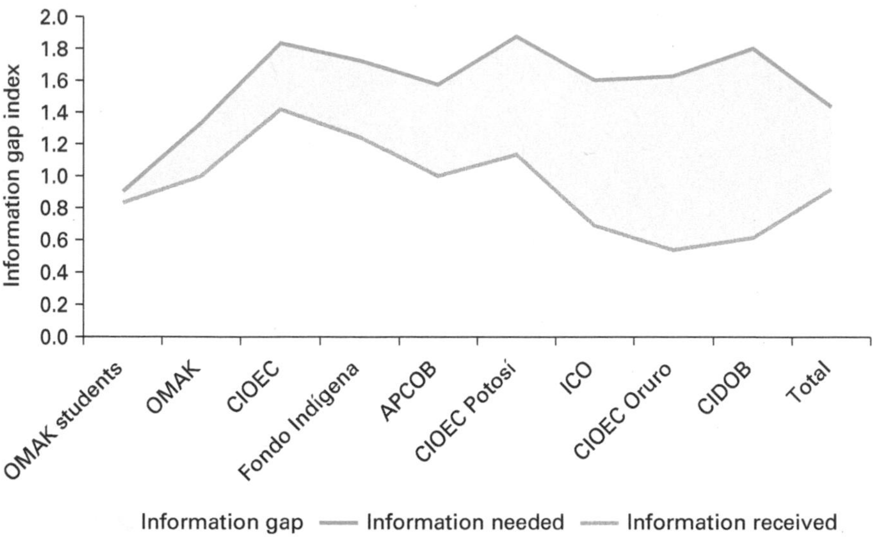

The demand for this type of information is dependent on the following variables: (a) educational attainment, (b) age, and (c) ICT program. First, in relation to educational attainment, the demand for more information on citizenship rights is the highest among respondents with the least education: 72 percent of respondents who have completed primary schooling or have less than five years of education said that their access is inadequate compared with 60 percent of the total sample (table E.17 in appendix E). The relationship between the demand for citizenship rights and educational attainment is statistically significant, as confirmed by the chi-square test ($\chi^2 = 15.612$, $p = 0.01 < 0.05$). Moreover, the Cramer V test—with a value of 0.180—indicates that the relationship is strong (table E.17, panel b, in appendix E).

Second, age is also an important factor. Older respondents (36 years and older) expressed a 1.8 times higher demand for information on citizenship rights than the younger cohort (35 years and under). Of those older than 36 years of age, 71 percent indicated a strong demand for improved access to information on their rights, compared with only 53 percent of the younger cohort (table E.18 in appendix E). The chi-square test confirms that this relationship is statistically significant ($\chi^2 = 10.359$; $p = 0.00 < 0.05$). The Cramer V test—with a value of 0.210—indicates that this is a relatively strong association (table E.18, panel b, in appendix E).

Third, the ICT program has a significant influence on respondents' ability to access the citizen rights information (figure 4.27). On the one hand, CIOEC program respondents said that access to information on citizenship rights is an important priority (INI = 1.83) and that they are generally satisfied with their access (IAI = 1.42). On the other hand, Fondo Indígena respondents expressed strong demand (INI = 1.73) but a greater level of dissatisfaction with their access (IAI = 1.24). This is noteworthy, as a main objective of both programs is to increase indigenous peoples' awareness of their rights. Similar to Fondo

Indígena, respondents from the other grassroots programs—CIOEC Potosí (INI = 1.88), CIDOB (INI = 1.80), and CIOEC Oruro (INI = 1.63)—expressed a strong desire for information on citizenship rights but insufficient access. The Kruskal-Wallis test confirms that the association between the demand for information on citizenship rights and ICT program is statistically significant (χ^2 = 55.73; p = 0.00 < 0.05). Moreover, the relationship between the level of information received is also associated with the program variable (χ^2 = 21.81; p = 0.01 < 0.05), as shown in table E.16 in appendix E.

Market Price Information

In terms of market prices, the information gap is relatively small (IGI = 0.39), with the majority of indigenous respondents considering such information to be a lower priority than access to information on education, agriculture, or citizenship rights. This small information gap may also be attributed to its supply, as market price information ranks as the second most received information (IAI = 1.03; figure 4.15). This reduces the demand for market price information (INI = 1.43), and respondents ranked it as the fifth most needed type of information (figure 4.14). A more detailed analysis shows an association between age and market prices.

As shown in figure 4.28, the group of respondents who are at least 36 years or older indicated a very high demand for this type of information (INI = 1.80). In comparison, the group of respondents who are 35 years or younger expressed an intermediate level of information needs in this area (IGI = 1.22).

Thus the information gap is 2.8 times larger for the older cohort than for the younger cohort of respondents. The Kruskal-Wallis test confirms this statistically significant association (χ^2 = 34.07, p = 0.00 < 0.05), as shown in table E.19 in appendix E. In relation to the differences by program, CIOEC respondents exhibited both the greatest demand—reaching the highest possible INI of 2.0—and the best access to market price information of all programs, indicated by an IAI of 1.80. Consequently, CIOEC respondents exhibited the smallest information gap (IGI = 0.20) for market prices.

FIGURE 4.28   Information gap regarding market prices, by age

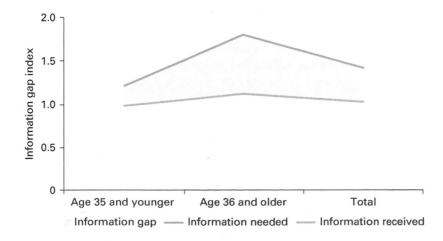

FIGURE 4.29   Information gap regarding market prices, by program

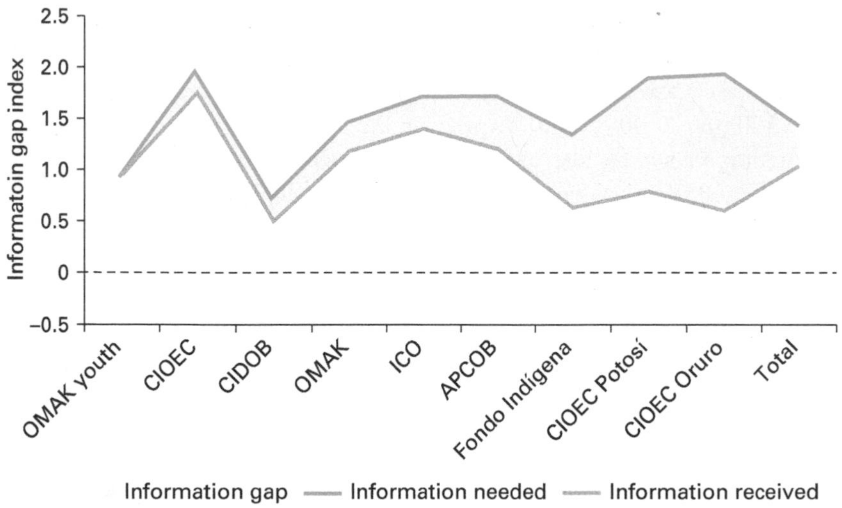

Information gap ——— Information needed ——— Information received

Second, both OMAK leaders and CIDOB respondents have small information gaps, but for different reasons (figure 4.29). In the case of CIDOB, the small information gap is attributable to the very low priority given to having access to market prices (INI = 0.71). While in the case of OMAK leaders, access to market prices is important (INI = 1.45) and adequately fulfilled through a fair amount of information (IAI = 1.18). The resulting small information gap is not as significant as in the case of the CIOEC program, since both the INI and IAI values represent an intermediate level of demand for and supply of information.

While the respondents from CIOEC Oruro (INI = 1.92) and CIOEC Potosí (INI = 1.88) said that having access to market price information is a high priority, they also expressed concern that they are unable to access sufficient information—evidenced by the very low information availability indexes for these programs (IAI = 0.60 and 0.79, respectively). As a consequence, both of these grassroots programs have large information gaps (for CIOEC Oruro, IGI = 1.32; for CIOEC Potosí, IGI = 1.10). ICO respondents rank third, in terms of having their market pricing needs met (IGI = 0.31). The Kruskal-Wallis test confirms a statistically significant association between the respondents' information needs in this area and the ICT program variable ($\chi^2 = 77.14$, $p = 0.00 < 0.05$) and between the availability of market price information and the ICT program ($\chi^2 = 47.26$, $p = 0.00 < 0.05$), as shown in table E.20 in appendix E.

Information on Government Policies

Finally, indigenous peoples do not consider improved access to information on government policies to be a high priority. They ranked this type of information—with an intermediate INI level of 1.22—as the seventh most important topic out of nine (figure 4.14). Only information related to job opportunities and festivities and religious events was rated lower. The small

FIGURE 4.30   Information gap regarding government policies, by program

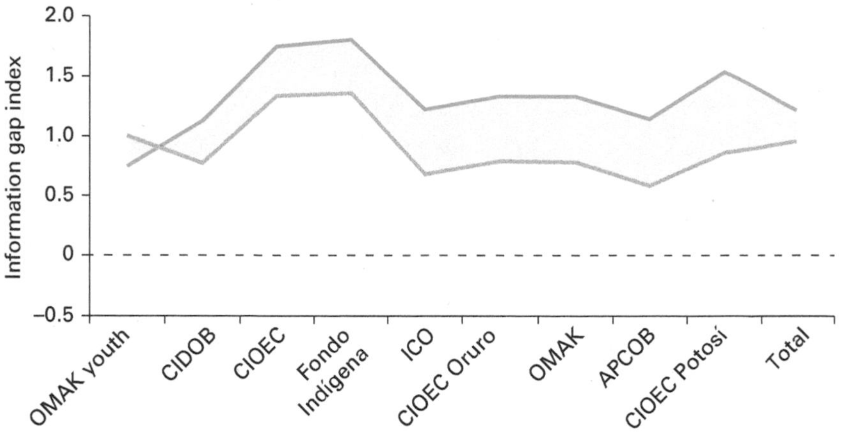

information gap of 0.27 can be attributed largely to low demand (figure 4.30). Government policies also have the second smallest information gap overall, ranking behind only festivities and religious events (IGI = 0.16).

There is a strong association between the need for government information and the intermediary organization ($\chi^2 = 53.60$, $p = 0.00 < 0.05$). The only groups expressing strong demand for this type of information were indigenous leaders who participated in the CIOEC and those who participated in Fondo Indígena programs, with a corresponding INI of 1.75 and 1.81, respectively. The high level of demand in both programs is noteworthy, since both organizations work to convince the Bolivian government of the need to incorporate indigenous peoples' interests and rights into its policy making. Better access to information on government policies and programs is one of the organizational mandates espoused by both organizations. Respondents from grassroots programs expressed a very low demand for information on government policies. For instance, participants in the CIDOB, APCOB, and OMAK programs perceived improved access to this topic as a low priority, as evidenced by low INI values (CIDOB = 1.13; APCOB = 1.14; OMAK leaders = 1.33).

The demand for government policies is even lower among respondents who predominantly speak an indigenous language in their households, particularly among women and youth. However, the greatest indifference toward government policies is among youth (INI = 0.75), producing a negative information gap (−0.25) that indicates information overload on this topic.

As shown in figure 4.31, female respondents have a lower demand for government information than male respondents (INI = 1.00 and 1.31, respectively). Consequently, they also have a significantly smaller information gap in this area than male respondents (IGI = 0.00 and 0.38, respectively). The relationship between information needs and gender is statistically significant ($\chi^2 = 6.95$, $p = 0.01 < 0.05$), but the association between the availability

FIGURE 4.31　**Information gap regarding government policies, by gender**

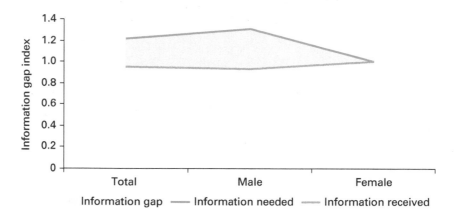

Information gap ——— Information needed ——— Information received

FIGURE 4.32　**Information gap regarding government policies, by ethnicity**

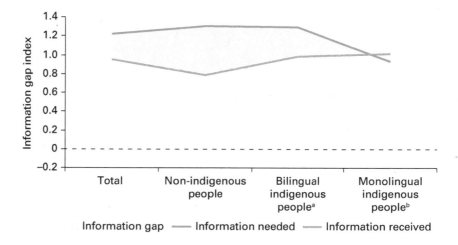

Information gap ——— Information needed ——— Information received

a. Respondents who speak Spanish and an indigenous language.
b. Respondents who only speak an indigenous language at home.

of information and gender is not statistically significant ($\chi^2 = 0.33$, $p = 0.057 > 0.05$), as shown in table E.22 in appendix E. This suggests that while women have relatively equal access to government information, they lack strong demand for this type of information.

Moreover, as shown in figure 4.32, there is a negative association between demand for information about government policies and ethnicity ($\chi^2 = 8.84$, $p = 0.01 < 0.05$), but no statistically significant relationship between perceived availability of government information and ethnicity ($\chi^2 = 2.47$, $p = 0.291 > 0.05$).

In the aggregate sample, respondents who primarily speak an indigenous language in their household expressed a very low demand for information about government policies (INI = 0.93), pointing to an information overload on government policy—evidenced by a negative IGI of –0.09. In contrast, non-indigenous respondents said that access to government policies is important, as indicated by a relatively high INI of 1.31 (table E.23 in appendix E).

FIGURE 4.33 **Information gap regarding government policies, by ethnicity (without youth cohort)**

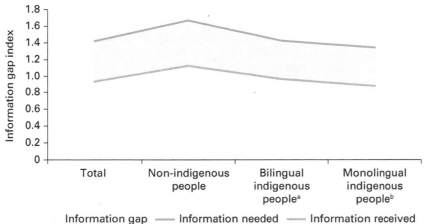

a. Respondents who speak Spanish and an indigenous language.
b. Respondents who only speak an indigenous language at home.

The more detailed data analysis shows that the Aymara expressed the lowest demand for information on government policies, with an extremely low INI for this topic (0.84), followed by the Quechua (INI = 0.91). As shown in table E.24 in appendix E, for both the Aymara and Quechua respondents, the information gap index is negative (IGI = –0.13 and –0.09, respectively).

In the following discussion, the youth cohort is excluded from the analysis, considering that this group has expressed an extremely low demand for information on government policies, which might distort the overall findings.

As shown in figure 4.33 and as expected, the overall demand for government information increases when the youth cohort is excluded (INI = 1.22 and 1.42, respectively), resulting in a somewhat larger information gap (IGI = 0.49). The association between demand for information on government policies and ethnicity is confirmed: the non-indigenous respondents have a significantly higher demand for information on this topic (INI = 1.67) than both the respondents who speak an indigenous language and Spanish (INI = 1.43) and those who speak primarily an indigenous language at home (INI = 1.35). The Kruskal-Wallis test confirms a statistically significant association between the respondents' need for government information and ethnicity (χ^2 = 1.399, p = 0.49 < 0.05), as shown in table E.25 in appendix E.

The data also point to the fact that it is significantly easier for non-indigenous respondents to access government information than for indigenous respondents. The IAI is considerably higher for the non-indigenous group (1.12) than for the group of respondents who primarily speak an indigenous language in their household (0.88). The Kruskal-Wallis test, however, shows no statistically significant association between the availability of government information and ethnicity (χ^2 = 2.42, p = 0.298 > 0.05). This can be explained by the relatively high IAI for the respondents who speak Spanish and an indigenous language (0.96). Despite the lack of a statistically significant association

TABLE 4.10   Exchange of information between the communities and the government (without youth cohort)

Answer	Number of respondents	% of respondents
Very good	27	9
Good	32	11
Normal	69	24
Bad	70	24
There are no information exchanges	84	29
Not applicable	7	2
Total	289	100

between the two variables, respondents who primarily speak an indigenous language encounter major barriers to accessing government information.

Table 4.10 supports these results. The majority of respondents (53 percent) said that the information exchanges between the government and indigenous communities are problematic. Indeed, almost a third said that there is no exchange of information between the government and indigenous peoples. Moreover, they are not interested in improving the exchange of information with government agencies.

The analysis of the communication needs of indigenous peoples reveals the problematic nature of vertical information and communication flows between indigenous communities and the formal institutions of the state as well as between the indigenous and non-indigenous populations. The most important priority of indigenous peoples, discussed next, is to improve the horizontal communications within the indigenous movement and not to enhance the vertical communications with government, international organizations, or NGOs.

Communication Flows

Respondents identified their top priority as being the need to improve the communication flows between local communities, followed by the need to enhance communication flows between indigenous organizations and local communities (figure 4.34).

The third priority is to improve communications between indigenous organizations at the national level. Notably, 55 percent of the respondents said that communicating with the government is of little or no importance. Those who primarily speak an indigenous language in their household are especially critical of the government, with more than 60 percent saying that this kind of information is of little or no value at all. Representatives of the Aymara and Quechua ethnic groups have a particularly negative view, with 36 and 42 percent, respectively, saying that no communication at all with the government is necessary, compared with only 22 percent of the overall sample. Furthermore, 37 percent of indigenous youth are very critical of the government. Respondents from CIOEC are the most interested in government information, with 91 percent of the respondents from Oruro and 92 percent of those from CIOEC Potosí saying that they would like to enhance communication with the central government. This demand is consistent with the program,

FIGURE 4.34   Most needed flow of communication, by type of organization

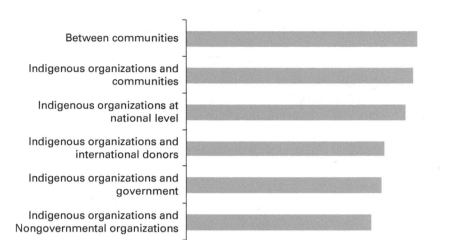

which seeks to enhance indigenous peoples' access to government information, in particular regarding bidding documents related to government programs.

Indigenous peoples have an even more critical opinion of NGOs, with 62 percent saying that they see little or no need at all to communicate with them. The only exceptions are participants in the programs of APCOB and ICO, where 89 and 73 percent, respectively, said that communicating between indigenous organizations and NGOs is very important. These findings are confirmed by the focus groups organized in Concepción and Comarapa. In both cases, participants said that local communities have a very good relationship with APCOB and ICO (focus group in Comarapa in July 2006; focus group in Concepción in April 2006). The respondents to the youth survey and the representatives of the national indigenous organization CONAMAQ are the most critical of NGOs, with 35 and 44 percent, respectively, indicating that there is no need at all to communicate with them.

Why do indigenous respondents show so little interest in improving their access to information and in enhancing communications with the government and outside organizations? Why do they prioritize enhancing information and communication flows among their communities? What underlying factors of the broader sociopolitical context have caused indigenous peoples to look inward instead of trying to improve information sharing and to enhance their relationship with the rest of Bolivian society?

The specific findings of the survey regarding information needs have to be seen in the broader sociopolitical context. Indigenous peoples seem to take a critical view of the traditional development approaches that have been applied by previous Bolivian governments and many NGOs. Based on the top-down nature of many development programs, the benefits for indigenous peoples have frequently been limited, and many programs have failed to have any significant positive long-term impact on their livelihoods (van Niekerk 1994). Consequently, indigenous peoples have become increasingly critical about this type of development intervention, which might explain their low interest in improving the flow of information and communication with the Bolivian government and NGOs.

Another aspect requiring further investigation is the gap that seems to be emerging within the indigenous movement between the grassroots leadership at the community level and the leadership at the regional, national, and international levels. In particular, indigenous women and youth continue to be excluded from leadership positions, and their participation is very low in national indigenous organizations.

Value and Expectations: The Potential of the Internet to Promote Indigenous Development

To what extent do indigenous peoples believe that ICTs can provide them with an instrument to overcome the information and communication gaps identified and enhance their relationships with government? This section addresses the following questions: (a) What is the value of ICTs to indigenous peoples and what are their expectations about the role that ICTs could play in enhancing their well-being? (b) Can ICTs facilitate a two-way flow of information between indigenous peoples and the non-indigenous population? Can ICTs, in the view of indigenous peoples, improve their relationships with the central government?

Usefulness of the Internet for Community Development

This section investigates peoples' perceptions regarding the usefulness of the Internet for their personal development and its potential to aid their communities' overall development. The survey included a specific question on how useful people thought the Internet would be for the development of indigenous peoples. As shown in figure 4.35, 85 percent said that the Internet might be either useful or very useful for their community, and 60 percent said that it would be very useful.

This is one of the most unexpected results of the survey. At the outset of my fieldwork, I expected that many indigenous peoples would have "ideological" concerns about the Internet as a medium developed in the United States and dominated by "Western cultures" that could pose a cultural "threat" to indigenous culture. In the Bolivian context, many indigenous peoples are skeptical about the influence of outsiders and *mestizos* on their communities, and I did not expect them to have a positive attitude toward new technologies (McNeish 2002; Gustafson 2009).

This unexpected finding was confirmed by the results of the focus groups, where indigenous elders and women, who had never used the Internet, expressed their strong support for this technology. For instance, in a focus group with regional leaders from CONAMAQ, an indigenous woman from a

FIGURE 4.35   Usefulness of the Internet for indigenous peoples' development

FIGURE 4.36   Interest by non-users in using the Internet in the future

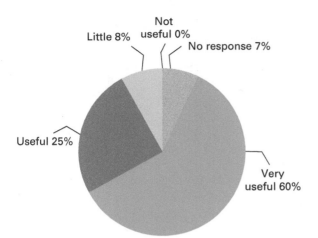

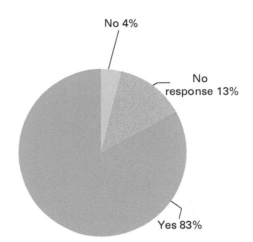

very small, remote community in Chuquisaca—one of the poorest regions of Bolivia—listened to the discussions for two hours and then stood up to express her opinion:

>> I come from a small community from Chuquisaca, where we basically have very little contact with technology; however, during one of my travels I saw children use the Internet in urban centers—I strongly believe that we as indigenous peoples cannot afford anymore to stay behind—we cannot allow that our children are once again excluded from this type of development. Our children have the same rights as children in the cities, and we cannot afford to sit here and discuss this forever. We need to take action and do everything possible so that our children will have the same kind of opportunities as children in the cities. We need to do something about this and pressure the government to provide our communities with access to this technology. (CONAMAQ focus group with indigenous leaders in La Paz, February 2006)   <<

Indigenous representatives expressed the same opinion in several focus groups. It was particularly striking to hear elders and indigenous women who had never had the opportunity to use a computer or the Internet encourage indigenous communities to engage with the new technology. They had seen this happen in the urban centers and felt that they could not afford to be excluded from the potential benefits of these technologies, as they had been excluded from many services in the past. Most women and elders said that their children should have access to these technologies in order to have the same opportunities as the non-indigenous children in urban centers.

The groups least likely to have had the opportunity to access the Internet (indigenous women and elders) have the most favorable views and high expectations of its potential benefit. As shown on table 4.11, 78 percent of the respondents who had never used the Internet nevertheless indicated that it might be a very useful instrument for indigenous development.

TABLE 4.11 **Differences in expectations about the Internet, by previous use**

Have you used the Internet before?	Very useful	Useful	Not very useful	Total
Yes	21.9	35.5	27.8	23.7
No	78.1	64.5	72.2	76.3
Overall (%)	69.1	25.0	5.9	100
Total number of responses	210	76	18	304

By comparison, only 21 percent of the respondents who had used the Internet before said that it is very useful, while almost 28 percent of the people who had used it said that it is not very useful. The other variables that influence indigenous peoples' views of the usefulness of the Internet are age, level of educational attainment, and ICT program. It is not surprising that the youth with a relatively high level of education said they have an even more favorable opinion of the Internet than the rest of the respondents; 71 percent of respondents who are under the age of 18 and have a high school degree said that the Internet is very useful. It is, however, quite surprising that the respondents from the two grassroots programs of CIOEC in Potosí and CIOEC in Oruro have the highest expectations for the Internet, with 94 and 89 percent of respondents, respectively, indicating that they believe that the Internet would be very useful to them. In particular, the finding that the participants of the program of CIOEC in Potosí are the most optimistic about the value of the Internet deserves a closer analysis, since the respondents come from the poorest communities in the entire sample, and the CIOEC program in Potosí was modest in size and the most recent initiative to promote ICTs among indigenous peoples (UNDP 2004b; interview with ICT program coordinator, CIOEC Potosí).

The focus group discussions with small farmers from Potosí confirm these findings. Several participants expressed their deep concerns about the future of their community; several community leaders stressed that the department of Potosí is being "forgotten" by the "powerful" elites in La Paz. One leader from the municipality of Llica, Potosí, expressed this view in the following way:

>> Many of my brothers and sisters from my community here in Potosí feel isolated from the rest of the country, and many times we have the impression that we are being forgotten by the people in power. Our communities have to find a way to make a living under extreme conditions, and we have the impression that it is always getting harder for us to feed our families and to get ahead in life. At the same time, we are convinced that a better future is possible for our children, and, for this to happen, we need to strengthen the bonds between our communities and indigenous organizations. CIOEC is supporting us in this effort, and I have heard about the Internet and I have great hope that if we stand together and learn how to use it well, we and in particular our children can benefit from it and thus

our communities will benefit as well. (CIOEC focus group with small producers in Potosí, August 2006)   <<

In addition, 83 percent of the respondents who had never used the Internet expressed their interest in doing so in the near future (figure 4.36). Finally, the large majority (80 percent) of the respondents said that the Internet would have more positive than negative effects on their community.

Only a small minority (14 percent) said that the Internet would have more negative than positive effects on the community. Nevertheless, there are notable variations based on differences in the intermediation process. The ICO program in Comarapa is unique because all of the respondents (100 percent) said that the Internet would have more positive than negative effects on their community. Among participants from CIOEC Potosí, 94 percent of the respondents said that the Internet would have mostly positive effects. The only group that was relatively skeptical about the potential impact of the Internet was participants from the national indigenous organization CONAMAQ, where only 58 percent of respondents said that the Internet would have more positive than negative effects on local communities, while 14 percent said that things would remain the same with the introduction of the Internet, compared with 1 percent of the overall sample.

The Internet and the Relationship between Indigenous Peoples and the Government

An important finding of the survey is that a significant majority (90 percent) of indigenous peoples have very high expectations about the role that the Internet might play in improving the relationship between the government and indigenous peoples (figure 4.37). Men (94 percent) are far more optimistic about this issue than women (77 percent), and this sentiment is also influenced by the degree of ethnicity of the respondents and by the intermediary. In Comarapa, for instance, where the large majority of participants were non-indigenous, all respondents (100 percent) said that the Internet has the potential to enhance

FIGURE 4.37 Expectations about the role of the Internet in improving the relationship between the government and indigenous peoples

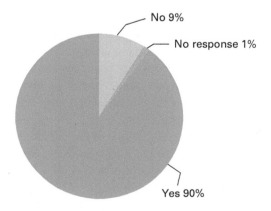

No 9%

No response 1%

Yes 90%

TABLE 4.12   Ways in which the Internet can change the relationship between the government and communities

Indicator	Number of respondents	% of respondents
Improve the access to information about social government programs	82	22
No answer or don't know	44	11
Improve the access to information about government policies	31	8
Improve health services	24	7
Facilitate government processes (birth certificate, identification card)	23	6
Improve education	19	5
Improve access to information about government programs and policies	11	3
Improve access to information and education	10	3
All of the above	121	33
Total	365	100

the relationship between the government and indigenous peoples. One possible interpretation of this number is that the ICO program had contributed to improving the relationship between the state and local communities. However, when people were asked to identify the manner in which the Internet could improve this complex relationship, most people said that they have only a very vague idea of how new technologies might improve government policies toward, and services for, indigenous peoples.

The only specific area that the participants could identify was government transparency, but only 8 percent thought that the Internet would enhance the transparency of government and improve the access of local communities to information.

As shown in table 4.12, respondents said that they have no concrete idea of how the Internet might help their communities; only a very small percentage said that it could improve their communities' education (5 percent) and health services (7 percent). A third of respondents indicated that the Internet would support their community in all of the aspects mentioned. The findings confirm that indigenous respondents have very high expectations about the role that this technology could play in improving their relationship with government.

In smaller groups, many indigenous peoples, particularly indigenous elders and women, said that they do not have a clear idea about how the Internet could improve their living conditions, but that they associate it with a vague concept of being able to overcome the remoteness and current forms of social and political exclusion of their community; they believe that this new medium might provide their children with new opportunities to improve their well-being. The discussion and the analysis of the survey data, however, suggest that many indigenous peoples have "unrealistic expectations" about the role that the Internet can play and that these expectations may not be fulfilled once the Internet reaches their community.

Conclusions

This chapter has used the results of a quantitative survey of non-ICT users, combined with focus groups and interviews, to describe indigenous peoples' information needs and communication patterns, as well as their awareness, use, and expectations of ICTs. The survey results were analyzed within the overall socioeconomic context of indigenous peoples in Bolivia, including the significant geographic differences between the highlands and lowlands in the country. The chapter also paid special attention to the extent to which intermediary organizations influence the information and communication patterns and levels of ICT awareness and capacities of indigenous peoples.

First, indigenous peoples have very low awareness of and access to the Internet and other forms of ICTs (cell phones). This finding raises a key question: What are the main reasons for this low awareness and the extremely low exposure to these technologies? A possible explanation could be that the overall level of Internet use is very low in Bolivia and that, as in developed countries, individual factors, such as age, gender, and educational attainment, are responsible for the internal "digital divide" within Bolivia.

The results from the survey, however, show that, for indigenous peoples, social contextual factors, like belonging to an indigenous group or living in a remote community, are decisive in terms of whether or not a person has the opportunity to access ICTs. In contrast to the situation in urban centers or in developed countries—where individual factors like age, gender, or educational attainment determine whether or not a person has access to and used ICTs—for indigenous peoples the main reason for not having access to ICTs is being indigenous. The socioeconomic profile of the participants has clearly shown that there is a very strong correlation between being indigenous and being poor and that the higher the degree of ethnicity in a municipality, the higher the level of poverty and the lower the likelihood of having access to ICTs. The analysis of the socioeconomic and demographic factors shows that rural communities that are less indigenous have a higher likelihood of having access to ICTs than those that are more indigenous (UNDP 2004a). In this sense, indigenous people's low awareness of, and access to, ICTs and the Internet reflect the overall economic and social exclusion of indigenous peoples in Bolivia.

Second, the chapter has highlighted that indigenous peoples continue to rely to a large extent on traditional mechanisms of information exchange and forms of communication, such as community meetings and personal visits, which are based on the community's existing social structures and strong kinship networks and are embedded in the strong oral tradition of indigenous cultures. This finding raises an important issue about the broader political economy of the information and communication flows in Bolivia. The results from the survey demonstrate that indigenous peoples communicate and exchange information mainly with each other and that there is an important "lack" of information exchange and communication with the *mestizo* population and with the government in urban centers. The communications profile and the information ecology described in this chapter demonstrate that indigenous peoples continue to use very different sources of information and communications channels than the non-indigenous population. While the urban *mestizo* population uses newspapers, television, and commercial radio stations as their main sources of information, indigenous peoples receive the core of their information from traditional sources of information based on their strong kinship-based social networks.

Third, indigenous peoples continue to face major information gaps: 52 percent of the respondents of the survey requested greater access to information than they currently have. The survey provided an indication about the type of information that indigenous communities need the most. The respondents expressed the strongest demand for (a) educational information, (b) information on agricultural practice, (c) citizenship rights information, and (d) health information. The most significant finding from the information needs analysis, however, is that most indigenous peoples are not interested in receiving more information from the central government. In particular, indigenous youth (50 percent) and women (34 percent) from grassroots organizations said that they have very little interest in receiving more information from the government. Many indigenous peoples are not very interested in being better informed about government policies and programs. The two indigenous groups that are the least interested in receiving government information are the Aymara and Quechua, who come from the highlands, live in conditions of extreme poverty, and face systematic social and political exclusion within Bolivian society.

However, a large majority of indigenous peoples (84 percent) identified improved access to information on citizenship rights as one of their most important priorities and 76 percent said that they currently do not receive any, or receive only very little, information about their rights. This discrepancy between people's need for information concerning their rights and the very low access to this information highlights the failure of "vertical accountability" within Bolivia. Vertical accountability, understood as the "accountability of the state to civil society," has at its center the "relationship between the state and individual citizens or groups of people." An important aspect of vertical accountability is the transparency of the state: politicians and state employees are obligated to make information about government policies and programs available to citizens who are affected by them. The concept of vertical accountability is best understood as the "mirror" of the right to information by its citizens—one is ideally reflected in the other (Blackburn 2000, 4). Bolivia has a tradition of weak vertical accountability, and the political culture has long been dominated by a caudillist, clientelistic, and patrimonial system (Blackburn 2000, 4). The country represents a "mosaic" of very different, almost separate, realities, characterized by poor interaction between the state and its citizens. The consequences are widespread frustration with the formal political system. The findings of the survey seem to indicate the coexistence of two different realities. Each segment of society continues to use its own information and communications channels, and there is little opportunity for truly intercultural communication and dialogue between the indigenous and non-indigenous populations.

Indigenous peoples are very critical of the role that NGOs play in their communities, with a large majority of respondents (62 percent) saying that they do not want to receive more information from NGOs. Many indigenous leaders believe that too many outsiders have benefited from "development" programs, with most of the benefits and resources going to government officials, NGO staff, and "development professionals" and very few going to the local communities. Indigenous elders, particularly, stressed the negative effects of NGO or government programs on community structures, in terms of the negative effects they have on the social coherence within indigenous communities.

Programs are frequently overly specialized in one particular sector, do not address the needs of indigenous communities holistically, and are often implemented in a top-down way. Such programs can lead to the fragmentation of communities and often result in unsustainable projects that are "imposed" on communities, instead of responding to their real needs. Indigenous peoples are frequently excluded from the planning and implementation of such programs. Consequently, the benefits for them are very limited, and the programs do not manage to reach financial and social sustainability (Butz, Lonergan, and Smit 1991).

Thus many indigenous leaders have concluded that it is necessary to form a new indigenous leadership of professionals and to seek a direct relationship between the government, international donors, and local communities (focus group with indigenous leaders in Tiwanaku, April 2005; CIOEC focus group with indigenous national leaders, August 2005).

In addition, a gap is emerging between the grassroots and the national indigenous leadership. National indigenous leaders are predominantly male, between the ages of 30 and 44, and well educated, while grassroots indigenous leaders are often small-scale farmers, women, and youth who are increasingly critical of the lack of "transparency," "accountability," and "inclusiveness" of the indigenous leadership, particularly at the national and international levels. As the analysis has shown, many respondents recognize that the indigenous movement faces problems similar to those of the central government, such as how to achieve an open and transparent flow of information, as well as problems related to providing grassroots groups with a voice in national policy dialogues and the ability to participate in national political decision-making processes. There are major constraints and barriers to overcoming these mutual "information and communication gaps": (a) the isolation of the traditional information and knowledge systems of rural communities from the formal information systems and communications channels of the urban centers; (b) the extremely limited access to ICT services within indigenous communities; (c) the low capacity of indigenous peoples to use new forms of ICTs; and (d) the difficulties of adjusting the content to the local information needs and cultural context, including the challenges of translating the information into indigenous languages.

Notes

1. The Kruskal-Wallis test is used to investigate one independent variable with two or more levels and an ordinal dependent variable (Scanlan 2002; Hollander and Wolfe 1999). In the analysis here, the ISI represents the

ordinal dependent variable, and the type of intermediary organization represents the different groups of the independent variable.

2. Amateur, or ham, radio is defined as a two-way means of communication carried out by amateurs as a leisure-time activity and individuals interested in radio technique solely with a personal aim and without pecuniary interest (McGraw-Hill 2002).

References

Albó, Xavier, and Victor Quispe. 2004. "¿Quiénes son indígenas en los gobiernos municipales?" Cuadernos de Investigación 55, Centro de Investigación y Promoción del Campesinado (CIPCA), Plural, La Paz.

Blackburn, James. 2000. "Popular Participation in a Prebendal Society: A Case Study of Participatory Municipal Planning in Sucre, Bolivia." Ph.D. diss., University of Sussex.

Butz, David, Steven Lonergan, and Barry Smit. 1991. "Why International Development Neglects Indigenous Social Reality." *Canadian Journal of Development Studies* 12 (1): 143–57.

Girard, Bruce, ed. 1992. *A Passion for Radio: Radio Waves and Community*. Montreal: Black Rose Books.

Gustafson, Bret. 2009. *New Languages of the State: Indigenous Resurgence and the Politics of Knowledge in Bolivia*. Durham, NC: Duke University Press.

Hall, Gillette, and Harry A. Patrinos. 2006. *Indigenous Peoples, Poverty, and Human Development in Latin America*. London: Palgrave.

Hollander, Myles W., and Douglas A. Wolfe. 1999. *Nonparametric Statistical Methods*. New York: John Wiley.

INE (Instituto Nacional de Estadísticas). 2001. *National Census of Housing and Population 2001*. La Paz: Ministry of Planning and Coordination, INE.

ITU (International Telecommunication Union). 2003. "World Telecommunication Development Report 2003: Access Indicators for the Information Society (Executive Summary)." ITU, Geneva.

McGraw-Hill. 2002. *Encyclopedia of Science and Technology*, 5th ed. McGraw-Hill.

McNeish, John A. 2002. "Globalization and the Reinvention of History and Community in Highland Bolivia." *Journal of Peasant Studies, Special Issue on Latin American Peasantry* 29 (3-4, April-July): 228–69.

Meier, Kenneth J., and Jeffrey K. J. Brudney. 1997. *Applied Statistics for Public Administration*. Belmont: Wadsworth Publishing.

Molina, Ramiro, and Xavier Albó. 2006. *Gama étnico y lingüística de la población boliviana*. La Paz: Sistema de las Naciones Unidas en Bolivia.

Norris, Pipa. 2001. *Digital Divide: Civic Engagement, Information Poverty, and the Internet Worldwide*. Cambridge, MA: Harvard University, John F. Kennedy School of Government.

O'Farrell, Claire, Patricia Norrish, and Andrew Scott. 1999. "Information and Communication Technologies (ICTs) for Sustainable Livelihoods." Preliminary study, Intermediate Technology Development Group, Agricultural Extension and Rural Development Department, London.

Psacharopoulos, George, and Harry Patrinos. 1994. *Indigenous Peoples and Poverty in Latin America: An Empirical Analysis.* Washington, DC: World Bank.

Scanlan, Craig. 2002. "Introduction to Nonparametric Statistics." University of Medicine and Dentistry of New Jersey, Newark. http://www.umdnj.edu /idsweb/idst6000/nonparametric_analysis.pdf.

Slater, Dan, Jo A. Tacchi, and Peter Lewis. 2002. "Ethnographic Monitoring and Evaluation of Community Multimedia Centres: A Study of Kothmale Community Radio Internet Project, Sri Lanka." DFID, London.

UNDP (United Nations Development Programme). 2004a. "Encuesta de Capacidades Informacionales" (ECADI)." Survey administered by UNDP in preparation for its *2004 National Human Development Report, Interculturalidad y globalización.* La Paz.

———. 2004b. *Human Development Report 2004: Cultural Liberty in Today's Diverse World.* New York: UNDP.

———. 2005. *Human Development Report 2005: International Cooperation at a Crossroads; Aid, Trade, and Security in an Unequal World.* New York: UNDP.

Van Niekerk, Nico. 1994. "Desarrollo rural en los Andes: Un estudio sobre los programas de desarrollo de organizaciones no gubernamentales." Leiden Development Studies, the Netherlands.

CHAPTER 5

Beyond Access to Meaningful Use

This chapter goes beyond the concept of "access to ICT (information and communication technology)" to identify the multiple factors that are associated with indigenous peoples' use of ICTs. It argues that the actual use of ICTs and not simply access to the Internet is critical for evaluating the impact of ICTs on peoples' well-being.

The chapter proceeds as follows. It begins by presenting empirical evidence from a user survey administered to indigenous peoples in rural Bolivia and then compares the results from my own user survey with those of the *Encuesta de Capacidades Informacionales* (ECADI)—a national household survey of people's perceptions and uses of ICTs conducted by the United Nations Development Programme (UNDP) in preparing its *2004 Human Development Report* (UNDP 2004). The section presents the main results from the logistical regression analysis—the statistical method chosen to analyze the UNDP data— outlining which characteristics (age, gender)—are best suited to explaining variations in the dependent variable of ICT use. Subsequently, the analysis uncovers some current barriers preventing indigenous peoples from using ICTs and provides insight into the role that intermediary organizations play in promoting peoples' ICT use so as to improve the economic, social, political, and cultural dimensions of indigenous peoples' lives. The final section draws conclusions from both the statistical analysis of my ICT user survey and the regression models applied to the UNDP data set, placing the findings in the broader context of the study.

A critical aspect of the theoretical approach described in chapter 1 is to apply a capability perspective to the study of ICTs and to move beyond the concept of "ICT access." Such an approach deemphasizes the role of technology per se and stresses instead the need to focus on understanding the interactions between technology and people. As such, the study seeks to unpack the conditions and processes under which ICTs can enhance indigenous peoples' human capabilities. In this context, the actual use of ICTs, rather than mere access, is a prerequisite for ICTs to improve the lives of indigenous peoples. The chapter addresses the following research questions:

- What individual and socioeconomic factors are associated with indigenous peoples' ICT use?
- To what extent do indigenous peoples' use of ICTs depend on their individual characteristics (for example, age, gender, education), ethnicity, or on the broader social and institutional context of indigenous communities in Bolivia?
- What structural barriers impede indigenous peoples from being able to use ICTs?

A central hypothesis posed in chapter 1 is that the use of ICTs has to be facilitated by an effective and local intermediary organization in order to have significant positive effects on indigenous peoples' well-being. Therefore, the chapter also addresses the following question:

- Can different outcomes in indigenous peoples' use or non-use of ICTs be explained by differences in the existing ICT programs?

Principal Findings

The main finding of this chapter, derived from the empirical analysis of my own user survey and the ECADI survey, is that indigenous peoples are disproportionately disadvantaged in terms of their Internet use relative to the non-indigenous population. The triangulation of my survey results with the statistical results obtained through a regression analysis of the national data sets of the ECADI survey allows me to generalize the findings to the national context in Bolivia and thus to make general statements about the conditions under which indigenous peoples can use ICTs.

First, only a weak association exists between ICT access and Internet use. As such, having access to the Internet is not sufficient to ensure its use. This is particularly true for indigenous communities, as the majority of indigenous peoples do not use the Internet despite having access to it. In fact, a significant majority of indigenous peoples (58 percent) who have access to the Internet don't use it. This finding confirms the central assumption made in the theoretical framework that mere access to ICTs neither ensures that people can enhance their individual well-being through these technologies nor automatically leads to meaningful benefits for the communities in which the access is being provided. To the contrary, the multiple structural barriers facing indigenous peoples, such as low educational attainment, impede them from using the Internet; for this reason, solely enhancing their access to ICT infrastructure is not sufficient to reduce digital inequalities in Bolivia. The findings confirm that a much more nuanced analysis is needed; the emphasis should be on issues related to the actual use of technology and on the extent to which people can derive real, meaningful value from their use.

Second, ethnicity (belonging to a specific group) and socioeconomic factors (literacy rates and poverty levels), not individual factors (age, gender, education, and income), are critical in explaining the stark digital inequalities within Bolivia. Indeed, the significant disparities among the various indigenous groups, and not only between the indigenous and non-indigenous populations, suggest the need to take a more differentiated approach to analyzing digital inequalities in Bolivia. For example, the data analysis finds that the Quechua are almost twice as likely as the Aymara to use the Internet—with 80 percent of the Aymara never having used it. The Aymara are the most disadvantaged group of ICT users within Bolivia.

The empirical results point to the fact that the described digital inequalities reflect the existing social exclusion of indigenous peoples within Bolivian society. In particular, due to the severe socioeconomic disparities between the indigenous and non-indigenous populations of Bolivia, the experiences of indigenous peoples are shaped by structural barriers. In this context, an unexpected finding is that while education is a key predictor of use, important differences between indigenous and non-indigenous peoples persist even when indigenous peoples acquire advanced levels of education. It is particularly noteworthy that these differences persist among indigenous peoples who have achieved tertiary education. Thus the chapter argues that it is not sufficient to enhance the individual human capabilities of indigenous peoples to overcome the multiple barriers to their Internet use; rather, to reduce their "digital exclusion" within Bolivia, it is necessary to address the underlying structural barriers that lead to their social exclusion, high levels of poverty, and lack of social opportunities.

Finally, the empirical evidence shows that intermediation through an ICT program plays a critical role in improving indigenous peoples' Internet use. The strongest association is between the variable "ICT program" and Internet use. Thus a critical finding is that targeted interventions benefiting indigenous communities, particularly in rural areas, can contribute significantly to reducing digital disparities within Bolivia. Such interventions support indigenous peoples in overcoming the multiple structural barriers they face in terms of their ability to make use of the Internet.

Data Analysis

This chapter presents the results from an ICT user survey administered to 148 respondents over a 12-month period between 2005 and 2006 in rural Bolivia. The survey focused on demographic, individual, and socioeconomic factors influencing indigenous peoples' ICT use and capabilities. It also captured their perceptions about the principal barriers to the effective use of ICTs in their communities and their views on the impact of ICTs on their well-being across different (economic, social, and political) dimensions of their lives.

In order to understand the relationship between intermediary organizations and indigenous peoples' ICT use, I divided the full data sample into six clusters distinguishing between six types of intermediary, as shown in table 5.1, by type of intermediation: (a) *nongovernmental organization* (NGO): Instituto de Capacitación del Oriente (ICO) and Apoyo para el Campesino Indígena del Oriente de Bolivia (APCOB); (b) *national grassroots organization*: Coordinadora de Integración de Organizaciones Económicas

TABLE 5.1   Affiliation of respondents, by clusters of intermediaries

Cluster	Name of organization	Number of respondents	% of respondents
NGO in the lowlands	ICO Comarapa, APCOB	19	13
National grassroots organization	CIOEC leaders	25	17
Grassroots organization in the highlands	CIOEC Oruro, CIOEC Potosí, and OMAK	43	29
National indigenous organization	CIDOB and CONAMAQ	26	18
Government organization	EnlaRed Municipal	11	7
International indigenous organization	Fondo Indígena	24	16
Total		148	100

Campesinas Indígenas y Originarias (CIOEC leaders); (c) *grassroots organization in the highlands*: CIOEC Potosí, CIOEC Oruro, and Organización de Mujeres Aymaras del Kollasuyo (OMAK); (d) *national indigenous organization*: Confederación de Pueblos Indígenas de Bolivia (CIDOB) and Consejo Nacional de Ayllus y Markas del Qullasuyu (CONAMAQ); (e) *government organization*: EnlaRed Municipal; and (f) *international indigenous organization*: Fondo Indígena.

The same logic—differentiating clusters by type of intermediary—guides the analysis of the non-user survey, which is the subject of chapter 4. The only difference between the user and non-user sample is that the user sample also includes data from EnlaRed Municipal, an ICT program coordinated by the Federation of Municipal Associations of Bolivia, a government organization that provides technical assistance and serves as an information clearinghouse for local government officials.

Role of Demographic and Socioeconomic Factors in Explaining ICT Use

The socioeconomic status of ICT users, like that of non-users, is highly representative of indigenous peoples' overall situation in Bolivia. As shown by comparing table 4.2 in chapter 4 with table 5.2, the only difference is that the proportion of ICT users who are indigenous is smaller than the proportion of non-users who are indigenous (65 and 81 percent, respectively), with primary language spoken in the household used as a proxy for ethnic identity.

The lower level of ethnicity in the user sample in comparison to that in the non-user sample can be attributed largely to the inclusion of EnlaRed, a government organization (figure 5.1). The majority of the surveyed participants from EnlaRed (56 percent) identified themselves as monolingual Spanish speakers, reflecting the fact that fewer indigenous than non-indigenous peoples work for the government. This inequality has begun to change,

TABLE 5.2   ICT users' ethnicity (principal language spoken in the household used as proxy for ethnic identity)

Principal language	Number of respondents	% of respondents
Spanish	53	35
Indigenous language (primary) and Spanish	80	54
Indigenous language, Spanish, and English	15	11
Total	148	100

FIGURE 5.1   Principal language spoken in the household, by type of organization

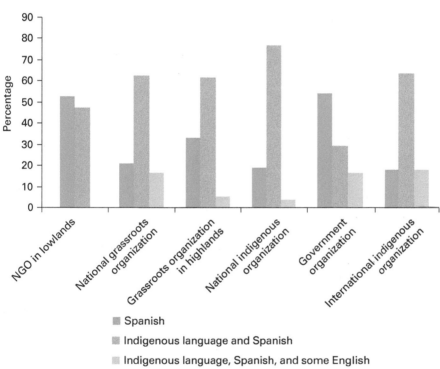

at least in the upper echelons of government, since Evo Morales came to power in 2006, but it will take much longer to trickle down to the local and municipal levels. Furthermore, the majority of respondents in the ICO Comarapa program predominantly speak Spanish, reflecting the fact that a relatively low percentage of the overall population in Comarapa is indigenous. A significant 11 percent of the overall sample is trilingual, speaking an indigenous language, Spanish, and English. This is an indication that the level of educational attainment of the user sample is significantly above the national average, since only a very small percentage of the Bolivian population speaks English.

Educational Attainment

It is evident from the data that education is a major factor in ICT use. As shown in figure 5.2, 40 percent of users have attained secondary schooling, 14 percent have advanced technical schooling, and about one-third (31 percent) have attended a university. Users are, by and large, better educated than non-users, with 15 and 35 percent, respectively, having attained only a primary education or less and 9 and 45 percent, respectively, having attained advanced technical schooling or a university degree. Secondary education is more of a leveling field, with users and non-users split fairly evenly (40 and 56 percent, respectively).

However, figure 5.2 also demonstrates the very high educational level overall of all respondents. About 15 percent of the respondents (both users and non-users) have a university education and 45 percent either have completed or at least have some secondary education. Comparing these figures with indigenous peoples' educational attainment at the national level, shown in table 5.3, reveals that the educational attainment of respondents is significantly higher than the

FIGURE 5.2   Use of ICTs, by level of education

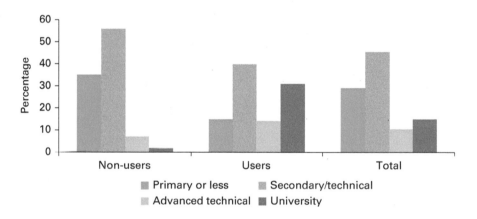

TABLE 5.3   Educational attainment in Bolivia, by ethnicity and gender, 2002

% of persons 15 years of age and older

| Indicator | Indigenous | | | Non-indigenous | | | |
	Male	Female	All	Male	Female	All	All
Still in school	9.4	7.3	8.3	18.6	16.9	17.7	12.6
If not still in school, highest educational achievement							
None	7.7	26.4	17.5	3.1	7.8	5.5	12.0
Incomplete primary	49.5	46.4	47.9	31.7	30.6	31.1	40.3
Incomplete secondary	15.8	9.1	12.3	20.9	19.6	20.2	15.9
Complete secondary	9.7	6.2	7.9	17.1	16.5	16.8	11.9
University	9.7	6.8	8.2	18.9	18.2	18.6	12.9

national average. Only 13 percent of Bolivians have a university degree, and about 27 percent have some secondary education (table 5.3). Furthermore, the educational attainment of both the users and non-users in the sample is quite high with regard to primary education; at the national level, 47 percent of Bolivians have only some level of primary education and 12 percent have no education at all, while about 29 percent of respondents have only primary education or less. However, neither the non-user nor the user survey included the category of no education as an option in the questionnaire.

An analysis of users' educational attainment by ethnicity points to the significant disparities between indigenous and non-indigenous peoples. In 2002, at the national level, school enrollment was significantly lower for indigenous school-age children than for non-indigenous children, with 18 percent of the non-indigenous population 15 years of age or older attending school, compared with only 8 percent of indigenous peoples (Jiménez Pozo, Landa Casazola, and Yáñez Aguilar 2006, 57). Low enrollment rates have important consequences for school achievement among adults. Nearly 18 percent of indigenous peoples ages 15 years and over had no education at all in 2002, compared with only 5.5 percent of the non-indigenous population. The difference is even more dramatic for indigenous women: 26 percent of indigenous women had no education, compared with 8 percent of non-indigenous women. Moreover, only 8 percent of indigenous peoples had reached university level, in contrast to 19 percent of non-indigenous people. These data show that there is a strong correlation between being indigenous and having less access to education in Bolivia. These national statistics are also confirmed by the empirical data from the user survey: of the respondents who have primary schooling or less, 78 percent are indigenous and only 22 percent are non-indigenous.

Unsatisfied Basic Needs and Human Development

Users fare better on the unsatisfied basic needs (UBN) index and the human development index (HDI) than non-users. As table 5.4 demonstrates, users tend to be better off economically than non-users. Almost half of the non-users (44 percent), compared with 15 percent of the users, are from municipalities with extreme poverty levels, categorized as levels IV and V in the UBN index.

TABLE 5.4 Poverty level of ICT users and non-users

UBN category	Non-users		Users	
	%	Cumulative %	%	Cumulative %
Satisfied basic needs (I)	4	4	14	14
At the threshold of poverty (II)	28	32	35	49
Moderate poverty (III)	24	56	21	70
Marginal population (IV)	35	91	8	78
Indigent population (V)	9	100	7	85
No response	0	0	15	100
Total	100	100	100	100

While 14 percent of users reside in municipalities with satisfied basic needs, a mere 4 percent of non-users enjoy the same comforts. The user sample had a no-response rate of 15 percent for municipality of residence, while the non-user sample had a 100 percent response rate in this regard. Nevertheless, even after adjusting the data to account for the no-response rate, users are much more likely to come from municipalities with satisfied basic needs than non-users.

Using the HDI to validate the correlation between income and ICT use confirms that the level of poverty is significantly higher among non-users than among users. Almost three-quarters of non-users (74 percent) are from municipalities with a low HDI index in the range of 0.5–0.6 compared with 50 percent of users (table 5.5). Furthermore, more than one-third of users (34 percent) are from municipalities with a relatively high HDI above 0.6.

Ethnicity

Another important socioeconomic variable affecting the impact of ICTs on indigenous peoples' well-being is their degree of ethnicity, as measured by the "ethnicity index" defined in chapter 3. An overwhelming majority of users (83 percent) are from municipalities with a high or medium level of ethnicity, and the sample was able to target indigenous peoples even when the spatial dimension of ethnicity, such as the low population of indigenous peoples in the lowlands, would normally bias their representation. In the highlands (table 5.6), 72 percent of users indicated that they live in municipalities with a high degree of ethnicity. In the lowlands, where the overall indigenous population is less than 10 percent, only about 7 percent of respondents live in areas

TABLE 5.5 Human development of ICT users and non-users
% of respondents

HDI	Non-users	Users
Less than 0.5	8	5
0.5–0.6	74	50
More than 0.6	17	34
No response	0	11
Total	100	100

TABLE 5.6 Ethnicity of respondents, by geographic region
% of respondents

Degree of ethnicity	Highlands	Valleys	Lowlands	Total
High	71.7	66.7	7.1	46.4
Medium	28.3	0.0	59.5	36.4
Low	0.0	33.3	33.3	17.3
Total number	76	15	42	110

characterized by a high degree of ethnicity, while two-thirds live in areas with a medium to high degree of ethnicity (67 percent). These figures indicate that the survey was successful in targeting indigenous peoples.

Municipal Illiteracy Rates

Finally, a significant majority of the respondents (63 percent) are from municipalities with high illiteracy rates. The respondents from the highlands are from municipalities with significantly higher illiteracy rates than those from the lowlands: 12 and 0 percent, respectively, are from municipalities with illiteracy rates above 25 percent (table 5.7). This indicator is particularly relevant for analyzing ICT use, since literacy is a key prerequisite for using the Internet (Heeks 1999). The high illiteracy rates in the respondents' municipalities are a major constraint on the use and diffusion of the Internet within their communities, particularly in rural areas.

Figure 5.3 illustrates the disparities in literacy between the indigenous and non-indigenous populations. The data reveal a strong correlation between a municipality's "degree of ethnicity," as measured by the ethnicity index developed by Xavier Albó (and described in appendix D), and the illiteracy rate. Albó uses three main criteria to differentiate between high, medium, and low degrees of ethnicity: (a) principal language spoken as a child being an

TABLE 5.7   **Illiteracy rate of respondents, by geographic region**
% of respondents, unless otherwise noted

Illiteracy rate	Highlands	Valleys	Lowlands	Total
Less than 10%	31.9	0.0	19.0	23.8
10–25%	56.5	40.0	81.0	62.7
More than 25%	11.6	60.0	0.0	13.5
Total number	46	15	42	103

FIGURE 5.3   **Illiteracy rates of municipalities, by degree of ethnicity**

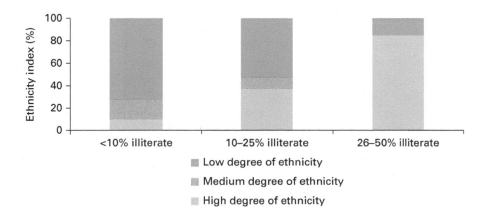

indigenous language, (b) being born in an indigenous community (rural or semi-urban area), and (c) self-identifying as being indigenous.

The ethnicity index is high when a majority of the population satisfies all three of these criteria; it is low when a majority of the population identifies itself as *mestizo* and does not speak an indigenous language. Using this ethnicity index, a strong correlation exists between the literacy rate and the ethnicity index of a municipality. Almost three-quarters of municipalities with a low illiteracy rate have a majority non-indigenous population, while almost 85 percent of municipalities with a high illiteracy rate also have a high ethnicity index (figure 5.3). These figures confirm the national statistics, attesting to the challenging situation with regard to education in indigenous communities (Hall and Patrinos 2005).

Role of Individual Factors in Explaining ICT Use

Most of the "ICT for development" literature analyzes the effects of individual factors on a person's capacity to use ICTs. Among the most frequently cited factors are gender, age, level of education, profession, and income. This section focuses on how these first three variables are correlated with ICT use. The survey did not include income in its questionnaire due to the sensitivity of indigenous peoples. This decision was guided by the results of a pilot survey—carried out prior to the main user survey—which indicated that indigenous peoples do not respond well to questions related to income (either they do not answer the question or they refuse to continue with the survey). Finally, there was also a very high no-response rate to the question about their profession, so it was decided not to include this variable in the analysis.

Gender

Given that both social and cultural factors shape access to and use of ICTs, I expected to find a strong association between gender and ICT use, as substantiated by much of the literature (Roman and Colle 2002; Martin-Barbero 1993; Richardson and Paisley 1998). Considering similar patterns in many of the other dimensions of economic and social life among indigenous peoples, I expected men and women to differ significantly in their ICT use. Indeed, in Bolivia, indigenous women are generally the most disadvantaged segment of society: the systemic, dual structures of prejudice against them—machismo and racism—predispose them to having significantly less access to education (more than four fewer years) and lower incomes than indigenous men and *mestizo* women, as well as an alarmingly high maternal mortality rate (Psacharopoulos and Patrinos 1994; Hall and Patrinos 2005).

So I was surprised when the comparative analysis between users and non-users found that gender is not a critical factor in ICT use. In both the user and non-user samples, the proportion of indigenous women is about the same—30 and 29 percent, respectively. As figure 5.4 shows, only a slightly higher percentage of men than women (38 and 34 percent, respectively) are Internet users.

FIGURE 5.4   Internet use of respondents, by gender

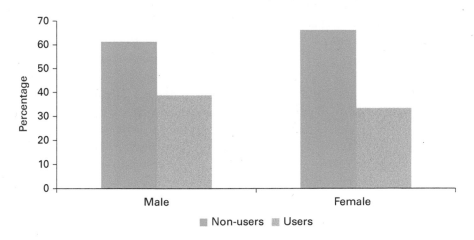

FIGURE 5.5   Educational attainment of respondents, by gender

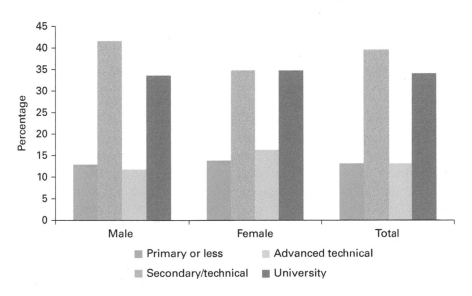

These results should be viewed with caution, considering the following: (a) there is a strong male bias in the survey, with only 30 percent of respondents being female, and (b) the women included are predominantly indigenous leaders. The absence of an association between gender and ICT use can be attributed to the fact that the relatively few women who do manage to attain positions of power within the indigenous leadership are highly educated and skilled (figure 5.5). In the user sample, more women than men (51 and 45 percent, respectively) have attended either an advanced technical school or a university. This particular set of indigenous women has managed to overcome many significant challenges, with education and access to leadership positions only two of many. It seems that the female indigenous leaders surveyed were able to overcome numerous barriers, through their individual agency, across many aspects of their lives, including in the educational and professional spheres. Therefore, they do not represent the average indigenous woman in Bolivia and thus have been able to overcome the ICT gap.

Age

The literature frequently mentions age as one of the key factors determining Internet access and use. The assumption is that younger people are more predisposed to embracing new technologies and tend to be savvier users by virtue of their youth (Livingstone 2003). Indeed, survey data from developed countries confirm the greater frequency of ICT use among youth than among the older generation (Norris 2001; Dertouzos and Gates 1998; Delgadillo, Gómez, and Stoll 2002). However, in the survey, three-quarters (75 percent) of indigenous youth are non-users, and the greatest prevalence of use (34 percent) is among persons between 19 and 35 years of age (table 5.8). This may be because rural schools do not have access to the Internet, making it difficult for indigenous youth to use it, and because complementary infrastructure, such as electricity and other basic services, is lacking. It may also be because indigenous leaders normally come of age in their early 30s, which falls within the age cohort of 19–35 years, and their Internet use is facilitated by the roles and responsibilities they fulfill as part of their leadership position—with access and training provided at no cost by the ICT programs evaluated in this study.

The statistical chi-square test confirms that there is a very weak association between age and ICT use (Kramer V = 0.075), and the results are not statistically significant ($p > 0.05$). This is illustrated in the cross-tabulations, as there is no clear trend between age and ICT use, with the exception of the youth cohort (younger than 18), who are undeniably the least advanced in their use.

To conclude, there is no statistically significant relationship between age and ICT use for indigenous peoples in Bolivia: thus age is not a critical factor vis-à-vis Internet use. It is, however, important to underscore that this survey was conducted only in rural areas and does not apply to the urban context.

Educational Attainment

Educational attainment is commonly considered a key factor influencing people's ability to take advantage of the information society (Campbell 2001; López and Villaseca 1996). It is no surprise that a certain degree of literacy is required to use the Internet. However, Heeks (2002) goes beyond the traditional literacy requirement acknowledged in the ICT literature; he stresses the

TABLE 5.8   Internet use of respondents, by age

% of respondents

Age (years)	Non-users	Users
Younger than 18	75.0	25.0
19–35	66.3	33.7
Older than 35	73.4	26.6
Total	70.4	40.6

Internet's inherent prerequisites, which in his view include not just literacy but also particular skills that allow users to access these technologies and to assess, evaluate, and apply the information provided by them. In his view, a substantial barrier to widespread, meaningful use of the Internet is that people need to have a certain level of capacity and "information literacy."

The survey findings confirm that education matters for ICT use. There is a substantial disparity in educational attainment between users and non-users, with the overall education level significantly lower for non-users than for users (table 5.9). About one-third of non-users (35 percent) said that they have a primary education or less, compared with 13 percent of users. Moreover, highly educated respondents who have attended an advanced technical school or a university make the most use of the Internet. Almost half of all users (46 percent), compared to a mere 7 percent of non-users, have a tertiary education.

The data indicate that a respondents' Internet use is strongly associated with his or her educational attainment. Someone with a university education is four times more likely to use the Internet than someone who has attained only primary education or less. However, a substantial majority of the high school graduates (67 percent) in the sample said they have not used the Internet (figure 5.6).

Strong Correlation between Being Indigenous and Lack of Access to Education

In order to contextualize these findings, which have substantiated the central role of education in Internet use, this section examines the implications of ethnicity

TABLE 5.9 Internet use of respondents, by education

% of respondents

Educational attainment	Non-users		Users	
	%	Cumulative %	%	Cumulative %
Primary or less	35	35	12	12
Secondary or technical	57	92	40	53
Advanced technical	5	97	13	66
University	2	98	33	98
No response	1	100	2	100

FIGURE 5.6 Internet use of respondents, by education level

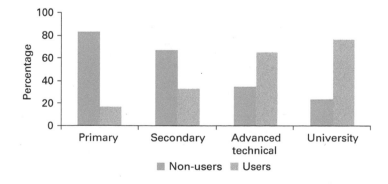

FIGURE 5.7   Educational attainment of respondents, by ethnicity

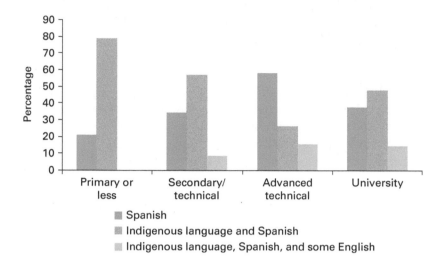

■ Spanish
▨ Indigenous language and Spanish
▨ Indigenous language, Spanish, and some English

for educational attainment in Bolivia. As discussed in chapter 3, there are severe disparities in schooling attainment between indigenous and non-indigenous peoples. The overwhelming majority (79 percent) of respondents who have attained only a primary education or less are indigenous; just 21 percent are non-indigenous (figure 5.7). A similar trend is evident among those with advanced technical schooling, which favors the non-indigenous at a ratio of 2:1—with 58 and 26 percent, respectively, of the non-indigenous and indigenous populations having attended a university or an advanced technical school.

On a more positive note, the figure also points to a burgeoning highly educated indigenous leadership; almost half of the indigenous respondents (48 percent) have attained a university education.

Weak Association between Internet Access and Use

The empirical evidence from my user survey reveals only a weak association between physical "access" to the Internet and its use. The access variable is here defined as the material or physical availability of the Internet (possession of computers and Internet connection or permission to use them) for respondents in the community in which they live (Van Dijk 2005, 21). This definition focuses solely on whether people have the possibility to use the Internet due to the physical availability of the necessary infrastructure either in private or in public, such as through a local school or a telecenter. Many authors have highlighted the limitations of using such a narrow definition of access and the need to define access in a much more holistic manner to include aspects such as motivational factor, ICT skills, and usage (Selwyn 2004; Van Dijk 2005; Warschauer 2003). This narrow definition also neglects the fact that many people are capable of using the Internet even though they have no physical access to ICTs in the community where they live (Heeks 2002). Despite these problems, I use such a narrow definition of access here in order to study whether or not physical access to the Internet is associated with indigenous peoples' use of this technology. Figure 5.8 shows that even if ICTs are available, this does not necessarily mean that people will use them.

The data also show that physical access improves the ability of indigenous peoples to use the Internet since the percentage of users is significantly higher (by 12.2 percent) in municipalities with access to the Internet than in municipalities with no access. However, this effect is much weaker than expected, and the majority of indigenous peoples (58 percent) do not use the Internet despite having access to it. This important finding is confirmed by the statistical chi-square test, which indicates a weak but statistically significant association between access and ICT use (phi = 0.14; $p > 0.05$).

Strong Association between Ethnicity and Internet Use

The data reveal a strong association between ethnicity and Internet use. As shown in chapter 4, there is a substantial negative correlation between being indigenous and having Internet access. In fact, three-quarters of all municipalities with a high degree of ethnicity (75 percent) have no access to Internet services, and the inverse is also true—78 percent of municipalities with a low degree of ethnicity have access.

Figure 5.9 demonstrates the digital inequalities between the non-indigenous (haves) and the indigenous (have-nots); half of the non-indigenous respondents

FIGURE 5.8 Internet use of respondents, by ICT access in the municipality

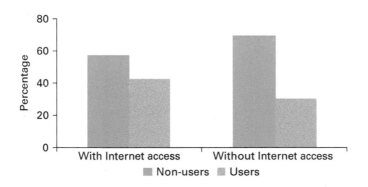

FIGURE 5.9 Internet use of respondents, by ethnicity

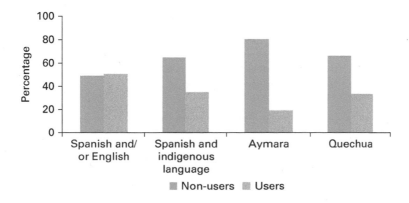

use the Internet, and two-thirds of the indigenous respondents do not. Moreover, it is noteworthy that there are significant differences in terms of Internet use among the various ethnic groups in Bolivia. The empirical findings reveal that 80 percent of the Aymara do not use the Internet and that they are the most disadvantaged group in terms of ICT use. In comparing indigenous groups, the Quechua have the highest percentage of Internet users (33 percent).

Role of Socioeconomic Factors in Explaining ICT Use

This section studies the association between ICT use and socioeconomic factors such as level of poverty, HDI, and illiteracy rates. It includes a detailed summary of the main empirical findings and considers the extent to which variations in indigenous peoples' ICT use are associated with their individual differences (age or gender) and their adverse socioeconomic conditions. It examines the following socioeconomic variables: poverty levels as measured by the UBN index, human development as measured by the HDI, illiteracy rates, and spatial or geographic location (highlands, valleys, and lowlands).

Poverty

A municipality's poverty level is associated with the respondent's ability to use the Internet. As shown in table 5.10, municipalities with lower poverty rates (higher index of satisfied basic needs) have a higher proportion of users: 27 percent of users are in extremely poor municipalities (UBN categories IV and V), 36 percent are in poor municipalities (UBN category III), and 41 percent are in municipalities with satisfied basic needs (UBN categories I and II).

Illiteracy

Furthermore, a municipality's illiteracy rate is associated with a person's Internet use (figure 5.10). In municipalities with high illiteracy rates (25–50 percent), more than two-thirds of the population does not use the Internet, while in municipalities with low illiteracy rates (less than 10 percent), the majority do use the Internet. Furthermore, the chi-square test verifies that the association between non-use and illiteracy is statistically significant (Kramer V = 0.24; p = 0.02).

Spatiality

As discussed in chapter 3, regional inequalities within Bolivia are severe. The Andean highlands is the most disadvantaged region, with extreme levels of

TABLE 5.10   Internet use and the unsatisfied basic needs index
% of respondents, unless otherwise noted

UBN index	Non-users	Users	Total number
Satisfied basic needs (I and II)	59.0	41.0	173
Moderate poverty (III)	63.8	36.2	116
Extreme poverty (IV and V)	73.2	26.8	179
Total	65.6	34.4	468

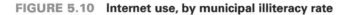

FIGURE 5.10   **Internet use, by municipal illiteracy rate**

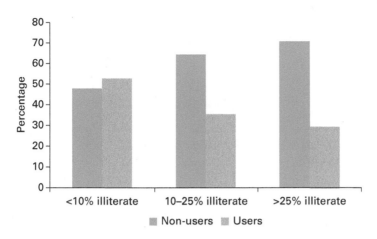

FIGURE 5.11   **Internet use, by geographic location**

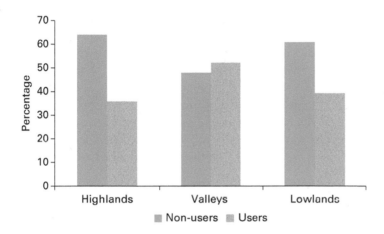

poverty, high maternal and child mortality, and lack of access to basic social services (education and health). The eastern lowlands is the most affluent region as a result of its recent strong economic growth rates, which can be attributed largely to the exploitation of natural resources. Because of these stark regional disparities, I expected spatiality to play a major role in influencing indigenous peoples' Internet use.

In fact, respondents from the Andean highlands are behind those from the lowlands and valleys in their Internet use: almost two-thirds of the highlands respondents, compared to more than 60 percent of the lowlands respondents, are non-users (figure 5.11). The highest Internet use in the sample is in the central valleys, where the majority (52 percent) of respondents indicated that they have used the Internet.

The low number of Internet users in the sample who are from the lowlands is due largely to the mechanisms used in the survey to target indigenous peoples in that region. This was done because these peoples are, by and large, excluded from the gains of the region's economic boom, continue to live in extreme poverty, and lack access to basic infrastructure services (telecommunication services, electricity). They therefore resemble indigenous peoples in the highlands. The statistical chi-square test indicates that this variable is statistically significant, but it has a relatively weak effect on people's ICT use (Kramer V = 0.084).

Role of Intermediary Organizations in Explaining ICT Use

The literature points to the importance of intermediary organizations in promoting the use of ICTs (Heeks 1999; McConnell 2000; Madon 2000). As shown in chapter 1, however, few studies investigate the role of intermediary organizations in depth (Nelson 1996; Madon 2004).

As shown in figure 5.12, a strong association exists between the type of intermediary organization and indigenous peoples' ICT use. While the empirical evidence indicates that the majority of indigenous peoples continue to be excluded from use of the Internet, the data also demonstrate that the intermediary process plays a critical role in supporting indigenous peoples in their use of this technology once they have access to it.

Figure 5.12 shows the significant differences in indigenous peoples' Internet use among the different ICT programs. Grassroots programs—in particular, the national CIOEC program, with 54 percent of users, and OMAK, with 52 percent of users—have the highest rates of Internet use. EnlaRed, a government-led program, and ICO Comarapa, an NGO program, are significantly less successful in promoting the use of the Internet. About two-thirds of respondents (66 percent) from both of these programs indicated that they have not used the Internet, and only one-third (33 percent) said they have used it.

FIGURE 5.12   ICT use, by program

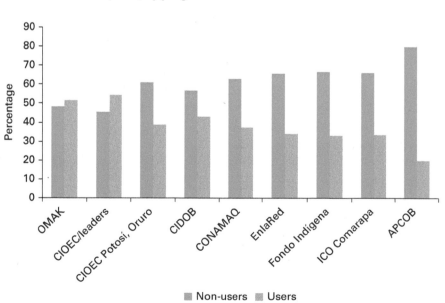

This high number of non-users indicates that these programs have encountered difficulties in their implementation. The results from the EnlaRed program are particularly notable since the majority of its participants are non-indigenous local government officials with high levels of educational attainment, who would be expected to have a much higher rate of Internet use. The case studies in chapters 8 and 9 analyze the multiple factors and ongoing processes that have led to the variations in ICT uses among participants in the different programs and the role that intermediary organizations play in the process of introducing ICTs to indigenous communities. The statistical chi-square test confirms a strong and statistically significant association between the ICT program variable—defined as (a) the ICT program works in the respondent's municipality and (b) no ICT program works in the respondent's municipality—and Internet use ($p > 0.05$). In fact, the correlation coefficient reaches the highest value of all independent variables (Kramer V = 0.43), indicating that the intermediary organization is the factor most strongly associated with indigenous peoples' ICT use.

Cross-Validation of the Results: Multivariant Regression Analysis of the ECADI National Household Survey

This section analyzes data from a national ICT survey conducted by the UNDP in 2004 (ECADI household survey) in order to complement and enrich my own research data. This analysis is useful for several reasons. First, the survey was conducted at the national level across 48 municipalities within all nine departments in Bolivia and had a significantly larger representative sample of 3,617 respondents than my own survey, which covered only 513 respondents. This adds important value to the study, as it allows me to assess indigenous peoples' uses of ICTs at a national scale. Second, it allows me to cross-validate my findings by comparing the two data sets with each other and to generalize some of the main findings from this chapter to a broader national scale. Third, the ECADI survey used a randomized sampling methodology, making it possible to carry out a multivariate regression analysis in order to determine not only the nature of the relationship between the dependent variable (ICT use) and the independent variables (gender and ethnicity), but also the magnitude of the interrelationship.

Finally, the regression analysis broadens its scope to include users from urban areas ($n = 916$). Considering that my survey focused only on rural indigenous areas, the urban perspective of the ECADI survey contributed to the research by providing a perspective on the degree to which spatiality matters in people's levels and uses of ICTs, and distilling some of the differences in use that can be attributed to either the urban or rural context. Finally, several key statistics are derived from the overall ECADI sample ($n = 3,617$), in order to understand the general ICT trends at the national level.

This section summarizes the main findings of several regression models using the data set of the ECADI survey in order to cross-validate the main results obtained from my own user survey and to investigate which of the independent variables can best explain variations in ICT use or non-use.

The section is divided into five subsections: (a) data analysis and the establishment of analogous subsamples; (b) assessment of individual factors that explain differences in indigenous peoples' ICT use; (c) investigation of ethnicity as a critical factor for ICT use; (d) overview of external socioeconomic factors that are associated with ICT use; and (e) influence of the presence of an intermediary organization on indigenous peoples' ICT use.

Data Analysis

In order to carry out an in-depth data analysis of the ECADI survey and to be able to compare its results to those from my own user survey, I established analogous subsamples. The first of these isolates ICT users in rural areas and non-users from communities with a majority indigenous population. As such, the ECADI subsample comprises 1,920 rural ICT users and non-users, while the subsample from my survey includes 513 respondents (including 148 users and 365 non-users).

In both user subsamples, the large majority of the population is indigenous—62 and 65 percent, respectively (table 5.11). In both surveys, primary household language is used as a proxy for ethnic identity, and the large majority of respondents from both subsamples come from municipalities with high illiteracy rates (table 5.12).

Table 5.13 illustrates the strong negative correlation between poverty and Internet use. In sum, the findings from both surveys confirm that ICT users are much better off economically than non-users and that while only one out of three users is poor, two out of three non-users are poor.

TABLE 5.11   Ethnicity of ICT users and non-users (ECADI vs. own user survey)

% of respondents

Primary language spoken	ECADI respondents	Own user survey respondents
Spanish	38	35
Indigenous language and Spanish	58	54
Indigenous language, Spanish, and English	4	11
Total	100	100

TABLE 5.12   Literacy of ICT users and non-users (ECADI vs. own user survey)

% of respondents

Illiteracy rate	ECADI cases	Own user survey
< 10%	31.9	23.8
10–25%	56.3	62.7
> 25%	11.8	13.5
Total	100	100

TABLE 5.13 Poverty rate of ICT users and non-users (as measured by unsatisfied basic needs)

% of respondents

UBN index category	Non-users		Users	
	ECADI	Own survey	ECADI	Own survey
Satisfied basic needs (I)	22	4	52	14
At the threshold of poverty (II)	18	28	14	35
Moderate poverty (III)	35	24	22	21
Marginal population (IV–V)	25	44	12	15
No response	0	0	0	15
Total	100	100	100	100

Degree to Which Individual Factors Explain Differences in ICT Use

This section describes the main findings from three regression models testing the extent to which individual factors such as age, gender, level of education, ethnicity, and socioeconomic status (as measured by income) affect Internet use or non-use (table 5.14). The three models demonstrate that both individual sociodemographic variables and socioeconomic status are relevant with respect to ICT use.

Age, Gender, and Education

Model 1 shows that age, gender, and education are influential factors determining people's ICT use. These variables, however, are able to explain only 19 percent of the model and thus can explain only to a certain extent indigenous peoples' ICT use.

Concerning the relationship between age and ICT use, the findings reveal that respondents in the 19- to 35-year-old cohort are the most likely to use the Internet; a member of this group is more than twice as likely to use the Internet as anyone else.

In contrast, those older than 35 are more than seven times less likely to use the Internet than persons 19–35 years of age—making them the most disadvantaged. The results also validate my own findings with respect to indigenous youth: persons under 19 are at a disadvantage, being 1.2 times less likely to use the Internet than the cohort 19 to 35 years of age.

Model 1 reveals that gender also plays an important role in determining use or non-use. Indeed, men are twice as likely as women to use the Internet. This finding differs from the results of my user survey, where men are only slightly more likely to use ICTs than women (38 and 33 percent, respectively). This discrepancy in the two surveys can be attributed largely to the higher level of educational attainment of the particular group of women in my sample, who are mostly indigenous leaders and part of the growing diversification of the indigenous leadership in Bolivia. As such, the women included in my sample are not representative of the average indigenous woman. Due to this particular bias in my survey, indigenous women respondents have a greater likelihood of using the Internet and of attaining advanced levels of use than

TABLE 5.14 **Three models regressing ICT use correlated with individual factors**

a. Model 1: ICT use regressed on individual factors (age, gender, profession, and education) as independent variables

Individual factor	Model 1		Model 2		Model 3	
	B	Exp (B)	B	Exp (B)	B	Exp (B)
Age						
Less than 19	0.54***	1.72	0.60***	1.83	0.604***	1.83
19–35	0.71***	2.02	0.76***	2.16	0.770***	2.16
Older than 35	−1.24***	0.28	−1.36***	0.25	−1.374***	0.25
Gender						
Male	0.43***	1.53	0.54***	1.71	0.535***	1.71
Female	−0.39***	0.67	−0.29***	0.79	−0.29***	0.79
Level of education						
Primary	−1.42***	0.24			−1.178***	0.31
Secondary education	−0.08***	0.91	−0.123***	0.87		
Advanced technical	1.51***	**4.52**			1.310***	3.71

b. Model 2: ICT use regressed on individual factors controlled by ethnicity

Ethnicity	B	Exp (B)	B	Exp (B)
Non-indigenous	0.674***	**1.95**	0.31***	**1.36**
Quechua	0.231***	1.26	0.22***	1.25
Aymara	−0.827***	0.43	−0.53***	0.59

c. Model 3: ICT use regressed on individual factors controlled by ethnicity and socioeconomic status (as measured by income)

Socioeconomic status	B	Exp (B)
High income	0.76***	2.14
Intermediate income	0.23***	1.25
Poor or extreme poverty	−0.98***	0.36

Note: n = 1,920. For model 1, pseudo R^2 = 0.190. For model 2, pseudo R^2 = 0.365. For model 3, pseudo R^2 = 0.398.
*** Regression coefficients are statistically significant, with $p < 0.05$.

those surveyed by ECADI. This finding suggests that the increasing number of indigenous women who, using their strong individual agency, are able to overcome barriers of racial and gender discrimination and to attain leadership positions in their communities have been able to overcome the digital divide as well. The first case study detailed in chapter 8 investigates the experiences of indigenous women and ICTs, providing a more detailed view of how they have managed to overcome paramount obstacles in their daily lives—the "digital divide" being only one among many.

The results from the regression model 1 also confirm that the level of education plays a crucial role in determining whether or not a person has the opportunity to use ICTs. The main finding from this model is that a person with an advanced education (university or technical schooling) is more than four times as likely as one with lower educational attainment to use the Internet. Moreover, there is a negative correlation between having either a primary or a secondary education and ICT use or non-use; the correlation coefficient reaches a steep negative of $B = -1.42$ with respect to primary educational attainment and a moderately negative value of $B = -0.082$ in relation to secondary education.

These findings confirm the results from my own user survey—particularly the finding of a strong association between education and ICT use. The data from my survey point to the particularly disadvantaged position to which indigenous peoples are relegated, as they need to acquire high levels of education (university degree or advanced technical schooling) in order to overcome one of many other barriers to using the Internet. Access to education in Bolivia is, on average, 3.2 years lower for indigenous than for non-indigenous peoples, and indigenous women have an average attainment of 4.2 years less (Psacharopoulos and Patrinos 1994). In this context, the results suggest that ICTs are not reaching the large majority of rural indigenous peoples, but only the new, highly educated, emerging leadership who have been able to overcome numerous other structural barriers.

Ethnicity

Model 2 adds the control variable of ethnicity in order to analyze the extent to which people's ICT use varies between different ethnic groups. The results from the regression model illustrate that ethnicity is essential to the analysis, as it is strongly correlated with the use or non-use of the Internet. Being indigenous is a disadvantage, with a non-indigenous person almost twice as likely as an indigenous person to use ICTs. The statistical log likelihood test further highlights that ethnicity is the dominant and most influential variable explaining differences in people's ICT uses: its inclusion in the model increases the model fit by 17.9 percent to reach 36.5 percent. Thus the statistical analysis of the national data set confirms the finding from my own user survey that there is a strong association between belonging to a certain ethnic group and a person's opportunities to use the Internet, irrespective of that person's individual characteristics, including age, gender, and education.

The model also reveals major disparities among and within the various indigenous groups. One of the most significant findings is that the Aymara are the most excluded in terms of ICT use. In fact, they are 2.9 times less likely than the Quechua to use ICTs and 4.5 times less likely than the non-indigenous population. These results point to the importance of taking a more nuanced approach when investigating "digital inequalities" within Bolivia and the value added in differentiating between the various indigenous groups (figure 5.13).

There are deeper disparities within indigenous groups as well. First, the most disadvantaged are persons who speak only an indigenous language, with 96 percent saying that they have never used the Internet. This group (88 percent females, 80 percent elders) consists largely of older indigenous women—a historically excluded segment of the population that has been characterized by the lowest socioeconomic indicators, including the

FIGURE 5.13   Use or non-use of ICTs, by ethnicity

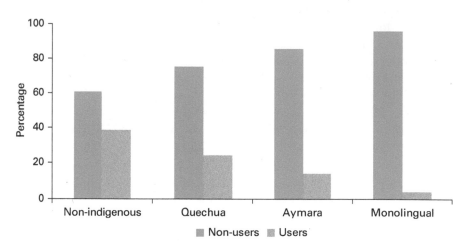

Note: *n* = 3,617.

least access to education and income in Bolivian society. Second, there are different degrees of exclusion between the Aymara and the Quechua, with only 15 percent of the Aymara and 25 percent of the Quechua indicating that they have used the Internet. Third, digital inequalities persist between the non-indigenous and indigenous populations, with nearly 40 percent of the non-indigenous population using ICTs. Finally, model 3 controls for differences in socioeconomic status (as measured by income) in order to confirm or refute the strong association between ethnicity and ICT use. The log likelihood test shows that income increases the model fit (R^2 = 39.8 percent), which confirms, as expected, the correlation between income and ICT use: high-income earners are more than twice as likely to use the Internet as anyone else. When income is included in the model, the magnitude of the effect of ethnicity diminishes, with a non-indigenous person being about 1.4 times more likely than an indigenous person to use the Internet, as opposed to twice as likely. In sum, there is a strong relationship between income and ICT use, and the strong correlation between ethnicity and ICT use exists irrespective of variations in income levels.

Why are there differences in ICT use or non-use among the various ethnic groups? The findings from the regression models suggest the importance of taking a more nuanced, multidimensional approach than is traditionally taken to assess digital inequalities in rural Bolivia. Not only are there important differences between the non-indigenous and indigenous populations, but there are also significant differences among the various indigenous groups. The rest of this section analyzes the main reasons for these digital inequalities in relation to ethnicity, specifically the differences in socioeconomic status and educational attainment between the Quechua and the Aymara and between them and the non-indigenous population.

Ethnicity and Socioeconomic Status

The first possible factor explaining these digital inequalities is the difference in socioeconomic status (as measured by income) between the non-indigenous, Quechua, and Aymara. The analysis of the ECADI survey reveals that there is a strong correlation between income and ethnicity. The large majority of indigenous peoples live in poverty or extreme poverty, but there are significant differences among indigenous groups (figure 5.14). While more than three-quarters (77 percent) of the Aymara live in poverty or extreme poverty, about half of the Quechua are poor.

There are also major differences between the indigenous and non-indigenous populations, with only about one-third (31 percent) of the non-indigenous population living in conditions of poverty and extreme poverty. Furthermore, as shown in figure 5.15, the differences in use or non-use are dependent on socioeconomic status. According to the ECADI survey, only about 4 percent of persons living in extreme poverty and 19 percent of those living in poverty use the Internet, compared with more than 62 percent of the high-income group (figure 5.15). The underlying reason for the existing digital inequalities in rural Bolivia seems to be differences in socioeconomic status. However, a more in-depth analysis of the ECADI data confirms the results from my own ICT survey and ascertains that ethnicity is the most significant factor determining differences in ICT use or non-use, even after controlling for differences in socioeconomic status.

As expected, there is a strong correlation between income and Internet use. On average, high-income earners enjoy a significantly higher rate of Internet use than their lower-income counterparts (figures 5.16 and 5.17). Among the Quechua, the incidence of ICT use decreases from about a third among the

FIGURE 5.14 Socioeconomic status, by ethnicity

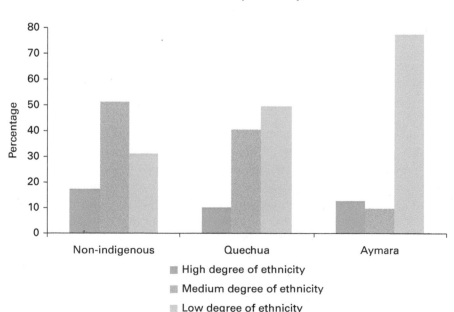

FIGURE 5.15   Simple ICT use, by socioeconomic status

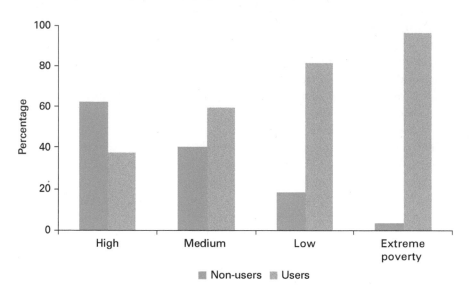

FIGURE 5.16   Simple ICT use of high-income earners, by ethnicity

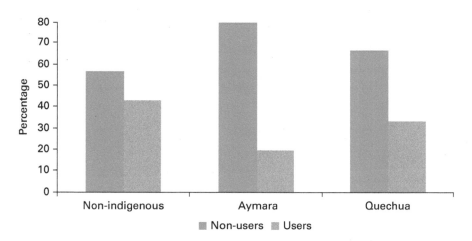

high-income group to less than a tenth among the low-income group (33.3 and 8.7 percent, respectively). More important, some digital inequalities can only be attributed to ethnicity and persist irrespective of income. Among the high-income cohort, 20 percent of the Aymara, 33 percent of the Quechua, and 43 percent of the non-indigenous population use ICTs. In other words, an Aymara—in spite of earning the same high income as a Quechua or a non-indigenous person—remains significantly disadvantaged in his or her opportunities to use ICTs. The same holds true for respondents in the low-income group: a non-indigenous person is 2.0 times and a Quechua is 1.6 times more likely than an Aymara to use ICTs (figure 5.17). Therefore, income cannot explain all of the disparities between ethnic groups.

Ethnicity and Educational Attainment

A second possible explanation for the digital inequalities among ethnic groups is educational attainment. An unexpected finding is that more of the Quechua achieve tertiary and advanced schooling levels than either the Aymara or the non-indigenous population. More than 28 percent of the Quechua have an advanced degree, compared with only 19 percent of the Aymara and 23 percent of the non-indigenous respondents (figure 5.18). Moreover, about two-thirds of the Aymara (67 percent) have a secondary education, compared with only about half of the Quechua and 48 percent of the non-indigenous population.

The level of educational attainment is indeed a key factor influencing whether or not a person uses ICTs (figure 5.19). While a mere 5 percent of persons with primary schooling have used the Internet, 50 percent of those with a tertiary education have used it. Having a secondary education is not sufficient to increase ICT use: only 20 percent of those with a high school education have used the Internet. Tertiary education has the strongest positive

FIGURE 5.17   Simple ICT use of low-income earners, by ethnicity

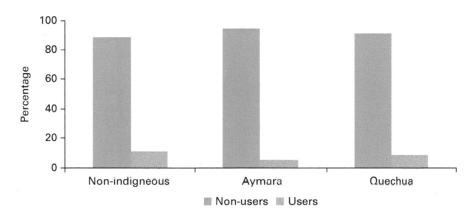

FIGURE 5.18   Educational attainment, by ethnicity

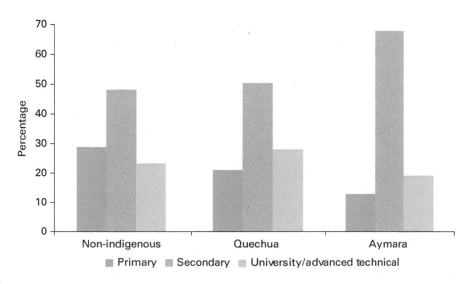

FIGURE 5.19   Use of ICTs, by educational attainment

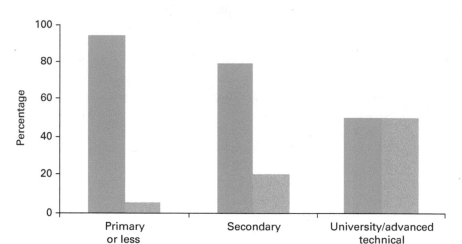

FIGURE 5.20   Use of ICTs among persons with tertiary education, by ethnicity

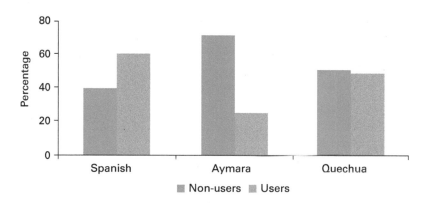

correlation with ICT use, which helps to explain the low percentage of the Aymara in rural Bolivia who use the Internet. The vast majority of the Aymara (67 percent) have a secondary education, but they continue to be excluded from using these technologies.

To what extent is an Aymara from rural Bolivia able to enhance her or his likelihood of using the Internet by pursuing an advanced education? Figure 5.19 demonstrates that significant disparities between the Aymara, the Quechua, and non-indigenous populations persist, even when controlling for differences in educational attainment. Whatever their level of education, the Aymara are the most disadvantaged in terms of Internet use.

In fact, of all respondents with an advanced or a university degree, only 28 percent of the Aymara, compared to 49 percent of the Quechua, said they have used ICTs (figure 5.20). Furthermore, ethnic disparities between indigenous and non-indigenous populations persisted, even among the highly educated indigenous. The only ethnic group in which a majority (60 percent) of the population has used ICTs is the non-indigenous population. In sum, despite a strong association between education and the respondent's use or non-use of ICTs, indigenous peoples—in particular the Aymara—continue to be significantly disadvantaged vis-à-vis Internet use, even when attaining the same advanced level of education as the non-indigenous population. Ethnicity remains the most important variable affecting ICT use.

Degree to Which Socioeconomic Factors Explain Differences in ICT Use

This section assesses the relationship between socioeconomic factors (poverty, illiteracy) external to a person's demographics (gender, age) and her or his human capabilities (level of education) with a person's ability to use ICTs. Model 4 includes municipalities' poverty levels as measured by the UBN index and illiteracy rates, in order to investigate the extent to which an association exists between both of these variables and indigenous peoples' ICT use. The regression analysis is applied to the subsample of all rural respondents ($n = 1,920$).

Poverty and Illiteracy

Model 4 shows that there is a significant relationship between both socioeconomic variables—the poverty and illiteracy rates—and ICT use (table 5.15). For instance, a person who lives in a municipality with satisfied basic needs is 2.5 times more likely to use the Internet than a person from a municipality with unsatisfied basic needs.

The log likelihood test shows that the UBN alone explains 15.5 percent of the model, indicating that it contributes significantly toward explaining a person's ability to use ICTs. Furthermore, the results highlight the importance of municipal illiteracy rates: municipalities with illiteracy rates under 5 percent have twice the ICT use as those with illiteracy rates of 15–30 percent. Illiteracy rates contribute significantly to the model: they are equivalent to 4 percent of the model fit, and R^2 reaches a value of 0.194.

TABLE 5.15 Model 4: Use of ICT regressed on socioeconomic variables (unsatisfied basic needs and illiteracy rates)

Socioeconomic variables	Model 4a		Model 4b	
	B	Exp (B)	B	Exp (B)
UBN category				
I	0.943***	2.568	0.719***	2.052
II and III	0.104***	1.109	−0.145***	0.865
IV and V	−0.332***	0.718	−0.286***	0.751
Illiteracy rate				
Under 5%	0.711***	2.035	0.427***	1.533
From 5–15%	0.236***	1.266	0.135***	1.145
From 15–30%	−0.295***	0.745	−0.482	0.617

Note: $n = 1,920$. For UBN category, pseudo $R^2 = 0.155$. For illiteracy rate, pseudo $R^2 = 0.194$.
*** Regression coefficients are statistically significant, with $p < 0.05$.

TABLE 5.16   Model 5: Use of ICT regressed on external socioeconomic variables controlled by geographic region

Geographic region	B	Exp (B)
Highlands	−0.383***	0.68
Valleys	0.174***	1.190
Lowlands	0.555***	1.742

Note: pseudo R^2 = 0.254.
*** Regression coefficients are statistically significant, with $p < 0.05$.

In model 5, both of these independent variables are controlled for geographic location in order to test the extent to which ICT use is dependent on geography (highlands, valleys, and lowlands). My own user surveys found a strong association between the external variable geography and indigenous peoples' use or non-use of ICTs. As shown in table 5.16, the regression model confirms the important relationship between these two variables. A person who lives in the lowlands is 1.7 times as likely to use the Internet as a person who lives in any other region in Bolivia. The model also confirms the main finding of my survey results: the highlands are the most disadvantaged region in Bolivia with regard to ICT use. A person from the lowlands is three times more likely to use the Internet than a person from the highlands. In the log likelihood test, R^2 increases to 0.254, indicating that this variable significantly improves the model and thus its capacity to explain variations in people's simple ICT uses.

However, adding geography to the model reduces the influence of the other two independent variables: UBN and illiteracy. For instance, the Exp (B) for municipalities with illiteracy rates under 5 percent is reduced from 2.0 to 1.5, indicating that the inclusion of the geographic variable reduces the influence of the other variables. Nevertheless, both UBN and illiteracy continue to have an important influence on determining whether or not a person lives in a municipality where the enabling environment is conducive to the use of ICTs.

Geographic Location

Is geography associated with indigenous peoples' ICT uses, and to what extent do digital inequalities between ethnic groups persist in urban areas? More than three-fourths of the Aymara live in rural areas, in contrast to 45 percent of the Quechua (figure 5.21). Moreover, 72 percent of the non-indigenous respondents live in urban rather than in rural remote areas.

As shown in figure 5.22, geography plays an important role in the existing digital inequalities between urban and rural dwellers. This particular data set includes the full ECADI sample (n = 3,616) and reveals stark digital inequality; the overwhelming majority—92 percent—of respondents from rural areas reported never having used the Internet before, compared with two-thirds (64 percent) of their urban counterparts.

Structural Barriers and Intermediary Factors

Is the intervention of ICT programs correlated with indigenous peoples' ability to use new forms of ICTs? To what extent can ICT programs overcome,

FIGURE 5.21   Spatiality and ethnicity (rural vs. urban populations)

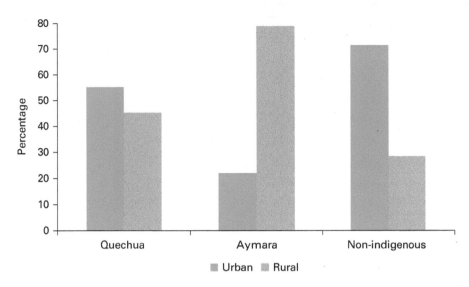

FIGURE 5.22   Geographic factors explaining differences in simple ICT use

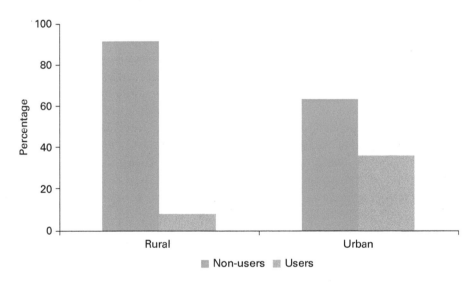

at least partially, the structural barriers to ICT use (high levels of poverty, illiteracy rates, and lack of basic infrastructure and social services)? Can such programs increase the likelihood of indigenous peoples being able to use ICTs, even under the adverse overall socioeconomic conditions in rural areas, particularly the Andean highlands?

Model 6 indicates that the provision of basic infrastructure services (access to Internet and electricity services) is positively correlated with indigenous peoples' ICT use (table 5.17). In addition to the "digital divide," there is a significant "electricity divide" in Bolivia; the use of ICTs is severely limited by

TABLE 5.17   Model 6: ICT use regressed on basic infrastructure services (access to ICTs and to electricity services)

Type of infrastructure	B	Exp (B)
Access to electricity		
Less than 25%	−0.668***	0.513
25–50%	−0.411***	0.663
50–75%	0.08***	1.008
More than 75%	1.071	**2.92**
Access to ICTs		
Municipality with Internet services	0.453***	1.573
Municipality without Internet services	−0.227***	0.797

Note: n = 1,920. For access to electricity, pseudo R^2 = 0.158. For access to ICTs, pseudo R^2 = 0.181.
*** Regression coefficients are statistically significant, with $p < 0.05$.

the fact that the large majority (70 percent) of rural, mostly indigenous, communities lack access to basic electricity services (VMEEA 2005). Having access to electricity services is a critical precondition for the ability to use ICTs. The model shows that a person who comes from a municipality where 75 percent or more of the population has access to electricity services is almost three times as likely to use the Internet as a person who comes from a municipality with a low percentage of access to electricity services. The opposite is true for a person who lives in a municipality where less than 25 percent of the population has access to electricity services, where the correlation coefficient indicates a strong negative relationship with ICT use (B = −0.668). The availability of electricity services, not the availability of Internet services, is the central determinant, since this independent variable alone can explain 15.8 percent of the model.

When the variable "access to ICTs" is added to the model, there is an important correlation between this variable and people's ability to use ICTs. The data show that a person who lives in a municipality with access to Internet services is almost twice as likely to use the Internet as a person who lives in a municipality without such access. However, including the "access to ICTs" variable enhances the model fit by a modest 2.3 percent (R^2 = 18.1 percent), indicating only a weak correlation between access and indigenous peoples' ICT use.

The findings from this model reveal the unexpected result that having access or not to electricity—rather than having access to ICTs—is the variable that explains, to a large extent, people's ability to use the Internet.

In model 7, ethnicity is introduced as a control variable, since the main findings from my own user survey indicate that belonging to a specific ethnic group—not pure access to ICTs—is the most significant variable determining a person's ICT use. The results reveal the unanticipated finding that once the model controls for variations in ethnicity, the variable "ICT access" does not enter into the model anymore. A person of Aymara origin remains 3.2 times less likely to use the Internet than a non-indigenous person and 1.9 times less likely than a Quechua, irrespective of ICT access. The major effect of the

TABLE 5.18 **Model 7: ICT use regressed on basic infrastructure services (access to ICTs and access to electricity services) controlled by ethnicity**

Infrastructure services	B	Exp (B)
Access to electricity		
Less than 25%	−0.687***	0.503
25–50%	−0.402***	0.669
50–75%	0.046***	1.047
More than 75%	1.043***	**2.83**
Access to ICTs*		
Municipality with Internet services		
Municipality without Internet services		
Ethnicity		
Non-indigenous	0.609***	1.838
Quechua	−0.047***	0.954
Aymara	−562***	0.570

Note: n = 1,920. For access to ICTs, pseudo R^2 = 0.158. For ethnicity, pseudo R^2 = 0.237.
*** Regression coefficients are statistically significant, with $p < 0.05$. * Regression coefficients are not statistically significant, with $p > 0.05$.

control variable of ethnicity is also confirmed by the log likelihood test, which shows that the inclusion of this variable significantly increases the model fit from 15.8 to 23.7 percent (table 5.18).

Thus the empirical results highlight that the reason for the high levels of non-use of ICTs among indigenous peoples in rural Bolivia is not the lack of access to ICT services but instead the group-level effect of belonging to a specific indigenous group. In other words, for an Aymara person, the reason for non-use of the Internet is her or his very ethnicity, which excludes her or him from a core set of economic and social opportunities—in this case, the use of new forms of ICTs. Not being able to use the Internet is only a reflection of the broader and deep-seated social exclusion of the Aymara within Bolivian society.

Model 7 also shows that having access to electricity services is an important determinant of indigenous peoples' Internet use. A possible explanation is that the variable "access to electricity services" reflects the low level of living conditions and the high level of extreme poverty in indigenous communities. Thus lack of access to electricity services can be used as an indication for both the remoteness and the overall degree of socioeconomic development of a community. Consequently, it is not surprising that this independent variable remains in the model, although its effect is slightly reduced because of the correlation between ethnicity and lack of electricity services. The findings are consistent with those obtained from the empirical analysis of my own user survey: lack of physical access to ICT services is not a key determining factor in the digital disparities in Bolivia.

The empirical evidence has demonstrated that other factors—specifically intermediary factors or the availability of local content and capacity-building activities—are essential to enabling indigenous peoples to make meaningful use of the Internet. Because of significant differences in the social and cultural context of indigenous peoples, it is crucial either to adopt existing content to their needs or to develop specific "culturally appropriate" content if indigenous communities are to derive real benefit from Internet use. It is also essential to support indigenous peoples in developing ICT capabilities so that they can meaningfully use these technologies and appropriate them to their own individual and collective realities.

Model 8 explores the extent to which this intermediary process is a critical factor for indigenous peoples' Internet use. This model studies whether or not ICT programs—which complement the provision of ICT infrastructure services with activities such as capacity building and content creation—can influence people's ability to use ICTs and the Internet. The findings confirm the main results from my own user survey: the presence of an ICT program is an essential factor for determining whether or not a person is able to use ICTs and the Internet (table 5.19).

The log likelihood test confirms the importance of this variable in explaining whether or not people are able to use ICTs: it can explain 14.2 percent of the overall model, indicating a high model fit. The data also show the strength of the positive correlation between the presence of an ICT program and people's ability to use the Internet.

A person who lives in a municipality with an ICT program is 3.6 times more likely to use the Internet than a person who lives in a municipality without such a program. This finding confirms the main finding from my user survey: for the question of whether or not indigenous peoples have the opportunity to use ICTs, the presence of an intermediary ICT program is essential.

TABLE 5.19 Model 8: ICT use regressed on intermediary factor (ICT program) controlled by access to ICTs and electricity services

Factor	B	Exp (B)
ICT program		
Municipality with an ICT program	1.294***	**3.649**
Municipality without an ICT program	−0.192***	0.147
Access to electricity		
Less than 25%	−0.687***	0.503
25–50%	−0.402***	0.669
50–75%	0.046***	1.047
More than 75%	1.043***	**2.83**
Access to ICTs*		
Municipality with Internet services		
Municipality without Internet services		

Note: $n = 1,920$. For access to electricity, pseudo $R^2 = 0.142$. For access to ICTs, pseudo $R^2 = 0.158$. *** Regression coefficients are statistically significant, with $p < 0.05$. * Regression coefficients are not statistically significant, with $p > 0.05$.

The correlation between the ICT program variable and the ability of indigenous peoples to use ICTs is much stronger than the association between the pure provision of access to ICTs and use. The results from models 7 and 8 show that although having access to ICT services is an important variable influencing people's opportunities to use ICTs, the presence of an "ICT program in a municipality" has more than twice the effect, with Exp (B) = 3.6, as mere "access to ICTs," with Exp (B) = 1.5. This finding confirms the results from my own user survey: the provision of ICT infrastructure, while necessary, is not sufficient for enabling indigenous peoples' Internet use.

Model 9 investigates the extent to which the influence of the independent variable "ICT programs" holds when controlling for differences in poverty levels (table 5.20). The results confirm a positive relationship between the presence of an ICT program and a person's ability to use ICTs, even in municipalities with high levels of poverty or extreme poverty.

The influence of the variable "ICT programs" on a person's ICT use is reduced, so that a person who lives in a municipality with an ICT program continues to be 1.8 times more likely to use ICTs than a person who lives in a municipality without such a program. Including in the model the control variable of the UBN index reduces this likelihood by about half, but the strong association between the variable "ICT program" and an indigenous person's use of ICTs remains. This result is confirmed by cross-tabulating the simple use of ICTs with the ICT program (figure 5.23).

There is a direct and positive relationship between the presence of an ICT program in a municipality and people's ICT use. Even in poor and extremely poor municipalities (UBN categories IV and V), the intervention by an ICT program can enable people to use ICTs.

In municipalities with high and extreme poverty rates, more than twice as many people use ICTs in municipalities with an ICT program

TABLE 5.20 Model 9: ICT use regressed on intermediary factors controlled by UBN and ethnicity

Factor	B	Exp (B)	B	Exp (B)
ICT program				
Municipality with an ICT program	0.584***	**1.792**	0.474***	**1.606**
Municipality without an ICT program	−0.292***	0.747	−0.474	0.622
UBN category				
I	1.028***	2.796	0.900***	2.459
II and III	0.360***	1.037	−0.10***	0.905
IV and V	−0.452***	0.636	−0.37***	0.534
Ethnicity				
Non-indigenous			0.473***	1.606
Quechua			−0.031	0.982
Aymara			−418***	0.658

Note: For UBN index, pseudo R^2 = 0.186. For ethnicity, pseudo R^2 = 0.199.
*** Regression coefficients are statistically significant, with $p < 0.05$.

FIGURE 5.23   Simple ICT use in poor and extremely poor communities (with and without an intermediary program)

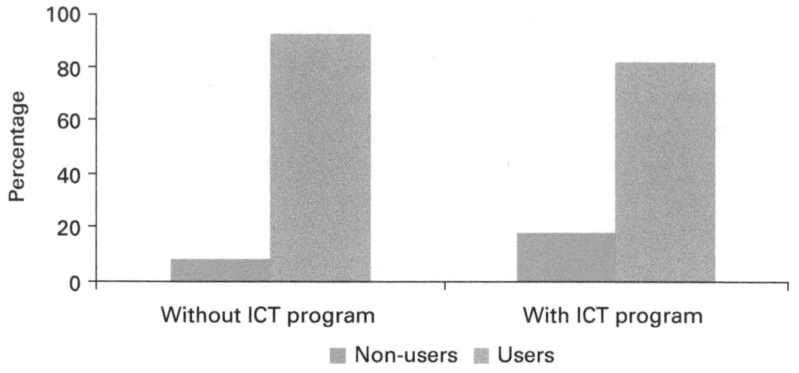

■ Non-users   ■ Users

as in municipalities without a program (18 and 8 percent, respectively). The number of users seems relatively low, but this is not suprising given the many structural barriers, such as the lack of basic infrastructure services and high illiteracy rates, that programs face in providing ICT services to rural indigenous communities. As the findings from the regression models have shown, people's ability to use ICTs is strongly dependent on the overall poverty levels in the municipality.

Finally, model 9 analyzes the extent to which the strong effect of the ICT program holds when controlling for differences in both poverty levels and ethnicity. Does the importance of the intermediary to people's Internet use hold irrespective of ethnicity? The main finding is that the magnitude of the influence of the ICT program variable is only slightly reduced; people who live in a municipality with an ICT program continue to be 1.6 times more likely to use ICTs than people who live in a municipality without such a program. This finding is also confirmed by figure 5.23. Furthermore, the log likelihood indicates the minor influence of this variable since the model fit is only increased by 1.3, reaching 19.9 percent. Regardless of ethnicity, the intermediary factor remains an essential variable explaining the differences in people's ICT use.

Conclusions

This chapter has shown that indigenous peoples are disproportionately disadvantaged in terms of their Internet use relative to the non-indigenous population. Table 5.21 summarizes the main statistical results.

There is a strong correlation between being poor and being indigenous and between poverty and ICT exclusion. Digital inequalities in Bolivia reflect the deep structural inequalities within Bolivian society; ethnicity is a major influence on whether a person is poor. The statistical analysis confirms the national poverty data, with approximately one-third of the non-indigenous population, two-thirds of the bilingual indigenous population, and almost three-quarters of the monolingual indigenous population being poor (Psacharopoulos and Patrinos 1994; Hall and Patrinos 2005). In relation to ICTs, there is a

TABLE 5.21   Summary of statistical results

dependent variable: Internet use

Independent variables	Chi-square	Phi (correlation)	Kramer V (correlation)	Asymptotic significance	Sample size (N)
Sociodemographic variables					
Gender	1.068	0.04		0.30	507
Age	2.945		0.075	0.22	445
Level of education	**87.605**		**0.42**	**0.00**	**505**
Socioeconomic variables					
Ethnicity (measured by language)	**73.963**		**0.36**	**0.00**	**501**
UBN index	13.285		0.164	0.01	491
HDI	6.109		0.172	0.04	497
Illiteracy rates	**24.222**		**0.24**	**0.02**	**491**
Population density	9.206		0.22	0.01	491
Geographic location	5.917		0.084	0.05	508
ICT infrastructure					
Access to ICTs	9.333	−0.14		0.02	484
ICT intermediary					
ICT program	**103.329**		**0.43**	**0.00**	**513**

Note: The bold numbers indicate independent variables that strongly influence indigenous peoples' ICT use.

significant correlation between poverty, ethnicity, and ICT use. Among the non-indigenous population, 33 percent live above the poverty line, and 40 percent use ICTs. Only 26 percent of the indigenous bilingual population lives above the poverty line, and only 14 percent use ICTs. The most disadvantaged is the monolingual indigenous population, with 86 percent living below the poverty line and only 4 percent using new forms of ICTs.

There is a weak association between access to and actual use of ICTs. The empirical evidence confirms the main assumption of the theoretical framework: it is necessary to go beyond the concept of mere access to analyze the effects of ICTs on indigenous peoples' well-being. The data reveal the unexpected finding that the high rates of non-use among indigenous peoples—approximately two-thirds of whom do not use the Internet—cannot be explained simply by lack of access to ICT services. The majority of indigenous peoples (58 percent) who have access to new forms of ICTs do not use this technology.

There are significant digital disparities among the various indigenous groups. These differences exist not only between the non-indigenous and indigenous peoples, but also among the different indigenous groups, with the Aymara being the most disadvantaged. An Aymara person is 2.8 times less likely to use the Internet than a Quechua and 4.5 times less likely than a non-indigenous person. These digital inequalities among ethnic groups persist even after controlling for individual factors such as age, gender, educational attainment, and socioeconomic status.

Schooling matters to ICT use, but digital disparities between the indigenous and non-indigenous populations persist, even when controlling for educational attainment. The statistical analysis confirms that disparities in educational attainment between the indigenous and non-indigenous peoples affect use. Given that only 5 percent of survey respondents with a primary education have used the Internet, compared to half of those with a tertiary education, and that indigenous peoples have generally low educational attainment—only 75 percent of indigenous males and 84 percent of indigenous females have completed only primary education or less—it is not surprising that a large majority are excluded from ICT use (Psacharopoulos and Patrinos 1994). Even when attaining the same advanced level of tertiary education as the non-indigenous population, they continue to be severely disadvantaged in terms of their Internet use, at a ratio of almost 2:1. As such, education, although important, is insufficient to overcoming the substantial barriers to Internet use.

The presence of an intermediary program is the most significant factor explaining indigenous peoples' ICT use. The intervention of an ICT program can play a significant role in supporting indigenous peoples' use of ICTs, in spite of the multiple structural barriers that have led to the "digital exclusion" of the vast majority of indigenous peoples. The empirical evidence from both my own user survey and the national ECADI survey highlights the strong association between the presence of an ICT program in a municipality and people's use or non-use of ICTs. In fact, a person who lives in a municipality with an ICT program is 3.6 times more likely to use the Internet than anybody else. This finding holds regardless of differences in spatiality, poverty levels, and access to ICT infrastructure.

References

Campbell, Tim. 2001. *The Quiet Revolution: The Rise of Political Participation and Leading Cities with Decentralization in Latin America and the Caribbean.* Pittsburgh, PA: University of Pittsburgh Press.

Delgadillo, Karen, Ricardo Gómez, and Klaus Stoll. 2002. "Community Telecentres for Development: Lessons from Community Telecentres in Latin America and the Caribbean." IDRC, Ottawa.

Dertouzos, Michael L., and Bill Gates. 1998. *What Will Be: How the New World of Information Will Change Our Lives.* HarperOne.

Hall, Gillette, and Harry A. Patrinos. 2005. *Indigenous Peoples, Poverty, and Human Development in Latin America.* London: Palgrave Macmillan.

Heeks, Richard. 1999. "Information and Communication Technologies, Poverty, and Development." Development Informatics Working Paper 5, University of Manchester, IDPM, Manchester.

———. 2002. "I-Development Not E-Development: Special Issue on ICTs and Development." *Journal of International Development* 14 (1): 1–11.

INE (Instituto Nacional de Estadística). 2002. *Encuesta de mejoramiento de las condiciones de vida [Survey on Improvement in Living Conditions].* La Paz: INE.

Jiménez Pozo, Fernando, Wilson Landa Casazola, and Ernesto Yáñez Aguilar. 2006. "Bolivia." In *Indigenous Peoples, Poverty, and Human Development in Latin America*, edited by Gillette Hall and Harry Patrinos. London: Palgrave Macmillan.

Livingstone, Sonia. 2003. "Children's Use of the Internet: Reflections on the Emerging Research Agenda." *New Media and Society* 5 (2): 147–66.

López, Enrique J., and May González Villaseca. 1996. "IT as a Global Economic Tool." In *Information Technology, Development, and Policy: Theoretical Perspectives and Practical Challenges*, edited by Edward M. Roche and Michael J. Blaine. Avebury: Ashgate.

Madon, Shirin. 2000. "The Internet and Socio-Economic Development: Exploring the Interaction." *Information Technology and People* 13 (2): 85–101.

———. 2004. "Evaluating the Developmental Impact of E-Governance Initiatives: An Exploratory Framework." *Electronic Journal on Information Systems in Developing Countries* 20 (5): 1–13.

Martin-Barbero, Jesus. 1993. *Communication, Culture, and Hegemony: From the Media to Mediations*. London: Sage.

McConnell, Scott. 1998. "Connecting with the Unconnected: Proposing an Evaluation of the Impacts of the Internet." In *The First Mile of Connectivity: Advancing Telecommunications for Rural Development through a Participatory Communication Approach*, edited by Don Richardson and Lynnita Paisley. Rome: FAO.

Nelson, Diane. 1996. "Maya Hackers and the Cyberspatialised Nation State: Modernity, Ethnostalgia, and a Lizard Queen in Guatemala." *Cultural Anthropology* 11 (3): 287–308.

Norris, Pipa. 2001. *Digital Divide: Civic Engagement, Information Poverty, and the Internet Worldwide*. Cambridge, MA: Harvard University, John F. Kennedy School of Government.

Psacharopoulos, George, and Harry Patrinos. 1994. *Indigenous Peoples and Poverty in Latin America: An Empirical Analysis*. Washington, DC: World Bank.

Richardson, Don, and Lynnita Paisley, eds. 1998. *The First Mile of Connectivity: Advancing Telecommunications for Rural Development through a Participatory Communication Approach*. Rome: FAO.

Roman, Raul, and Royal D. Colle. 2002. "Themes and Issues in Telecentre Sustainability." Development Informatics Working Paper 10, University of Manchester, IDPM, Manchester.

Selwyn, Neil. 2004. "Reconsidering Political and Popular Understandings of the Digital Divide." *New Media and Society* 6 (3): 341–62.

UNDP (United Nations Development Programme). 2004. "Encuesta de capacidades informacionales (ECADI)." Survey administered by UNDP in preparation of its *2004 National Human Development Report: Interculturalidad y globalización*. UNDP, La Paz.

Van Dijk, Jan A. G. M. 2005. *The Deepening Divide: Inequality in the Information Society*. Thousand Oaks, CA: Sage Publications.

VMEEA (Vice Ministerio de Electricidad y Energías Alternativas). 2005. *Anuario estadístico del sector eléctrico boliviano*. La Paz: VMEEA.

Warschauer, Mark. 2003. *Technology and Social Inclusion: Rethinking the Digital Divide*. Cambridge, MA: MIT Press.

CHAPTER 6

Beyond Use: Expanding Poor People's ICT Capabilities

As chapter 5 has shown, to understand the impact of information and communication technologies (ICTs) on indigenous peoples' well-being, it is necessary to move beyond the concept of "access" and instead analyze the different factors that enable people's uses of ICTs within their socioeconomic, political, and cultural context. This chapter takes the analysis one step further, arguing that to assess the role that ICTs can play in the well-being of indigenous peoples, it is not sufficient to evaluate factors that influence whether people have the opportunity to use ICTs. It is essential, as described in the theoretical framework in chapter 1, to investigate people's "ICT capabilities"—the extent to which a person is proficient in using computers and the Internet in a meaningful manner (appendix F). The notion of ICT capabilities draws on the conceptualization of ICTs by Michel Menou (2002) and Mark Warschauer (2003), who have emphasized that the impact that ICTs have on people's lives depends on the *level of ICT use* rather than on mere access and use.

In order to address the main research question of the study—whether and under which conditions ICTs can enhance indigenous peoples' well-being—it is critical not only to understand whether people are using ICTs but to analyze their proficiency or capabilities to use them. As laid out in the theoretical framework, it is necessary to unpack people's ICT capabilities—their abilities to transform the range of information and communication options made available to them into actual or realized functionings (Garnham 1997, 32)—before assessing their impact on people's well-being. The chapter studies the multiple factors that influence indigenous peoples' ICT capability, since the ability to use ICTs in a meaningful way constitutes an important precondition for people to enhance their well-being in multiple dimensions of their lives. Specifically, the chapter addresses the following questions:

- Which variables (individual or socioeconomic) are associated most strongly with a person's ICT capabilities?

- To what extent can differences in indigenous peoples' ICT capabilities be explained by differences in existing ICT programs?
- Which structural barriers impede indigenous peoples from reaching advanced ICT capabilities?

The chapter presents the empirical evidence of indigenous peoples' ICT capabilities based on micro-level data gathered through the same user survey used in the previous chapter. In the first section, I use these data to investigate the relationship between a person's ICT capabilities and variables such as individual characteristics (age, gender, and ethnicity) and socioeconomic factors. In the second, I cross-validate the results by applying a series of regression models to the data from the *Encuesta de Capacidades Informacionales* (ECADI)—the national household survey of people's perceptions of ICT administered by the United Nations Development Programme (UNDP), which is described in chapter 5. At the center of this analysis is an assessment of the magnitude of the effects that various factors have on people's ICT capabilities.

Principal Findings

The principal findings from this chapter indicate that the main underlying reasons for the existing digital inequalities are deeply rooted in the systematic inequalities that indigenous peoples face as a group in Bolivian society.

The empirical evidence demonstrates that the ICT capabilities of an indigenous person from a rural community are independent of his or her individual sociodemographic characteristics, including age, gender, and education. They do, however, depend on the broader economic, social, political, and cultural factors that have led to the social exclusion and marginalization of indigenous peoples in Bolivia. The most significant and unexpected result—that indigenous peoples' ICT capabilities are independent of their educational attainment—points to the fact that, irrespective of a person's sociodemographic characteristics, individual agency, or both, belonging to a marginalized group is a critical determinant of the extent to which a person can realize the potential benefits of ICTs.

Both the statistical analysis of my survey data and the regression models confirm that ethnicity and ICT capabilities are strongly associated with each other. This analysis illustrates that the process of improving individual human capability through ICTs cannot be separated from the more structural socioeconomic, political, and cultural impediments of indigenous peoples' development in Bolivia.

Another central finding of the chapter is that there are important digital inequalities among the various indigenous groups and not only between the indigenous and non-indigenous populations. Indeed, the empirical evidence from both my user surveys and the national ECADI survey indicates that the Quechua as a group are almost twice as likely as the Aymara to attain a high level of ICT capabilities as the non-indigenous, suggesting that the Aymara are the most disadvantaged group in terms of ICT use. These results are consistent with the overall socioeconomic situation of the Aymara; the analysis of poverty presented in chapter 3 shows that this group has the highest incidence of extreme poverty and lives in the most remote areas of Bolivia. Their low capacity in terms of ICT use is consistent with their high levels of poverty and social exclusion within Bolivia.

A more in-depth analysis of the digital inequalities among various indigenous groups, however, shows that these disparities disappear in urban areas, indicating that the migration of indigenous peoples to urban centers can improve an indigenous person's opportunities to reach advanced ICT capabilities, irrespective of the indigenous groups to which she or he belongs. The statistical data indicate that while the differences between the non-indigenous and indigenous populations are slightly reduced, important digital disparities between those groups persist even in urban areas. Another important finding from the analysis of the urban data set is that, as opposed to rural areas, improvements in an indigenous person's educational attainment are related with more advanced ICT capabilities. Thus it seems that the multiple structural impediments for indigenous peoples to make meaningful use of ICTs are less significant in urban areas.

Moreover, the chapter stresses that people's physical "access" to ICTs is not associated with advanced ICT capabilities. In fact, the findings point to the absence of any statistically significant relationship between these two variables, suggesting that the provision of ICT infrastructure services in rural areas does not have its intended effect and de facto fails to enable people to make meaningful use of the Internet. This finding suggests that while the provision of ICT services is certainly an important enabling factor for indigenous peoples to derive benefits from the use of ICTs, it is insufficient, by itself, to improve their well-being. This is a central finding, since the majority of government programs in Bolivia have focused their attention solely on improving access to ICT services in rural areas. The findings suggest an explanation for the limited impact of these existing government programs: they do not recognize the need to go beyond mere access to meaningful use.

Finally, the chapter demonstrates that, in spite of the multiple structural impediments mentioned above, a strong association exists between effective local intermediation through an ICT program and a respondent's ICT capabilities. The data reveal that the intermediary factor is the strongest determinant influencing a person's ICT proficiency. Respondents who have participated in grassroots-level ICT programs have relatively high ICT capabilities. For instance, in the program implemented by the grassroots organization CIOEC (Coordinadora de Integración de Organizaciones Económicas Campesinas Indígenas y Originarias), 28 percent of participants have attained advanced and 61 percent have attained intermediate ICT capabilities. This result is noteworthy, since the CIOEC program is implemented in Oruro and Potosí in the Andean highlands—the poorest and most economically depressed region in Bolivia. In fact, more than half of the indigenous respondents who participate in an ICT program and live in a community with extremely high rates of poverty and illiteracy nevertheless have been able to attain very high ICT capabilities. A second unexpected finding is that while almost two-thirds of indigenous peoples from the highlands are unable to use the Internet, the overwhelming majority of the one-third who do use ICTs have reached a very high level of proficiency. These findings indicate that indigenous peoples can overcome considerable socioeconomic constraints and significantly improve their ICT capabilities through the ongoing support of an effective local intermediary.

Both of these findings—(a) that the ICT program is the variable that has the strongest association with ICT capabilities and (b) that the participants of grassroots organizations have attained the most advanced ICT capabilities—are central to the study. The first confirms the hypothesis established in the theoretical framework—that the intervention of an ICT program is an essential factor determining whether indigenous peoples can enhance their well-being through ICTs. The second finding—that grassroots organizations are most effective in supporting indigenous peoples to make meaningful use of ICTs—confirms the second hypothesis, which stresses the importance of indigenous peoples' local appropriation of the technology. The empirical results indicate that in the CIOEC program, in particular, local indigenous leaders have "appropriated" ICTs and attained high ICT capabilities through their use.

Data Analysis

The chapter analyzes the multiple factors that influence indigenous peoples' ICT capabilities and thus their ability to make meaningful use of ICTs. The empirical evidence presented in this chapter is based on data from my user survey and the ECADI, which were used in chapter 5 to analyze people's use of simple ICTs. The demographic and socioeconomic profile of respondents is the same as described in chapter 5. The only difference is that this chapter includes users from urban areas ($n = 916$). The sample was expanded to examine the degree to which spatiality matters in indigenous peoples' level of ICT use, distilling some of the differences in use that can be attributed to either the urban or rural context.

In order to capture differences in ICT capabilities, the chapter establishes three levels of proficiency for use: basic, intermediate, and advanced. The definition of ICT capability is based on question 22 of the user survey: For which activities have you used the Internet? Appendix F describes in more detail how this variable and its categories were constructed.

Empirical Results of ICT User Survey

This section is divided into five subsections analyzing (a) the overall ICT capabilities in the sample; (b) the role of individual factors, including gender, age, and education; (c) the role of ethnicity; (d) the role of socioeconomic factors, including poverty, human development, illiteracy, and spatiality; and (e) the role of intermediary factors (the presence of an ICT program) in influencing people's ICT capabilities.

Overall ICT Capabilities in the Sample

The majority of the overall sample (59 percent) said that they have attained an advanced or intermediate ICT use (figure 6.1). This is surprising, considering

FIGURE 6.1   ICT capabilities of users

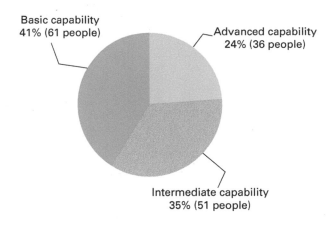

Basic capability
41% (61 people)

Advanced capability
24% (36 people)

Intermediate capability
35% (51 people)

Note: n = 148.

that many individual and socioeconomic factors significantly limit the ability of indigenous peoples to use new ICTs.

Although the user sample represents only about a third (29 percent) of all respondents, the vast majority (71 percent) of people who participated in the survey indicated that they do not use ICTs. At the same time, the finding that the one-third minority representing the user sample has attained considerable experience in using ICTs is important, given that a certain level of meaningful use is a prerequisite for people to derive benefits from ICTs.

Role of Individual Factors in Explaining ICT Capabilities

The ICT literature frequently points to the role of individual sociodemographic factors, such as gender, age, education, and ethnicity, in explaining differences in people's ICT uses (DiMaggio and Hargittai 2001; Warschauer 2003). This section explores the role of individual factors in determining the extent to which indigenous peoples can make meaningful use of ICTs.

Gender

Indigenous peoples' ICT capabilities are gender-neutral. About a quarter of all users (24 percent) are advanced, and this proportion is almost the same for men and women (24 and 26 percent, respectively). Indeed, the level of ICT use is slightly higher among women than among men (figure 6.2). Applying the chi-square test confirms that there is no substantive relationship between gender and ICT capabilities (Kramer V = 0.341) and that this association is statistically insignificant ($p = 0.053 > 0.05$).

As discussed in chapter 5, these results should be viewed with caution. There is a strong male bias in the survey, with only 30 percent of respondents being female, and the women who were included are predominantly indigenous leaders. The absence of an association between gender and ICT capabilities can be attributed to the fact that the relatively few women who

FIGURE 6.2   ICT capabilities and gender

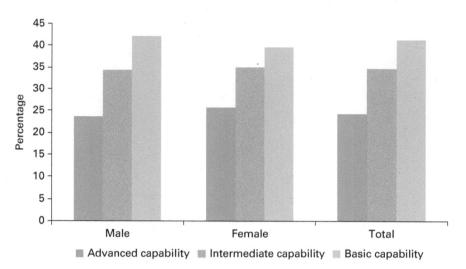

■ Advanced capability   ■ Intermediate capability   ■ Basic capability

do manage to attain positions of power within the indigenous leadership are highly educated and skilled, contesting the norm and beating the odds confronting the average indigenous woman. As such, this particular set of indigenous women leaders have managed to overcome numerous significant challenges, including education, access to leadership positions, as well as ICT capabilities.

Age

The users with the most advanced ICT capabilities are not indigenous youth, but persons 25–44 years of age (figure 6.3). The data indicate that while 25 percent of the overall sample and 30 percent of persons older than 25 are advanced users, only 10 percent of youth (more than 25 years old) are advanced users. In fact, almost half of youth (46 percent) are only basic users, while about two-thirds (65 percent) of persons between the ages of 25 and 44 are either advanced or intermediate users.

The statistical chi-square test confirms a very weak association between age and ICT use (phi = 0.21), and the results are not statistically significant ($p > 0.05$). The cross-tabulations show no clear trend in age and ICT use, with the exception of youth, who overwhelmingly have only basic or intermediate levels of use.

The finding of no statistically significant relationship between age and ICT capabilities is unexpected because surveys in developed countries have found that youth have a strong affinity for new technologies and much higher ICT capabilities than older persons (Livingstone 2003). However, for indigenous youth in rural Bolivia, structural barriers, rather than individual character-istics, are the greatest impediment to ICT use. My qualitative fieldwork con-firmed that rural indigenous youth are among the most disadvantaged groups in Bolivian society: they are denied opportunities in many dimensions of their

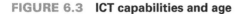

FIGURE 6.3   **ICT capabilities and age**

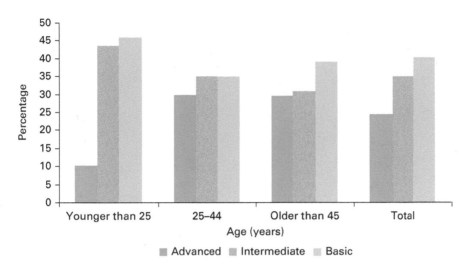

lives such as education, where absentee rates of teachers, grade repetition, and dropout rates of students are especially high in rural indigenous communities (Jiménez Pozo, Landa Casazola, and Yáñez Aguilar 2006). Thus the low level of ICT capabilities among indigenous youth reflects their overall disadvantaged situation within Bolivian society.

Education

Not all structural barriers deny access equally: educational attainment, for example, does not play a key role in shaping people's ICT capabilities. As shown in figure 6.4, 44 percent of the overall sample, compared with 43 percent of respondents with a tertiary education, have basic ICT skills. Higher levels of education influence ICT capabilities only slightly. The difference between persons with primary, secondary, and advanced technical education is minor. For example, the share of advanced ICT users is about the same among those who have attained a primary or a secondary education (17 and 14 percent, respectively). The greatest differential is between users with a university degree and those with only a primary education; the former are almost twice as likely as the latter to have advanced ICT capabilities.

In sum, while educational attainment plays a central role in determining whether a person is able to use the Internet, as shown in chapter 5, it does not have a significant influence on a person's ICT capabilities. This finding is confirmed by the chi-square test, which establishes that the relationship between respondents' schooling and their ICT capabilities is neither substantial (phi = 0.226) nor statistically significant ($p = 0.189 > 0.05$).

Role of Ethnicity

Once the analysis includes data on ethnicity, clear differences between the indigenous and non-indigenous respondents emerge. A large majority of

FIGURE 6.4   **ICT capabilities and educational attainment**

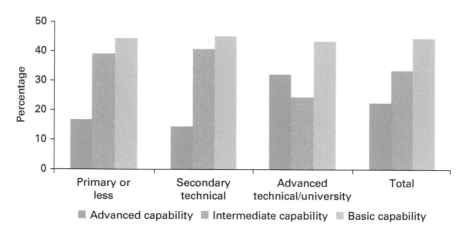

FIGURE 6.5   **ICT capabilities and ethnicity**

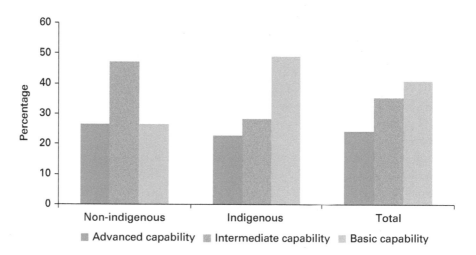

non-indigenous respondents (74 percent) have either advanced or intermediate ICT capabilities; a relatively small proportion (26 percent) have only basic ICT capabilities (figure 6.5). The chi-square test confirms this finding, showing a substantial relationship between ethnicity and ICT capabilities (phi = 0.279) that is statistically significant $(p = 0.06)$. Although p is slightly above the 5 percent mark, the relatively large sample of 148 respondents justifies this as an acceptably significant relationship.

What difference does education make for indigenous peoples in terms of expanding their ICT capabilities? The data analysis shows that even highly educated indigenous people do not have the same opportunities to use the Internet or to attain advanced ICT capabilities as their non-indigenous counterparts (figure 6.6). While the number of indigenous respondents who have advanced ICT capabilities significantly increases from 22.8 percent

FIGURE 6.6 **ICT capabilities and ethnicity (persons with advanced educational attainment)**

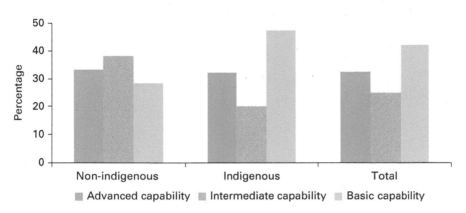

(all indigenous respondents) to 32.3 percent (indigenous respondents with tertiary education), it is notable that a significant gap between the indigenous and non-indigenous respondents remains. In fact, almost 48 percent of indigenous respondents with an advanced technical or university education have only basic ICT capabilities, compared with 29 percent of non-indigenous respondents.

These results suggest that advanced educational attainment is not sufficient for indigenous peoples to attain advanced ICT capabilities. These findings underscore that the group-level variable of ethnicity, and not a person's individual sociodemographic variables, such as gender, age, and education, is the most influential variable determining indigenous peoples' ICT capability. Thus the next section considers the extent to which variations in indigenous peoples' ICT capabilities can be explained not by differences in their individual characteristics but by the socioeconomic conditions they face in their daily lives.

Role of Socioeconomic Factors in Explaining Differences in ICT Capabilities

This section studies the association between socioeconomic factors and ICT capabilities. In particular, it studies the influence of (a) poverty levels as measured by the unsatisfied basic needs (UBN) index, (b) human development as measured by the human development index (HDI), (c) illiteracy rates, and (d) spatiality or geographic location (highlands, valleys, and lowlands) on people's ICT capabilities.

Poverty

The municipality's poverty level is a key factor defining indigenous peoples' ICT capabilities. Thus figure 6.7 demonstrates the clear relationship between the unsatisfied basic needs index and ICT capabilities.

The proportion of advanced ICT users is twice as high in municipalities with relatively low levels of poverty (UBN categories I and II) as in municipalities with relatively high levels of poverty (UBN categories IV and V): 31 and

FIGURE 6.7   ICT capabilities and municipal level of poverty

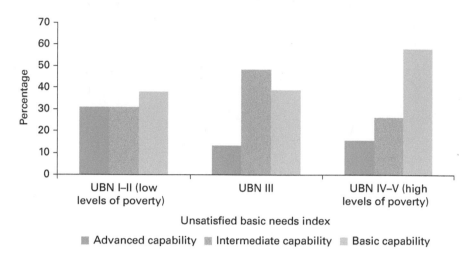

Unsatisfied basic needs index

■ Advanced capability   ■ Intermediate capability   ■ Basic capability

16 percent of respondents, respectively. Moreover, while a significant majority (58 percent) of users from extremely poor municipalities have low ICT capabilities, only a little over a third (39 percent) of those from better-off municipalities (UBN categories I and II) have such low proficiency. The chi-square test reveals that the relationship between unsatisfied basic needs and a person's ICT capabilities is statistically significant ($p = 0.095$) and that there is a moderate association between the two variables (phi = 0.28).

Human Development

In order to understand the relationship between ICTs and human capabilities, it is critical to go beyond the traditional measures of poverty that have focused on income poverty. Therefore, I use the HDI—a more comprehensive measure that takes into account the nonincome dimensions of poverty such as life expectancy and education— in my analysis. This section examines whether a municipality's HDI is correlated with the ICT capabilities of respondents. It includes data from a national assessment carried out by the UNDP that measures Bolivia's HDI at the municipal level (UNDP 2003).

The correlation between a municipality's HDI and a person's ICT capabilities is negative (figure 6.8). In municipalities with a significantly low HDI—equivalent to less than 0.5—half (50 percent) of the respondents said that they have basic ICT capabilities and only 13 percent said that they have advanced capabilities. In comparison, almost twice as many respondents in the overall sample (24 percent) indicated that they are advanced users.

Furthermore, in municipalities with a high HDI—equivalent to more than 0.6—almost a third (28 percent) of respondents indicated that they are advanced users. This is a sizable proportion, considering that the overall level of ICT use is relatively low. However, the chi square test shows that the correlation between these two variables is relatively weak (phi = 0.22) but statistically significant ($p = 0.027 < 0.05$).

FIGURE 6.8 **ICT capabilities and municipal level of human development**

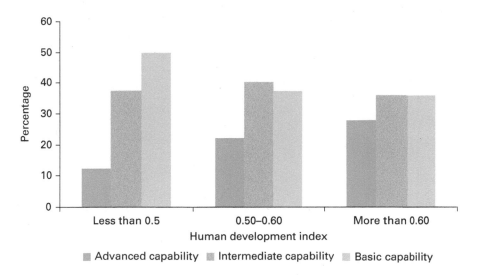

Illiteracy

The ICT literature frequently refers to illiteracy as a key impediment to Internet use (Heeks 1999; Kenny 2003). The large majority of the poor are de facto excluded from the meaningful use of the Internet because they are illiterate. In this section, I use empirical findings to determine the extent to which high illiteracy rates among indigenous communities in rural Bolivia constitute a significant barrier to attaining advanced ICT capabilities.

Municipal illiteracy rates significantly influence an indigenous person's ICT capabilities. This finding points to a fundamental aspect of the capability approach: a person's individual capabilities and the extent to which she or he can realize the valued functionings are intertwined with the presence or absence of social opportunities (Sen 1999). High illiteracy rates are a critical external constraint for indigenous peoples in Bolivia, preventing many from using or deriving benefits from the Internet. While the correlation between illiteracy rates and peoples' ICT capabilities is negative, it is remarkable that the majority of indigenous peoples living in municipalities with high illiteracy (53 percent) are nevertheless able to attain either intermediate or advanced ICT capabilities (figure 6.9).

Also significant is the fact that in municipalities with low illiteracy rates, only a slightly larger proportion (64 percent) indicated that they have either an intermediate or an advanced ICT capability. Although not derivable from the statistical data, it seems that a substantial number of indigenous peoples, in spite of considerable educational constraints—low schooling attainment and high illiteracy rates—have overcome these barriers and achieved relatively high ICT capabilities. The case studies in chapters 8 and 9 document in greater detail the process by which indigenous peoples channel their agency to expand their human and collective capabilities.

FIGURE 6.9   **ICT capabilities and municipal illiteracy rates**

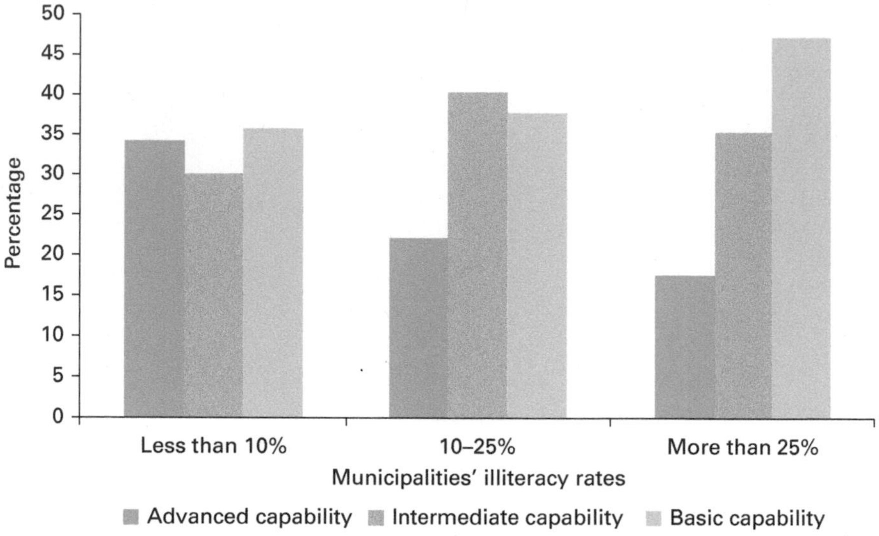

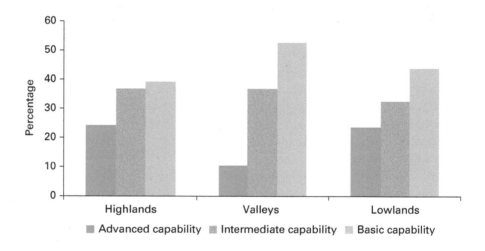

FIGURE 6.10   **ICT capabilities and spatiality**

Spatiality

Due to Bolivia's stark regional disparities, I expected spatiality to play a major role in influencing indigenous peoples' ICT capabilities. It is remarkable that respondents from the highlands who use the Internet have attained the highest level of ICT capabilities; 61 percent of them said that they are either advanced or intermediate users (figure 6.10). Among respondents from the valleys, who constitute the greatest proportion of Internet users, more than half (53 percent) said that they have low ICT capability, compared with 44 percent of respondents from the lowlands.

These results are counterintuitive, given Bolivia's deep-seated spatial inequalities. It is particularly striking that users from the highlands have attained the highest ICT capabilities, in spite of the multiple structural barriers that indigenous peoples confront in this region. In particular, the empirical findings from my user survey seem to conflict with national ICT statistics, which point to greater Internet diffusion in the lowlands and central valleys than in the highlands (Government of Bolivia 2005). A possible explanation may be that the user survey was targeted to municipalities with ongoing ICT programs, potentially linking the high levels of ICT capabilities with the effectiveness of these programs. In order to shed light on this issue, the next section investigates the role of ICT programs in influencing indigenous peoples' Internet use and the extent to which they affect their ICT capabilities.

Role of ICT Programs

There is a strong association between the ICT program variable and indigenous peoples' ICT capabilities (figure 6.11). Among the independent variables, the variable "ICT program" has the strongest influence on peoples' ICT capabilities: the Kramer V coefficient reaches a value of 0.48, representing a strong association between the two variables. Furthermore, the chi-square test also indicates that this association is statistically significant ($p = 0.001$).

Figure 6.11 displays the significant variation in terms of indigenous peoples' ICT capabilities by program. A comparative analysis of the different programs reveals that the respondents, who have participated in grassroots ICT programs (CIOEC, Organización de Mujeres Aymaras del Kollasuyo [OMAK], and Consejo Nacional de Ayllus y Markas del Qullasuyu [CONAMAQ]) have relatively more advanced ICT capabilities than those who have participated in either a nongovernmental organization (NGO) or a government program. For instance, almost one-third (28 percent) of the participants in CIOEC's program in Oruro and Potosí indicated that they have advanced ICT proficiency, and more than half (61 percent) said that they have intermediate ICT capabilities.

FIGURE 6.11 ICT capabilities and ICT program

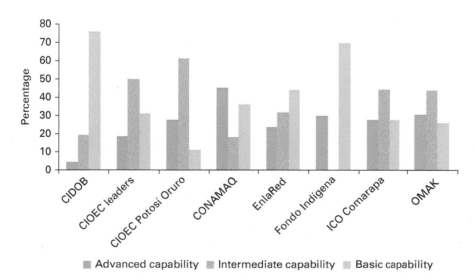

The empirical findings also show that 30 percent of OMAK respondents in Tiwanaku have advanced capabilities and 43 percent have intermediate ICT capabilities. The only other program where the participants have comparable advanced ICT capabilities was implemented by the Instituto de Capacitación del Oriente (ICO) in Comarapa: 28 percent of ICO participants have an advanced and 44 percent have an intermediate level of ICT use. In the case of CONAMAQ—a national grassroots organization—the results are more mixed, with 45 percent of respondents attaining advanced ICT capabilities, and 36 percent having only basic proficiencies to use ICTs. It is notable that participants in the EnlaRed Municipal program have relatively low ICT capabilities, with almost half (44 percent) having only basic and 24 percent having advanced ICT capabilities.

These findings are more revealing when considering the socioeconomic context in which the programs were implemented (figure 6.12). The analysis has shown that indigenous peoples' ICT capabilities are highly dependent on socioeconomic factors, such as municipal poverty and illiteracy rates. The respondents from the CIOEC (Oruro and Potosí) and OMAK (Tiwanaku) programs—68 and 65 percent of whom, respectively, are from the poorest and most economically depressed regions of Bolivia—nevertheless indicated that they have advanced ICT capabilities.

In contrast, respondents affiliated with ICO who have attained similarly advanced ICT capabilities come from municipalities with significantly lower poverty levels. In fact, only 5 percent of the EnlaRed participants are from municipalities with high rates of extreme poverty (UBN categories IV and V). In sum, while socioeconomic factors matter, the overriding variable shaping ICT capabilities seems to be the ICT program or intervention.

Furthermore, all of the grassroots-level projects have a substantial proportion of indigenous peoples among their participants. Both of these factors— project location in extremely poor municipalities and high degree of ethnicity

FIGURE 6.12   Municipal poverty level and ICT program

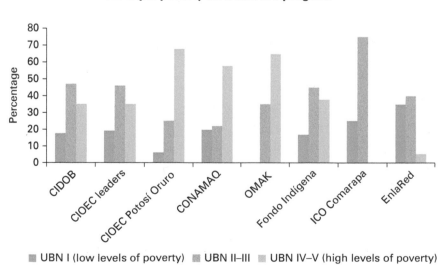

■ UBN I (low levels of poverty)   ■ UBN II–III   ■ UBN IV–V (high levels of poverty)

among participants—have considerable, negative effects on Internet use and ICT capabilities (figure 6.13).

The CIOEC program in Oruro and Potosí has the second highest proportion of indigenous beneficiaries among all of the programs, with three-quarters of its participants coming from municipalities with a high degree of ethnicity. In contrast, EnlaRed has a very low proportion of indigenous beneficiaries; more than half of its participants (56 percent) self-identified as monolingual Spanish speakers, and its geographic reach is primarily in non-indigenous municipalities (55 percent of municipalities are characterized by low or medium ethnicity). In the case of the ICO program, all of its participants have an intermediate level of ethnicity, pointing to the highly mixed character of Comarapa residents.

The empirical evidence shows that participants in the CIOEC and OMAK programs have ICT capabilities similar to those of respondents from the ICO program and slightly higher than those of respondents from EnlaRed. This finding is remarkable, since the overall socioeconomic conditions in Potosí, Oruro, and the Department of La Paz are much worse than those in Comarapa, where ICO implements its program, and in the municipalities where EnlaRed works. Furthermore, the participation of indigenous peoples is significantly higher in the CIOEC and OMAK programs than in either the ICO or EnlaRed programs. In the case of EnlaRed, the external socioeconomic conditions are much more favorable than those for CIOEC and OMAK. The program is implemented in 15 municipalities, some of which have relatively low poverty rates (Coroico, Cobija, Trinidad), and its participants are mostly non-indigenous. Nevertheless, the EnlaRed participants have relatively low levels of ICT capabilities.

FIGURE 6.13 Ethnicity and ICT program

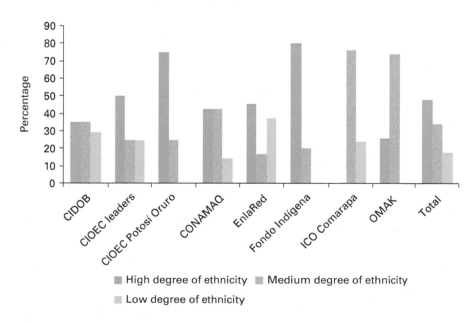

FIGURE 6.14   Educational attainment and ICT program

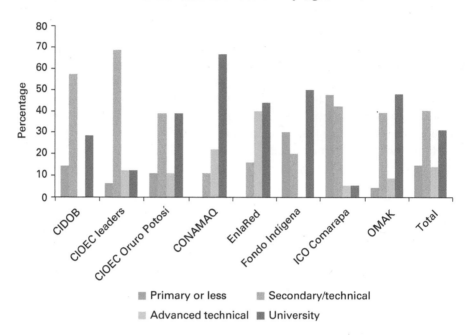

Primary or less ■   Secondary/technical ■

Advanced technical ■   University ■

Moreover, the ICT programs differ significantly in terms of their participants' educational attainment (figure 6.14). For instance, while 84 percent of the mostly non-indigenous participants from EnlaRed have an advanced education, almost half of those from ICO (47 percent) have only a primary education or less—the lowest educational attainment of all programs. This reflects the grassroots nature of ICO's program, which targets local farmers rather than government officials (EnlaRed) or indigenous leaders (Fondo Indígena).

The results for CIOEC Oruro and Potosí are quite surprising. Only 11 percent of CIOEC beneficiaries have primary education or less, and half (50 percent) have an advanced technical or university education. This high level of educational attainment can be attributed to the presence of a new, highly educated indigenous leadership working at the grassroots community level, not just in large urban centers. The numbers are nonetheless unexpected, given the extremely high poverty rates in Oruro and Potosí.

In order to understand the underlying reasons for these significant differences between the various ICT programs, the case studies in chapters 8 and 9 investigate the approaches and processes of the different intermediary organizations, focusing on grassroots and government programs.

Multivariate Regression Analysis of the ECADI National Household Survey

This section cross-validates the main findings obtained from the data analysis of my own user surveys by developing a series of multivariate models using the data set of the ECADI. The analysis seeks to verify the extent to which variations in the dependent variable of ICT capabilities can be explained by independent variables, such as individual factors (age, gender, profession, income, and ethnicity), socioeconomic attributes (poverty levels), and intermediation process (access to an ICT program). An important strength of regression analysis is that it allows us not only to determine whether or not there is a statistically significant relationship between the independent and dependable variables, but also to measure the magnitude of the interrelationship between these variables.

This section is divided into the following subsections: (a) data analysis based on the establishment of analogous subsamples; (b) the role of individual factors, including age, gender, and profession; (c) the role of ethnicity; (d) the role of spatiality; (e) external socioeconomic factors; and (f) intermediary factors in determining differences in ICT capabilities.

Data Analysis

As in the previous chapter, the data analysis established analogous subsamples to compare the results from my own user survey with those obtained from the ECADI survey. The first of these subsamples isolates ICT users living exclusively in rural areas, with the ECADI subsample totaling 255 and the subsample from my user survey totaling 148. The respondents in both of these subsamples share the following characteristics: they use the Internet on a regular basis, they live in rural areas, they are predominantly indigenous, and they live in municipalities with similar levels of poverty. These criteria allow me to focus the analysis on the key dependent variable detailed in this chapter—ICT capabilities—and to target the rural indigenous population.

First, the large majority or respondents in both subsamples are indigenous: 59 percent (ECADI) and 61 percent (my survey) using primary household language as a proxy for ethnic identity (table 6.1).

Second, the poverty level in both subsamples is similar: 34 percent of ECADI respondents and 36 percent of my survey respondents live in municipalities characterized as poor or extremely poor (table 6.2).

Finally, the large majority of respondents from both subsamples come from municipalities with high illiteracy rates: 54 and 72 percent, respectively (table 6.3).

Degree to Which Individual Factors and Ethnicity Can Explain Differences in ICT Capabilities

What is the influence of individual factors on indigenous peoples' ICT capabilities? This section analyzes the extent to which these factors make a difference

TABLE 6.1 Ethnicity of rural ICT users (ECADI vs. own survey)

% of respondents

Primary language spoken	ECADI	Own survey
Spanish	41	39
Indigenous language (primary) and Spanish	51	49
Indigenous language, Spanish, and English	8	12
Total	100	100

TABLE 6.2   UBN of rural ICT users (ECADI vs. own survey)
% of respondents

Unsatisfied basic needs category	ECADI	Own survey
Satisfied basic needs (I)	52	14
Population living at the threshold of poverty (II)	14	35
Moderate poverty (III)	22	21
Marginal population (IV–V)	12	15
No response	0	15
Total	100	100

TABLE 6.3   Literacy of rural ICT users (ECADI vs. own user survey)
% of respondents

Illiteracy rate	ECADI	Own survey
<10%	35.8	27.8
10–25%	44.6	61.8
>25%	9.8	10.4
Total	100.0	100.0

not only for people's access to ICT but also, and more important, for their capabilities to use these technologies.

Age, Gender, Profession, and Education

Model 1 incorporates individual sociodemographic characteristics, including age, gender, profession, and education, in order to investigate the extent to which these factors influence a person's ICT capabilities. None of these characteristics affects the outcome of the model, as none has a statistically significant relationship with a person's ICT capabilities (table 6.4).

One of the key premises found across the "digital divide" literature is that education is a key independent variable that explains differences in people's ICT capabilities (López and Villaseca 1996; Heeks 1999; Warschauer, 2002). However, an important finding—which contests the conventional beliefs of this literature—is that the variable "level of education" does not enter into the model: there is no statistically significant relationship between educational attainment and ICT capabilities in rural Bolivia ($p = 0.32 > 0.05$).

Ethnicity and ICT Capabilities

Unlike the other independent variables, ethnicity does enter the model, and the regression coefficient indicates that the correlation between this variable and ICT capabilities is strong and statistically significant. Thus a person's capability to use ICTs strongly depends on her or his ethnicity, with a significant correlation between being indigenous and having low ICT capability. A non-indigenous person is 1.5 times more likely to have advanced or intermediate ICT use than an indigenous person. As such, irrespective of individual characteristics, including educational attainment, ethnicity represents the determining

TABLE 6.4 Model 1: ICT capabilities regressed over individual factors (including age, gender, profession, education, and ethnicity)

Variable	B	Exp (B)	Significance
Age*		0.81	
Less than 20			
20–35			
Older than 35			
Gender*		0.56	
Female			
Male			
Profession*		0.67	
Farmer			
Self-employed			
Wage laborer			
Level of education*		0.32	
Primary			
Secondary			
University or technical			
Ethnicity*			
Non-indigenous	0.401***	1.49	0.035
Quechua	0.302***	1.35	0.041
Aymara	−0.71***	0.48	0.007

Note: n = 255; pseudo R^2 = 0.295.

*** Regression coefficients are statistically significant, with $p < 0.05$. * Regression coefficients are not statistically significant, with $p > 0.05$; thus there is no value for beta.

factor affecting people's ICT capabilities: both the regression coefficients and the log likelihood test indicate that it is the only independent variable that is statistically significant, with a high 29.5 percent model deviance (log likelihood test: R^2 = 0.295). These results confirm the main finding from my own user survey: ethnicity is the most important variable explaining ICT capabilities, as opposed to any other factors—even education.

Notably, there are important disparities within and among the various ethnic groups in rural Bolivia. The Aymara are particularly disadvantaged in their level of ICT use compared with the rest of the population. While the Quechua have almost the same proficiency of ICT use as the non-indigenous population, the Aymara are 2.8 and 3.0 times less likely than the Quechua and the non-indigenous, respectively, to reach either advanced or intermediate ICT capability—independent of age, gender, and profession (figure 6.15).

The ECADI data highlight the severe digital disparities among the various indigenous groups in Bolivia; only 5 percent of the Aymara have advanced ICT capabilities, compared with approximately one-fourth of the Quechua. In particular, the advanced levels of ICT proficiency among the Quechua are unexpected. This remarkable finding is analyzed further in a later section of this chapter.

In model 2, the magnitude of the correlation between ethnicity, individual factors, and ICT capabilities is controlled for socioeconomic status, as

FIGURE 6.15   **ICT capabilities, by ethnic group**

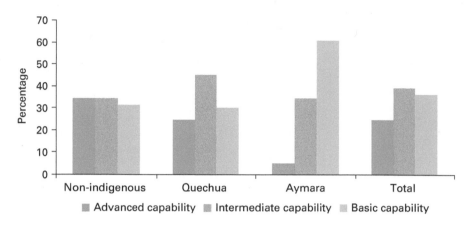

TABLE 6.5   **Model 2: ICT capabilities regressed over ethnicity and individual factors controlled by socioeconomic status (as measured by income)**

Variable	B	Exp (B)	Significance
Age*		0.81	
Gender*		0.56	
Education*		0.32	
Ethnicity			
Non-indigenous	0.409***	1.51	0.032
Quechua	0.294***	1.34	0.013
Aymara	−0.70***	0.49	0.007
Socioeconomic status			
High income*		0.21	
Intermediate income*		0.75	
Poor or extreme poverty*		0.65	

Note: pseudo R^2 = 0.295.
*** Regression coefficients are statistically significant, with $p < 0.05$. * Regression coefficients are not statistically significant, with $p > 0.05$; thus there is no value for beta.

measured by differences in household income (table 6.5). The "digital divide" literature indicates that both income and literacy are critical independent variables affecting people's level of ICT use (Gómez, Hunt, and Lamoureaux 1999; Richardson and Paisley 2000; Gómez and Casadiego 2002).

The results from model 2 illustrate the limitations of this premise: in the Bolivian context, a person's ICT capability is independent of income. When controlling for socioeconomic status, there is no statistically significant relationship between ethnicity and ICT use. In addition, the inclusion of this control variable does not significantly alter the strong correlation between ethnicity and ICT capabilities or the fact that none of the individual factors enters the model. Furthermore, the log likelihood test confirms that the control variable does not improve the fit of the regression model (R^2 remains constant at 0.295).

Ethnicity, Educational Attainment, and ICT Capabilities

To what extent does education explain the persistent disparities among Bolivian indigenous groups with regard to ICT capabilities? The empirical data reveal that significant discrepancies in ICT capabilities remain between the Quechua and Aymara, even when controlling for educational attainment. Even within the sample of respondents who have attained a tertiary education, half of the Aymara have not reached advanced or even intermediate ICT capabilities (figure 6.16). There are also disparities between the non-indigenous and indigenous populations: the overwhelming majority (83 percent) of non-indigenous peoples have attained a meaningful and informed level of use, compared with 63 percent of the Quechua.

These data confirm the findings from the second regression model: indigenous peoples—in particular, the Aymara—continue to be disadvantaged with regard to their ability to make efficient use of the Internet independent of their educational attainment. Digital inequalities persist even among those with a university or similarly advanced level of education.

Socioeconomic Status, Ethnicity, and ICT Capabilities

What is the relationship between ethnicity, socioeconomic status, and ICT capabilities? Within the high-income cohort, differences between the various ethnic groups persist (figure 6.17), with the majority of the non-indigenous and Quechua respondents (72 and 59 percent, respectively) having either advanced or intermediate ICT capabilities, compared with a little over one-third of the Aymara respondents (37 percent). These results confirm the findings of the second regression model: ICT capabilities are independent of a person's income level. Thus differences in socioeconomic status do not explain the prevailing digital inequalities between the Aymara and Quechua or between the indigenous and non-indigenous respondents.

FIGURE 6.16 ICT capabilities and ethnicity (persons with tertiary educational attainment)

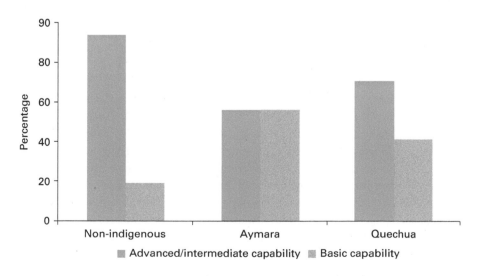

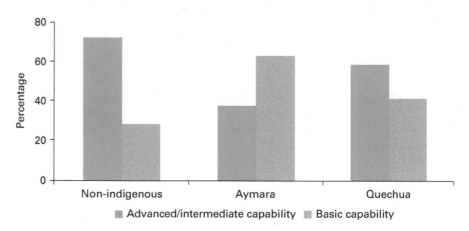

FIGURE 6.17   ICT capabilities and ethnicity (high-income cohort)

This more in-depth data analysis thus highlights the persistence of significant digital inequalities among the various ethnic groups in Bolivia independent of demographic factors (age, gender, education, or profession) or socioeconomic status as measured by income. Belonging to the Aymara ethnic group in and of itself places individuals at a disadvantage compared with other ethnicities. This important finding has been cross-validated by the results of the ECADI and those of my own user survey.

Degree to Which Spatiality Can Explain Differences in ICT Capabilities

This section applies the same regression models as above, but uses a broader sample that includes urban respondents ($n = 916$) to analyze the extent to which variations in the results can be attributed to spatiality. Model 3 is similar to model 1. It incorporates demographic factors, including age, gender, education, and profession, to examine how these affect the level of use (table 6.6). The only difference is the scope of the sample (model 1 focuses on rural dwellers; model 3 includes urban dwellers). In contrast to model 1, once the data from urban users are included, the individual factors of age and gender enter the model.

There is a negative correlation between age and ICT capabilities, with the regression coefficient reaching $B = -0.51$ for the cohort older than 35. A person under 20 years of age is more than twice as likely as a person over 35 to report having an advanced or intermediate ICT capability. The youngest cohort (persons under 20 years of age) also includes relatively more advanced ICT users ($B = 0.21$) than the 20–35 age group ($B = 0.29$). Furthermore, this model establishes that gender plays an important role in predicting level of use in urban areas, with men approximately twice as likely as women to be proficient in the use of ICTs. However, the influence of these two variables is not as substantial as the results would suggest: gender and age only account for 3 percent of the variation in ICT capabilities (log likelihood, $R^2 = 0.029$).

With respect to educational attainment, neither primary nor secondary schooling significantly influences the ICT capabilities of urban dwellers. However, in contrast to the model for the rural sample, there is a minor positive

TABLE 6.6 Model 3: ICT capabilities regressed over individual factors (age, gender, profession, and education) as independent variables

Variable	Model 3		Model 4 (controlled with ethnicity, socioeconomic status)	
	B	Exp (B)	B	Exp (B)
Age				
Less than 20	0.29***	1.34	0.21***	1.24
20–35	0.21***	1.24	0.17***	1.18
Older than 35	−0.51***	0.68	−0.38***	0.60
Gender				
Male	0.17***	1.19	0.19***	1.21
Female	−0.37***	0.70	−0.36***	0.69
Level of education				
Primary	−0.39*	0.74		
Secondary education	−0.21*	0.81		
Advanced technical or university	0.25***	1.28		
Profession*				
Self-employed				
Wage laborer				
Salesperson				

Note: $n = 916$; pseudo, age: $R^2 = 0.029$; level of education: $R^2 = 0.094$.
*** Regression coefficients are statistically significant, with $p < 0.05$. * Regression coefficients are not statistically significant, with $p > 0.05$.

relationship between tertiary educational attainment and advanced ICT use. Persons who have an advanced university degree are approximately 1.2 times more likely than others to be proficient ICT users. This is substantiated by the log likelihood test, as the inclusion of education improves the model fit only 6.5 percent ($R^2 = 0.094$). This finding supports the empirical results of my own user survey and confirms that while education is a critical independent variable with respect to use or non-use, its association with proficiency in ICT use is negligible.

In model 4, ethnicity and socioeconomic status are controlled for the various individual factors (table 6.7). In contrast to model 1, which only incorporates data from rural areas, model 4 includes urban data, and ethnicity no longer enters into the model. Instead, it is overshadowed by socioeconomic factors. The odds ratios indicate that the higher-income group is almost three times more likely to have advanced levels of ICT use than the poor or extreme poor. This strong correlation between ICT capabilities and socioeconomic status is also confirmed by the negative regression coefficient of $B = -0.50$ for respondents from the lowest socioeconomic group. The data shown in table 6.8 also indicate that adding this socioeconomic variable does not change the original relationship between age, gender, and the dependent variable. However, it does eliminate the minor influence that education has on ICT capabilities in model 3.

TABLE 6.7   Model 4: Individual factors controlled by ethnicity and socioeconomic status

Variable	B	Exp (B)
Ethnicity*		
Non-indigenous		
Quechua		
Aymara		
Socioeconomic status		
High income	0.494***	1.639
Intermediate income	0.009***	1.009
Poor/extreme poverty	−0.503***	0.604

Note: pseudo R^2 = 0.165.
*** Regression coefficients are statistically significant, with $p < 0.05$. * Regression coefficients are not statistically significant, with $p > 0.05$.

FIGURE 6.18   ICT capabilities and ethnicity (high-income cohort) in urban areas

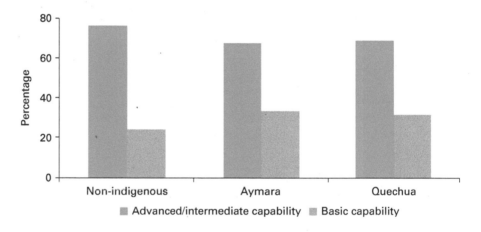

■ Advanced/intermediate capability   ■ Basic capability

Considering that this variable has only a minor effect on a person's level of ICT use, its elimination does not have significant consequences for the model fit. The log likelihood test even indicates that, with pseudo R^2 = 0.165, adding the socioeconomic status variable to the model significantly improves the model fit from about 9 percent to 16.5 percent, underlining its importance in explaining variations in people's ICT capabilities in urban areas.

Ethnicity, however, plays a minor role in explaining differences in urban dwellers' ICT capabilities (figure 6.18). In fact, differences between the Aymara and Quechua basically disappear once persons migrate to cities. The figure reflects the significantly smaller differences between the non-indigenous and indigenous populations, with the non-indigenous at only a slight advantage with regard to their ICT capabilities.

The main findings from regression models 3 and 4 highlight important differences between Bolivia's rural and urban areas. While in models 1 and 2 ethnicity is the most important variable determining ICT capabilities in rural areas, in models 3 and 4 ethnicity loses its importance in urban areas and is overshadowed by variations in educational attainment and socioeconomic status.

Degree to Which External Socioeconomic Factors Can Explain Differences in ICT Capabilities

To what extent do external socioeconomic variables (poverty, illiteracy rates, and geographic region) influence a person's ICT proficiency? The findings from my user survey show that there is a statistically significant relationship between these variables and ICT capabilities. Model 5 confirms that a person's ICT capabilities are dependent on the poverty level of the municipality and the geographic region in which she or he lives. As table 6.8 illustrates, a person who lives in a municipality with satisfied basic needs (UBN categories I or II) is twice as likely to have an advanced or intermediate ICT capability as a person who lives in a municipality with high levels of poverty (UBN categories IV or V). The log likelihood test confirms that the poverty level of the municipality in which a person lives significantly influences a person's ICT capability, explaining 14.2 percent of the model. Moreover, ICT capability is independent of overall literacy rates in the municipality, since this variable does not enter the model. While the literacy rate of a municipality is a key factor in whether individuals have the opportunity or not to use ICTs, it does not influence their level of ICT use.

In model 6, I include the spatial variable of geographic region to test the extent to which people's level of ICT use is affected by poverty level, regardless of geographic location (table 6.9). In this model, the municipality's poverty level continues to influence a person's ICT capabilities, although its effect is slightly reduced. For instance, the likelihood that a person is an advanced or intermediate user is only 1.2 instead of 1.3 times higher for a person who lives in a municipality with satisfied basic needs than for a person who lives in a municipality with unsatisfied basic needs. Moreover, ICT capability depends

TABLE 6.8 **Model 5: ICT capabilities regressed on external socioeconomic variables (unsatisfied basic needs and illiteracy rates)**

	Model 5		Model 6 (controlled by geography)	
Variable	B	Exp (B)	B	Exp(B)
UBN category				
I and II	0.291***	1.338	0.216***	1.241
III	0.216***	1.241	0.168***	1.183
IV and V	−0.429***	0.651	−0.353***	0.702
Illiteracy rates*				
Under 5%				
5–15%				
15–30%				

Note: $n = 1,920$; pseudo $R^2 = 0.142$.
*** Regression coefficients are statistically significant, with $p < 0.05$. * Regression coefficients are not statistically significant, with $p > 0.05$.

TABLE 6.9   Model 6: ICT capabilities regressed on external socioeconomic variables controlled by geographic region

Variable	B	Exp (B)
Geographic region		
Highlands	−0.383***	0.68
Valleys	0.174***	1.190
Lowlands	0.555***	1.742

Note: Pseudo R^2 = 0.254.
*** Regression coefficients are statistically significant, with $p < 0.05$.

on the region in which the person lives; a person from the lowlands is two and a half times more likely than a person from the highlands to be an advanced or intermediate user of ICTs.

The log likelihood test confirms the importance of this variable, since the model fit increases more than 11 percent to reach 25.4 percent. The findings from models 5 and 6 underline the role of external socioeconomic variables (poverty levels and geographic region) in explaining differences in the level of a person's ICT use. These results confirm the findings from my user survey; however, the strength of these relationships is significantly stronger in my survey than in the regression models described here.

In sum, external socioeconomic variables are crucial for indigenous peoples' ICT capabilities. As the analysis has shown, a structural problem exists—that is, people in the highlands continue to be excluded from many basic infrastructure services, such as electricity, water and sanitation, and ICTs—that can explain the digital inequalities that persist within Bolivia. The geographic region with the highest levels of extreme poverty, highest illiteracy rates, and lowest access to basic infrastructure services also has the lowest ICT use and capabilities. The Andean highlands has the highest percentage of indigenous peoples among the population and is home to the Aymara, who—as the regression models have demonstrated—are the most excluded from participation in the information society in Bolivia.

Degree to Which Intermediary Factors Can Explain Differences in ICT Capabilities

Do intermediary factors (access to ICT infrastructure and ICT program) have a statistically significant effect on people's ICT capabilities? My own user survey finds that while the provision of ICT infrastructure does not improve people's ICT capabilities, the presence of an ICT program is a key variable for determining people's capabilities to use ICTs.

The results from model 7 confirm these results only partially (table 6.10). First, the regression analysis shows that the variable "access to ICTs" does not enter the model, indicating that indigenous peoples' level of ICT use is independent of the provision of access to ICTs. Thus the mere provision of access to ICT services does not enable indigenous peoples to make meaningful use of this technology. The regression analysis confirms the key finding from my user survey that merely providing access to ICTs will not improve indigenous peoples' ICT proficiencies.

TABLE 6.10 **Model 7: ICT capabilities regressed on intermediary factor (access to ICT infrastructure, ICT program)**

Variable	B	Exp (B)	Significance
Access to ICTs*			0.37
Municipality with Internet services			
Municipality without Internet services			
ICT program			
Municipality with an ICT program	0.264***	1.303	
Municipality without an ICT program	−0.437***	0.645	

Note: n = 255; pseudo R^2 = 0.198.

*** Regression coefficients are statistically significant, with $p < 0.05$. * Regression coefficients are not statistically significant, with $p > 0.05$.

Second, the independent variable "ICT program" enters the model, but the relationship between this variable and a person's level of ICT use is relatively weak. A person who lives in a municipality with an ICT program is 1.3 times more likely than a person who lives in a municipality without such a program to have advanced or intermediate ICT capabilities. The log likelihood test shows that this variable explains 19.8 percent of the model—confirming that the presence or absence of an "ICT program" in a municipality is an important independent variable explaining differences in people's level of ICT use. The model fit achieved by this one independent variable is quite high, confirming the finding from my own user survey that the ICT program variable is a central factor in determining whether or not people can enhance their well-being through the use of ICTs.

One possible explanation for the relatively weak relationship between the two variables is that the ECADI survey was conducted in mid-2003, while most of my survey work was carried out in mid-2005. The time difference is important. Most ICT programs in rural Bolivia were initiated in 2001 or 2002, and the effects of the programs were not yet evident in 2003. Furthermore, the cluster sampling method of my survey focused on obtaining a sample from a small number of very active ICT programs. The UNDP survey, through its national reach and random sampling methods, was more representative of the rural areas in Bolivia. At the same time, it was less targeted than my survey, which reached respondents from municipalities with an active ICT program.

Both of these aspects are important for explaining why model 7 finds a relatively weak relationship between the presence of an ICT program and a person's level of ICT use. Nevertheless, the model confirms the main finding from my user survey: there is a positive relationship between the independent variable "ICT program" and "ICT capabilities." Moreover, there is a strong relationship between the presence of an ICT program and the use of ICTs. The work of ICT programs significantly increases indigenous peoples' opportunities to use ICTs and helps them to attain more advanced ICT capabilities, even in rural communities with high levels of poverty. Therefore, ICT programs are able to overcome, at least partially, the many barriers that indigenous peoples face in obtaining access to ICT services. The findings point to the significant possibilities that a proactive ICT policy offers for enhancing indigenous peoples' opportunities to use ICTs, improve their ICT proficiencies, and use these new technologies to improve their well-being.

Conclusions

This section summarizes the main conclusions arising from the empirical evidence presented in this chapter. Table 6.11 provides summary statistics.

Ethnicity and ICT use (or non-use) and *ICT capabilities are strongly associated with each other.* There are severe digital inequalities between indigenous and non-indigenous populations, particularly in rural Bolivia. Even after controlling for individual factors, such as educational attainment, gender, age, and profession, non-indigenous individuals have a 50 percent greater chance of having advanced ICT capabilities than their indigenous counterparts. Indeed, the association between two variables remains strong, even after controlling for variations in socioeconomic status, as measured by income. In fact, a non-indigenous person is 1.5 times more likely to have advanced ICT capabilities than an indigenous person, after controlling for socioeconomic status. Thus being indigenous may be an important barrier to the attainment of ICT capabilities and the meaningful use of ICTs.

An indigenous person's ICT capabilities are independent of her or his educational attainment. Furthermore, this chapter has confirmed that

TABLE 6.11   Summary of statistical results with ICT capabilities as the dependent variable

Independent variable	Chi square	Phi (correlation)	Kramer V (correlation)	Asymptotic significance	Sample size (n)
Sociodemographic					
Gender	0.341		0.053	0.843	145
Age	5.477	0.21		0.484	143
Level of education	6.137	0.23		0.189	143
Profession					
Socioeconomic					
Ethnicity (measured by language)	**9.060**	**0.27**		**0.060**	**145**
UBN index	**7.921**	**0.28**		**0.095**	**124**
HDI	5.112	0.22		0.027	130
Illiteracy rates					
Population density	2.353	0.15		0.061	124
Geographic location	1.826	0.12		0.768	141
ICT infrastructure					
Access to ICTs	3.127		0.17	0.209	127
ICT intermediary					
ICT program	**35.641**		**0.48**	**0.001**	**148**

Note: The bold numbers indicate independent variables that significantly influence indigenous peoples' ICT capabilities.

indigenous peoples' ICT capabilities are independent of their level of education. This finding is applicable to rural Bolivia, but it does not hold in urban settings. Although not discernible from the data, both the low quality of education and the lack of access to ICTs in rural schools may be contributing to the lack of correlation between educational attainment and ICT capabilities in rural areas.

Socioeconomic factors, irrespective of individual characteristics, are strongly associated with ICT capabilities. Socioeconomic variables (geography, municipal poverty levels, literacy rates, and ethnicity index), irrespective of variations in individual characteristics, are key to explaining the existing digital inequalities evident in Bolivia. Thus individual variations in ICT capability are embedded in Bolivia's broader socioeconomic inequalities. This finding emphasizes that for an indigenous person, the individual agency to improve her or his well-being through the use of ICTs is significantly limited by structural barriers, such as the poverty level of the municipality in which she or he lives. Moreover, the structural impediments vary significantly for the various ethnic groups, with the Aymara being the most negatively affected, as they reside largely in the Andean highlands—a region characterized by generally abysmal living conditions, an overwhelming absence of basic social services, and extreme levels of poverty.

Spatiality is strongly associated with ICT capabilities. The statistical results demonstrate that the region in which a person lives and the use (or non-use) of ICTs and ICT capability are strongly associated with each other. First, there are significant differences between the highlands and the lowlands, with a person living in the lowlands having a 74 percent greater likelihood of using ICTs than a person living elsewhere. Second, cities are equalizers, with a stark reduction in digital inequalities among ethnic groups residing in urban areas. In fact, when controlling for ethnicity and socioeconomic status, the differences between the Aymara and Quechua almost disappear, with a rate of 31 and 32 percent of ICT use, respectively, among high-income groups. The unexpected finding is that in urban areas individual factors (age, gender, educational attainment, and income) exert the most influence on ICT capabilities, overriding socioeconomic factors. Thus the individual characteristics and agency of urban dwellers are decisive in shaping their ICT use.

Indigenous peoples' ICT capability is independent of improvements in their access to ICTs. The regression analysis validates the finding from my own user survey: indigenous peoples' ICT capabilities are independent of their physical access to ICT services. In fact, no statistically significant relationship exists between these two variables, suggesting that providing ICT infrastructure services in rural areas does not enable people to make meaningful use of the Internet. Thus ICT service provision is a necessary, but insufficient, condition for improving indigenous peoples' well-being.

There is a strong association between variations in ICT programs and indigenous peoples' ICT use and capabilities. The statistical results confirm the study's main hypothesis: the presence of an intermediary organization promoting the effective use of ICTs is strongly associated with the meaningful use of ICTs. Both my user survey and the ECADI survey find that a proactive public policy that designs and implements targeted ICT programs in rural indigenous communities could play an important role in reducing the digital disparities between the indigenous and non-indigenous populations in Bolivia. Although this finding shows only a correlation and no causation between the two variables, it nevertheless validates the hypothesis by establishing that the ICT program is central to meaningful ICT use.

Thus, when analyzing the conditions under which indigenous peoples can enhance their well-being through the use of ICTs, it is important to take into account individual factors, group belonging and identity, and the broader socioeconomic context.

References

DiMaggio, Paul, and Eszter Hargittai. 2001. "From the 'Digital Divide' to Digital Inequality: Studying Internet Use as Penetration Increases." Working Paper 15, Princeton University, Center for Arts and Cultural Policy Studies, Princeton, NJ.

Garnham, Nick. 1997. "Amartya Sen's 'Capability' Approach to the Evaluation of Welfare: Its Application to Communications." *Javnost—The Public* 4 (4): 25–34.

Gómez, Ricardo, and B. Casadiego. 2002. "Letter to Aunt Ofelia: Seven Proposals for Human Development Using New Information and Communication Technologies." Publication based on the results of a workshop held at Cajamarca, Peru, March 2002. International Development Research Centre, Ottawa.

Gómez, Ricardo, Patrik Hunt, and Emmanuelle Lamoureux. 1999. *Telecentre Evaluation: A Global Perspective.* Ottawa: International Development Research Centre.

Government of Bolivia. 2005. *Estrategia nacional de TIC para el desarrollo en Bolivia (ETIC).* A.D.S.I.B., Vice Presidencia de la República, La Paz.

Heeks, Richard. 1999. "Information and Communication Technologies, Poverty, and Development." Development Informatics Working Paper 5, University of Manchester, IDPM, Manchester.

Jiménez Pozo, Wilson, Fernando Landa Casazola, and Ernesto Yáñez Aguilar. 2006. "Bolivia." In *Indigenous Peoples, Poverty, and Human Development in Latin America*, edited by Gillette Hall and Harry Patrinos. London: Palgrave Macmillan.

Kenny, Charles. 2003. "The Internet and Economic Growth in LDCs: A Case for Managing Expectations." *Oxford Development Studies* 31 (1): 99–113.

Livingstone, Sonia. 2003. "Children's Use of the Internet: Reflections on the Emerging Research Agenda." *New Media and Society* 5 (2): 147–66.

López, Enrique J., and May González Villaseca. 1996. "IT as a Global Economic Tool." In *Information Technology, Development, and Policy: Theoretical Perspectives and Practical Challenges*, edited by Edward M. Roche and Michael J. Blaine. Avebury: Ashegate.

Menou, Michel. 2002. "Information Literacy in National Information and Communications Technology (ICT) Policies: The Missed Dimension, Information Culture." White Paper prepared for UNESCO, Information Literacy Meeting of Experts, Prague, Czech Republic.

Richardson, Don, and Lynnita Paisley, eds. 2000. *The First Mile of Connectivity: Advancing Telecommunications for Rural Development through a Participatory Communication Approach.* Rome: FAO.

Sen, Amartya. 1999. *Development as Freedom.* New York: Knopf Press.

UNDP (United Nations Development Programme). 2003. "Índice de desarrollo humano en los municipios de Bolivia, 2003." UNDP, La Paz.

———. 2004. "Encuesta de capacidades informacionales (ECADI)." Survey administered by UNDP in preparation for its *2004 National Human Development Report: Interculturalidad y globalizacón.* UNDP, La Paz.

Warschauer, Mark. 2002. "Reconceptualizing the Digital Divide." *First Monday* 7 (7).

———. 2003. *Technology and Social Inclusion: Rethinking the Digital Divide.* Cambridge, MA: MIT Press.

CHAPTER 7

Poor People's Perspectives vis-à-vis the Internet's Impact on Their Well-Being

Chapter 6 emphasizes that indigenous peoples' capabilities to make meaningful use of information and communication technology (ICT) are a prerequisite for ICTs to have a positive impact on their well-being. This chapter expands the analysis even further by investigating the conditions under which meaningful ICT use not only strengthens a person's ICT capabilities but also enhances her or his informational capabilities. The alternative evaluation framework developed in chapter 1 stresses the critical role that information can play in development processes and argues for including a person's "informational capital" as an additional asset in his or her set of livelihood resources. The framework underscores that people's improved access to information and expanded informational capabilities can play a critical role in enhancing a person's human well-being, because informational capital can play an "agency role" in strengthening capital in the economic, political, and social dimensions of a person's life. Thus information, not ICTs per se, stands at the center of the analysis.

Within this process, factors such as gender and ethnicity influence whether or not, and the extent to which, an association exists between the attainment of advanced informational capabilities and improved human and social capabilities. Due to the indirect nature of the effects of ICT, the theoretical framework stresses the importance of understanding people's perceptions of this process and the way their use of ICT can affect their lives. This approach is based on the existing information systems literature, which has shown that perception-based measures of the effects of ICTs can be used as adequate proxies for actual benefits derived from the use of ICTs (Molla and Heeks 2007; Grover, Fiedler, and Teng 1997; Mirani and Lederer 1998; Saarinen 1997). Based on this theoretical background and research approach, the chapter analyzes the following question:

- Which factors influence people's perceptions of the impact that ICTs have on the various aspects of their lives?

The effects of this process will vary in different spheres. Consequently, the chapter analyzes the spheres in which advanced informational capabilities can expand a person's human and social capabilities. Specifically, it addresses the following question:

- In which dimensions (economic, political, social, and cultural) do people perceive ICTs to have the greatest impact on their well-being?

Finally, it is important to differentiate the informational capabilities of individuals from the collective or social capabilities and to analyze the impact that ICTs can have, not only on individuals but also on communities overall. This is especially true for indigenous peoples who generally hold strong community values. Thus the chapter also addresses the following questions:

- To what extent does the use of ICTs enhance indigenous peoples' human capabilities and improve their well-being?
- To what extent does the use of ICTs enhance indigenous peoples' social capabilities and improve the well-being of their communities?

The chapter analyzes the quantitative results from my impact user survey to assess indigenous peoples' informational capabilities and the mechanisms by which use of the Internet can have a significant impact on their well-being in various dimensions of their lives. It is divided into six subsections: (a) a summary of the chapter's principal findings, (b) a brief introduction to the survey design and data analysis, (c) a summary of the demographic and socioeconomic characteristics of the sample, (d) an examination of the perceptions of indigenous peoples vis-à-vis the impact of the Internet on their well-being, (e) key factors influencing people's perceptions of the Internet's impact on their well-being, by type of ICT program and ethnicity, and (f) differences in indigenous peoples' perceptions in various spheres of well-being, dependent on their level of informational capabilities.

Principal Findings

The principal finding of the chapter is that the attainment of advanced informational capabilities is the most critical factor determining the impact of ICTs on indigenous peoples' well-being. The chapter demonstrates that significant differences exist between a person's ICT capabilities and her or his informational capabilities. The concept of ICT capabilities encapsulates a person's ability to make efficient use of computer hardware, software, and ICT tools; the concept of informational capabilities is an information-centric approach, deemphasizing the role of technology in a person's ability to use these tools in multiple dimensions of his or her life. It includes four components: (a) ICT capability, (b) information literacy, (c) communication capabilities, and (d) content capabilities. In fact, the chapter demonstrates that a person's ability to use ICTs plays a significantly smaller role in improving her or his well-being than the ability to find, process, and evaluate information; to generate and share one's own knowledge; and to communicate effectively with others.

Second, the chapter shows that a positive association exists between a person's informational capabilities and her or his human and social capabilities in each of the six well-being dimensions (personal, economic, political, social, organizational, and cultural) investigated in this study. At the same time, the empirical findings also reveal important variations in the extent to which people's advanced informational capabilities translate into the expansion of human and social capabilities, whereby the positive impact on people's lives seems to be the strongest in the social and organizational dimensions. For instance, in the social dimension, the empirical data reveal a strong relationship between advanced informational capabilities and improved access to educational services in indigenous communities. However, the impact of ICTs on people's political and economic well-being is somewhat limited.

Finally, a central finding of the chapter is that the intermediary organization is the most critical factor for indigenous peoples to attain advanced informational capabilities and thus for ICTs to have a positive impact on their well-being. The empirical evidence demonstrates that variations in the respondents' informational capabilities depend largely on differences between the ICT programs.

Survey Design and Data Analysis

The analysis is based on results from the extended user survey ($n = 91$), which was designed to solicit indigenous peoples' perceptions of the impact of the Internet on their daily lives. In order to differentiate this survey from the simple user survey described in chapter 5, I refer to it as the impact survey. The impact survey questionnaire included, in addition to questions about people's sociodemographic background and level of ICT use, questions regarding their views about the impact of the Internet. To capture the multidimensional character of the impact that the Internet has had on people's lives, the questionnaire differentiated between the following six dimensions: economic, social, political, organizational, cultural, and personal. The survey questionnaire included two types of questions within the economic, social, and political dimensions. The first set asked about the effect of the Internet on the respondent's individual human capabilities. Specifically, it asked the following question with regard to each of the six dimensions: To what extent has the Internet helped you to achieve the following? In the economic, social, and political dimensions, the questionnaire also included a question related to the perceived impact of the Internet on their community's social capabilities: To what extent has the Internet helped your community to achieve the following? Based on the answers to these two questions, I developed a perceived impact index (PII) based on a population-size-weighted average using the following three-point rating scale, with values from 0 to 2. Specifically, the values were employed in the following way:

- 0: respondents perceive that the Internet does not have any impact on this dimension of their lives
- 1: respondents perceive that the Internet has only somewhat of an impact on this dimension of their lives

- 2: respondents perceive that the Internet has a strong impact on this dimension of their lives.

In order to be able to interpret the resulting data in terms of differences in people's perceived impact of the Internet on a specific dimension of their lives, I created four categories:

- Minimal impact: 0.0–0.49
- Low impact: 0.5–0.99
- Intermediate impact: 1.0–1.49
- High impact: 1.5–2.0.

The *minimal-impact* category (0.0–0.49) indicates that respondents believe that the Internet has had only a negligible impact on this dimension of their lives. Such a low value shows that they either perceive the Internet to be unimportant in this dimension or believe that it has not affected their lives in any substantial manner due to the potential barriers (for example, lack of access to information) they face in applying it in a meaningful way.

The *low-impact* category (0.5–0.99) indicates that respondents believe that the Internet has had a minor impact on this dimension of their lives.

The *intermediate-impact* category (1.0–1.49) indicates that respondents believe that the Internet has had a relatively high impact on this dimension of their lives.

The *high-impact* category (1.5–2.0) indicates that respondents believe that the Internet has had a critical impact on this dimension of their daily lives.

The data sample was then broken into clusters by type of intermediary to analyze once again the hypothesis that intermediary organizations are essential to explaining the extent to which ICTs can affect a person's well-being. The analysis uses the same clusters as are used in previous chapters, grouping organizations—Instituto de Capacitación del Oriente (ICO) Comarapa; Coordinadora de Integración de Organizaciones Económicas Campesinas Indígenas y Originarias (CIOEC) leaders; CIOEC Oruro and Potosí; Organización de Mujeres Aymaras del Kollasuyo (OMAK); and EnlaRed Municipal—into four clusters rather than the six analyzed previously (table 7.1). This is because the survey results do not include data from the Confederación de Pueblos Indígenas de Bolivia (CIDOB), a national indigenous organization, and the Fondo Indígena, an international indigenous organization.

TABLE 7.1   Affiliation of respondents, by cluster of intermediaries

Cluster	Name of organization	Number of respondents	% of respondents
NGO in lowlands	ICO Comarapa	19	20
National grassroots organization	CIOEC leaders	13	16
Grassroots organization in the highlands	CIOEC Oruro, CIOEC Potosí, and OMAK	39	42
Government organization	EnlaRed Municipal	20	22
Total		91	100

Demographic and Socioeconomic Profile of Respondents

The impact survey forms the second part of the user survey described in chapter 6. For this reason, the socioeconomic profile of respondents is the same as that described in chapter 6.

Males make up 72 percent of respondents to the impact survey and 74 percent of respondents to the user survey. Approximately one-fourth of respondents to both surveys are under 25 years of age (22 percent in the user survey and 24 percent in the impact survey), and 1 in 10 is older than 45 (13 and 11 percent, respectively).

However, there are also important differences between the two samples. Respondents to the user survey are slightly more educated than respondents to the impact survey, with 30 and 22 percent, respectively, having attained a university education. At the same time, more respondents to the impact survey have completed secondary schooling than respondents to the user survey (48 and 39 percent, respectively).

The most significant difference between the two samples is that the impact survey has fewer indigenous respondents than the user survey (61 and 71 percent, respectively). This can be attributed to the fact that the impact survey excluded respondents from two organizations with a high percentage of indigenous participants (Fondo Indígena, CIDOB). The respondents from these two organizations had self-identified as 100 percent indigenous, and thus their exclusion lowers the share of indigenous peoples in the impact survey. Another important difference is that the level of ICT use is slightly higher among respondents to the impact survey, 62 percent of whom said that they are either advanced or intermediary ICT users, compared with 53 percent of those in the user survey (figure 7.1). A possible explanation for this higher level of ICT use in the impact survey is the higher percentage of non-indigenous respondents in the sample, since ethnicity is a key factor influencing levels of ICT use.

For assessing the perceived impact of the Internet on people's well-being, it is very helpful that the level of ICT use is quite high among the respondents to the impact survey, since it is important that those surveyed understood

FIGURE 7.1   ICT capabilities of respondents to the impact survey

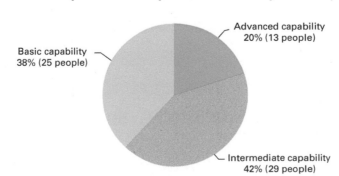

the possibilities and limitations of the technology so that they could make informed value judgments.

Indigenous People's Overall Perceptions of the Internet's Impact on Well-Being

Indigenous peoples perceive the Internet to have had a relatively high impact on all dimensions of their lives, as evidenced by the perceived impact index, which reaches an intermediate impact level for each dimension (figure 7.2). However, important differences are evident between the six dimensions. For instance, the respondents have a more positive view of the Internet's impact on the social dimension (education and health) of their lives (PII = 1.44) than on the other dimensions. They perceive that the Internet has had an intermediate effect on the personal dimension of their lives (PII = 1.41) and a relatively strong effect on the strengthening of indigenous organizations (PII = 1.33). This is the only dimension where all of the respondents indicated that the Internet has had an effect on their lives, with a very low no-response rate of 17 percent.

The respondents have a less positive view of the Internet's impact on the economic, cultural, and political dimensions of their lives than on the social, personal, and organizational dimensions. In particular, the perceived impact on the political dimension is a relatively low intermediate (PPI = 1.16), which indicates that a relatively large number of respondents said that the Internet has had a less positive impact on their lives in this dimension than in other ones.

In the economic dimension, respondents expressed a wide range of views. The respondents overall rated the impact as intermediate (PII = 1.27), but almost one-quarter (24 percent) either believe that the Internet has had no impact on the economic dimension of their lives or did not respond to the question.

FIGURE 7.2 **Perceived impact of the Internet on various dimensions of well-being**

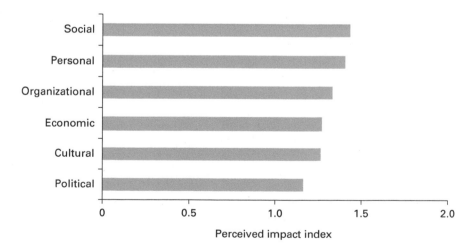

Perceived impact index

In the cultural dimension, respondents also rated the effects of the Internet as intermediate (PII = 1.26), supporting the findings from the qualitative field-work[1] that indigenous peoples are not very concerned that Internet technology might negatively affect their culture.

Key Factors Influencing People's Perceptions

What factors influence indigenous peoples' perceptions of the Internet's impact on their individual and social well-being? To answer this question, I begin by addressing the extent to which people's perceptions of the Internet's usefulness are influenced by sociodemographic variables. Are women more critical of the Internet than men, as other surveys assessing development interventions have found? Do older and younger persons have significantly different opinions of the Internet's impact? Are people's opinions shaped by their educational attainment, with persons who have attained a higher educa-tion responding more positively because they are better informed about the Internet's possible role in development? Finally, does a person's profession influence his or her views on the Internet's impact—that is, does a community leader expect more from the Internet than an indigenous farmer who might not see its immediate value?

The statistical analysis reveals that none of the sociodemographic variables significantly influences people's perceptions of the Internet's impact. Regardless of differences in gender, age, level of education, and profession, indigenous peoples have a generally positive view of the Internet. The elderly, in particular, are very open to this new technology and believe that it can have a positive impact on themselves and their communities.

Despite the overall positive view, respondents' perceptions of the Internet's impact vary significantly depending on (a) the intermediary organization responsible for the ICT program, (b) their own informational capabilities, and (c) their ethnicity. As in the analysis of levels of ICT use, people's perceptions of the Internet's impact are influenced primarily by socioeconomic variables (ethnicity) and the type of intermediation provided (grassroots- or community-led ICT program versus government-led program).

Program Affiliation

The variable "ICT program" is the most influential in people's perceptions of impact. The first overall finding is that respondents from grassroots-led ICT programs hold more favorable views regarding the Internet's impact on their well-being than respondents from the overall sample across all dimensions (figure 7.3). They rated all three dimensions as having a high impact on their well-being: social (PII = 1.69), economic (PII = 1.62), and cultural (PII = 1.54). In the economic dimension, respondents from national grassroots-led programs hold more positive views than respondents from the overall sample (PII = 1.62 and 1.27, respectively), as was shown in figure 7.2. Respondents from grass-roots programs also have more positive views than the overall respondents with respect to the social dimension, rating impact as high (PII = 1.69), while the overall sample rated it as intermediate (PII = 1.44).

FIGURE 7.3 **Perceived impact of the Internet on various dimensions of well-being (respondents from national grassroots-led programs)**

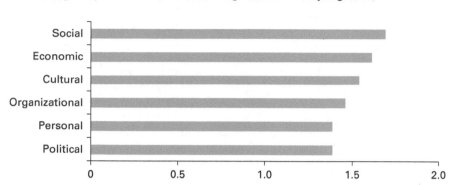

The second overall finding is that a strong association exists between the variable "ICT program" and indigenous peoples' perceptions of the impact of the Internet on their well-being in four (social, economic, organizational, and personal) out of six dimensions. This finding is confirmed by the Kruskal-Wallis test, which shows a statistically significant relationship between the perceived impact score and the ICT program variable in the social (χ^2=11.96; $p = 0.018 < 0.05$), economic ($\chi^2 = 15.55$; $p = 0.004 < 0.05$), organizational ($\chi^2 = 9.930$; $p = 0.042 < 0.05$), and personal ($\chi^2 = 9.423$; $p = 0.048 < 0.05$) dimensions.

Indeed, comparing the responses of participants in the EnlaRed, CIOEC, and ICO programs reveals significant differences of opinion. For instance, participants in the EnlaRed program have relatively negative views about the Internet's impact on their well-being. In the economic dimension, the large majority of EnlaRed respondents indicated that the Internet has not had any positive impact on their well-being, giving it a very low score for perceived impact (PII = 0.53). In the social dimension, the score was only 0.88, indicating that the EnlaRed respondents perceive the Internet to have had a minor effect on this dimension of their well-being. EnlaRed participants were primarily local government officials and thus may have understood the limited ability of the Internet to enhance the government's capacity to deliver social services to its citizens. The case study of the EnlaRed program presented in chapter 9 examines in more detail the underlying reasons why its participants perceive the Internet to have had such a low impact on indigenous peoples' well-being.

ICO participants have a generally more positive view of the Internet's impact, but they are still more critical of it than persons associated with CIOEC. For instance, while ICO participants believe that the Internet has had an intermediate impact on the provision of social services to indigenous communities (PII = 1.37), CIOEC respondents indicated that the Internet has had a high impact on their well-being in this dimension (PII = 1.69). To unpack how intermediary organizations influence people's perceptions of the Internet's impact on their well-being, the subsequent section examines the six dimensions of well-being more closely.

Impact on Economic Dimension of Well-Being

Respondents' views of the Internet's impact on their economic well-being vary significantly from program to program (figure 7.4). Respondents from grassroots programs have a more favorable view of the economic impact, with the majority of OMAK respondents (PII = 1.52), CIOEC leaders (PII = 1.50), and CIOEC Oruro and Potosí respondents (PII = 1.50) indicating that the Internet has had a high impact on the economic dimension of their lives.

In contrast, the majority of EnlaRed respondents indicated that the Internet has had no economic impact, as evidenced by the very low score (PII = 0.53). One reason for this disparity could be that the programs varied significantly and each intermediary stressed different dimensions of human development in its objectives. For instance, while CIOEC—an umbrella organization of local associations of indigenous producers—focused its program primarily on economic and organizational aspects, the EnlaRed program emphasized the political, as opposed to the economic, aspects of development. However, differences in the programs' objectives alone cannot explain differences in the perceived impact on this dimension. For instance, although the OMAK program had a political rather than an economic objective—its goal was to strengthen indigenous women's organizations at the local level—the majority of respondents indicated that ICTs have improved their economic opportunities, with a high score for impact (PII = 1.52). Chapter 8 investigates the OMAK program and analyzes the process and reasons why indigenous women perceive that ICTs are an effective instrument for improving their livelihoods.

While the ICO Comarapa program's objectives were purely economic—to strengthen the productive and economic activities of small-scale farmers—the respondents' opinions regarding the Internet's impact on their economic well-being are more negative, with an intermediate score for perceived impact (PII = 1.26). To understand the important differences in people's perceptions that are dependent on variations in the intermediary organization, I disaggregated the empirical data and looked into the set of indicators defining each dimension. The first analysis focuses on investigating the underlying reasons

FIGURE 7.4 Perceived impact of the Internet on economic well-being, by program

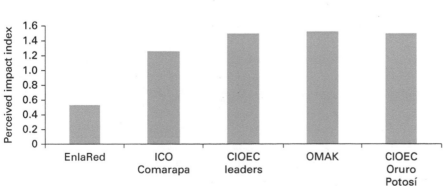

for people's distinct perceptions of the Internet's impact on their economic well-being.

Figure 7.5 illustrates that there are significant differences in people's perceptions of the Internet's impact within different aspects of the economic dimension. First, the respondents perceive the Internet to have had a relatively strong positive impact on both strengthening their community's producer organizations (PII = 1.42) and improving their access to market prices (PII = 1.35). However, they have less positive views regarding the Internet's impact on improving people's income (PII = 0.92) and reducing their travel costs (PII = 1.13). In particular, the low score for perceived impact on income is noteworthy, since income is such a critical aspect of economic well-being.

The results also highlight the importance of investigating not only the individual economic benefits that have been reaped from Internet use, but also whether the collective "economic capabilities" of the entire community have been strengthened. For indigenous peoples, community organizations are essential not only for conducting political decision making but also for increasing economic opportunities and their well-being (figure 7.5).

Most rural indigenous communities continue to rely on traditional forms of agricultural production (*mink'a*), which are based on collective rather than individual methods of production. In such a cultural setting, ICT programs need to be proactive and to respond to the community's existing social structures, emphasizing the value that the Internet can have for strengthening community-based organizations and for facilitating effective horizontal communications between producer organizations. Thus a more conventional approach to ICTs, which frequently focuses on an individualistic approach and emphasizes the role that the Internet can play in enhancing peoples' personal opportunities, is inadequate in the setting of indigenous communities. Although the majority of respondents have a favorable view of the Internet's impact on their collective capabilities (strengthening producer organizations), they believe that it has had a limited effect on their individual well-being (increasing household income).

FIGURE 7.5   **Perceived impact of the Internet on various components of economic well-being**

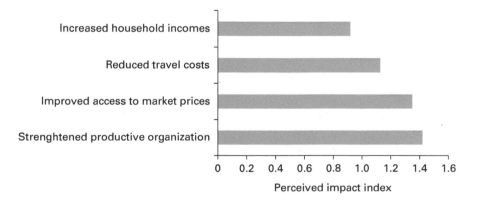

Perceived impact index

Impact on Social Dimension of Well-Being

Respondents' views of the Internet's impact on the social dimension of their well-being differ substantially depending on the program with which they are affiliated (figure 7.6). EnlaRed respondents have a more negative opinion about the Internet's effect on their well-being in this dimension than respondents from the other programs, giving it a low score for perceived impact (PII = 0.88). In contrast, the large majority of respondents associated with CIOEC Oruro and Potosí (PII = 1.50), CIOEC leaders (PII = 1.69), and OMAK programs (PII = 1.70) believe that the Internet has had significant positive effects on the social dimension of their lives.

A more detailed analysis breaking down the specific indicators in the social dimension reveals that respondents believe that the Internet has had a strong positive impact on improving access to education in indigenous communities (PII = 1.53), but a weaker impact on improving peoples' access to health services (PII = 1.32) (figure 7.7). It is noteworthy that the respondents agreed

FIGURE 7.6 Perceived impact of the Internet on social well-being, by program

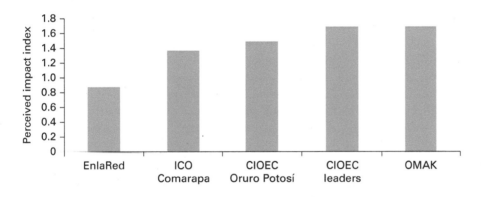

FIGURE 7.7 Perceived impact of the Internet on the social dimension of well-being

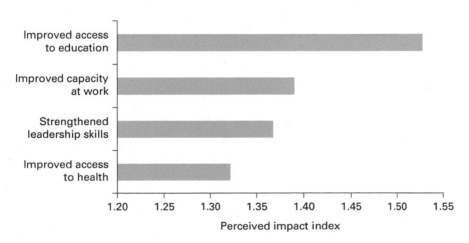

that ICTs have improved their human capabilities, with the perceived impact on leadership skills (PII = 1.37) and improved capacity in the workplace (PII = 1.39) both indicating intermediate impact. Participants affiliated with the CIOEC Oruro and Potosí program overwhelmingly (94 percent) believe that the Internet has improved their community's access to education, compared with only 42 percent of the participants affiliated with the ICO Comarapa program. This result is notable because CIOEC implements its programs in a region with extreme levels of poverty and lack of access to social services.

The empirical evidence presented here suggests that the people who have the least access to social services (CIOEC respondents) are the most convinced that the Internet has provided them with new opportunities to enhance their capabilities. The generally high scores for CIOEC respondents may be attributed to the program's grassroots focus on providing Internet services to local communities. CIOEC combined the Internet with traditional media (that is, community radios and printed newsletters) in order to reach a large audience of small-scale producers at the community level with both educational and market price information.[2]

Impact on Political Dimension of Well-Being

Regardless of program affiliation, respondents have a less positive view of the Internet in relation to the political dimension of their well-being. For three out of five ICT programs, they indicated that the Internet has had a limited impact on the political dimension of their lives (figure 7.8).

For participants in the EnlaRed, ICO, and CIOEC Oruro and Potosí programs, the score for perceived impact is low. In particular, the low value of PII = 0.94 for the EnlaRed program is notable, since this program's main objective was to enhance the transparency and accountability of local governments.

Only respondents in OMAK (PII = 1.39) and the CIOEC leaders program (PII = 1.44)—which specifically aimed to strengthen indigenous peoples' political participation in government—indicated that the Internet has had an intermediate impact on their political well-being. Respondents from both of these programs also have more positive views about the Internet's impact on other dimensions. For instance, the OMAK participants believe that the

FIGURE 7.8 Perceived impact of the Internet on political well-being, by program

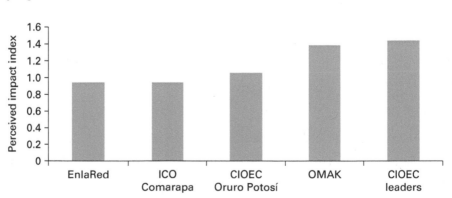

Internet has had a substantial impact on the social and economic dimensions of their well-being, with a high score for perceived impact (PII = 1.70 and 1.52, respectively). Similarly, the CIOEC leaders program gave higher scores for perceived impact on the economic, social, and organizational dimensions of well-being (PII = 1.50, 1.69, and 1.53, respectively) than on the political dimension (PII = 1.44).

Impact on Organizational Dimension of Well-Being

Respondents' views on the organizational dimension reflect differences in the priorities of the various intermediary organizations. Participants in grassroots programs have different views than participants in NGO programs (figure 7.9). The majority of participants from the CIOEC leaders (PII = 1.53), CIOEC Oruro and Potosí (PII = 1.44), and OMAK (PII = 1.48) programs indicated that the Internet has played an important role in strengthening the organizational capacity of their communities or networks.

The EnlaRed respondents indicated that they have a relatively favorable view of the impact on organizational well-being (PII = 1.35), which is noteworthy because they are the most skeptical of the impact on all of the other dimensions. This may mean that the EnlaRed program has primarily been successful in strengthening its network of local government officials in Bolivia's nine departments. Chapter 9 provides a case study of this government program, in order to investigate in more depth the impact of this program on indigenous peoples' well-being in the different dimensions of their lives. That chapter also sheds more light on the extent to which the program has indeed strengthened local government's organizational capabilities.

The data reveal that both of the CIOEC programs have improved indigenous peoples' well-being in the organizational dimension. Their main objective has been to strengthen producer organizations and rural cooperatives at the local level and to create a national network of small indigenous producers.

FIGURE 7.9 Perceived impact of the Internet on organizational dimension, by program

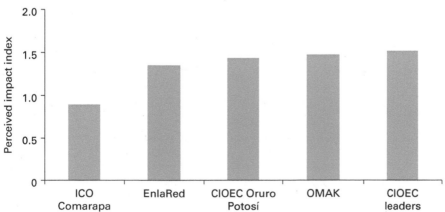

It is notable that participants from these programs think that the use of the Internet has been effective in this area, with two-thirds of participants in the CIOEC leaders program and more than half of those in the CIOEC Oruro and Potosí programs emphasizing that the Internet has had a substantial impact on strengthening their networks.

In contrast, respondents affiliated with the ICO Comarapa program have a much more negative view of the impact of the Internet on the organizational dimension, giving it a low score of 0.89. This value indicates that the program has had an insignificant impact on strengthening the participants' local small producer organizations and improving their networks with other producer organizations in Bolivia. The program's primary weakness was its centralization: it was initiated by a national nongovernmental organization (NGO) and not a direct result of local demand. The qualitative data confirm that while ICO efficiently provided services to indigenous communities, it did not transfer any of the decision-making processes to them.[3] The program was, in effect, paternalistic rather than participatory. Activities were defined and implemented largely by the NGO's technical staff, with only minor participation from local communities.[4] Another shortcoming in their approach was that the ITC program targeted individuals rather than local social organizations. Unlike the CIOEC program, it did not prioritize the strengthening of small producer organizations. Interviews with community leaders in Comarapa confirmed this point: indigenous leaders highlighted that the program was implemented in parallel with their existing social organizations and that all of the planning and implementation of the program's activities (training workshops) were carried out by the technical staff of ICO, with very little participation by the communities.

Impact on Personal Dimension of Well-Being

How does the impact of ICTs on the personal dimension of people's well-being vary by program? As shown in figure 7.10, the predominantly indigenous

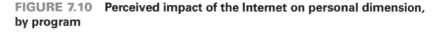

FIGURE 7.10 Perceived impact of the Internet on personal dimension, by program

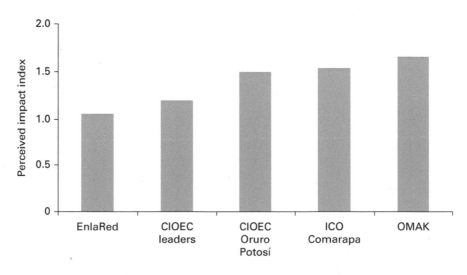

participants from the CIOEC leaders program rated this dimension as the least affected by the Internet, giving it a low score for perceived impact (PII = 1.19). This low score is noteworthy given their generally positive attitude of the impact on all of the other dimensions of well-being. In contrast, participants in the ICO Comarapa program indicated that the Internet has had a high impact on their personal development (PII = 1.53). The OMAK participants also believe that the use of the Internet has made a critical contribution to their personal development, as evidenced by the elevated PII value of 1.65. This finding seems to be consistent with the program's main objective of empowering indigenous women. Finally, the EnlaRed participants, similar to other dimensions, indicated that the Internet has had only a minor impact on their personal development (PII = 1.06). The case studies of OMAK and EnlaRed in chapters 8 and 9 examine the underlying reasons for these variations in the quantitative analysis and aim to understand the processes that led to these results.

Impact on Cultural Dimension of Well-Being

A majority of respondents think that the Internet has had a positive impact on the cultural dimension of their communities—for example, by preserving traditional indigenous knowledge or supporting programs that strengthen indigenous peoples' cultural identities. However, once again, there are substantial differences in the views of participants in the various programs (figure 7.11). Similar to the other dimensions, CIOEC participants hold the most favorable perception, saying that the Internet has had a substantial impact on their culture (PII = 1.56). In contrast, the majority of respondents affiliated with the ICO Comarapa (PII = 1.16) and EnlaRed (PII = 1.06) programs indicated that the Internet has had only a minor effect on the cultural dimension of their lives. This may be attributed to the purely technical nature of these two programs, in which cultural aspects played only a minor role.

FIGURE 7.11 **Perceived impact of the Internet on cultural dimension of well-being, by program**

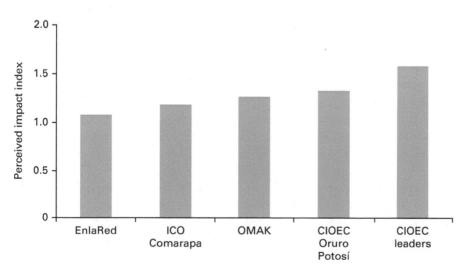

FIGURE 7.12   **Perceived impact of the Internet on peoples' cultural identity**

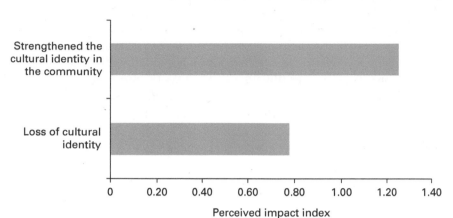

In light of indigenous peoples' long history of struggle and resistance against outside powers (Scott 1985), I expected them to have a much more negative perception of the Internet's impact, viewing it as yet another outside threat to their culture. However, this proved not to be the case (figure 7.12). The majority of respondents believe that the Internet has positively contributed to strengthening their cultural identity and gave this dimension a score indicating intermediate impact (PII = 1.25). Furthermore, they indicated that the Internet has not posed a significant threat to their cultural identity, giving it a score indicating low negative impact (PII = 0.78).

Ethnicity

With regard to ethnicity, indigenous respondents have a generally more positive view of the Internet's impact than non-indigenous respondents, except in relation to the provision of social services. Overall, the poorer indigenous participants are substantially more positive than their more wealthy non-indigenous counterparts about the Internet's impact on their economic and organizational well-being. Indigenous peoples from the highlands and from departments with the highest incidence of poverty (Oruro and Potosí) expressed the most favorable views of the Internet's role in enhancing their economic well-being. In focus group discussions with participants in the CIOEC program in Oruro and Potosí, most of whom were indigenous leaders of communities of subsistence farmers, the majority expressed frustration with the development policies of the central government and emphasized that they feel "abandoned" by the rest of the country.[5] They expressed the general sentiment that government and nongovernmental programs do not reach their communities and that they urgently need support in their daily struggle for survival.

Why are persons who face the most severe challenges in meeting their daily needs the most convinced that the Internet has helped them? In the focus groups, community leaders stressed that, being the most impoverished and excluded group in Bolivia, they potentially have the most to gain from this new media. One possible explanation for this finding is that indigenous peoples who live in the poorest and most remote areas of Bolivia have the most unrealistic expectations of the Internet, as they do of other development programs in their communities. However, many indigenous peoples from impoverished regions are extremely skeptical about "development" interventions that originate outside their own community and have frequently organized to prevent the implementation of such programs in their communities. Another possible explanation for the overwhelmingly positive views of this group is that the grassroots organization CIOEC was proactive in promoting the use of the Internet in this region and combined the Internet with other forms of ICTs (such as community radio), enabling it to reach local communities.

Perceptions about the Impact of the Internet on People's Well-Being and Informational Capabilities

Are variations in people's perceptions of the Internet's impact on the various dimensions of their lives associated with variations in their informational capabilities? This section analyzes the spheres (economic, political, cultural) in which people do and do not perceive a relationship between the attainment of advanced informational capabilities and improvements in their human and social capabilities. It also seeks to understand the factors that influence people's perceptions of the impact of the Internet on various dimensions of well-being and to analyze in more depth their perceptions about its impact in specific areas such as market access or vertical accountability.

First, there is a strong association between a person's informational capabilities and the ICT program variable. The chi-square test verifies that the relationship between these two variables is statistically significant ($p = 0.04$). The participants in grassroots-level programs (CIOEC and OMAK) have attained advanced informational capabilities, while those in NGO programs (ICO) and national governmental organizations (EnlaRed) have not. In fact, approximately two-thirds (67 percent) of the respondents from the CIOEC Oruro and Potosí program, compared with less than half (46 percent) of the overall survey respondents, said they have advanced informational capabilities (figure 7.13). More than half of the participants from OMAK (60 percent) have obtained advanced levels of informational capabilities.

At the same time, the empirical data show that respondents from the NGO program implemented by ICO have attained relatively low informational capabilities; only about one-third (32 percent) acquired advanced informational capabilities. Similarly, only 18 percent of the respondents from the government-led program EnlaRed have attained advanced informational capabilities, and almost one-third (29 percent) have reached only basic informational capabilities.

Comparing these results with data on the effects of ICT programs on people's ICT capabilities is revealing. The participants of the CIOEC program in Oruro and Potosí have attained both advanced levels of informational and ICT capabilities—with 68 percent of respondents reaching an advanced level of informational capability and 89 percent reaching either advanced or intermediate levels of ICT capacity. In contrast to the CIOEC program, the

FIGURE 7.13   Informational capabilities, by type of intermediary

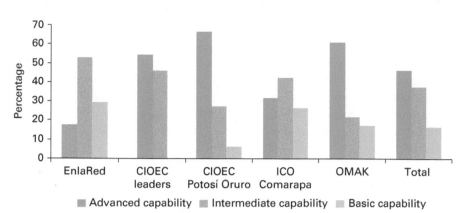

■ Advanced capability   ■ Intermediate capability   ■ Basic capability

FIGURE 7.14   Perceived impact of the Internet on various dimensions of well-being (respondents with an advanced level of informational capabilities)

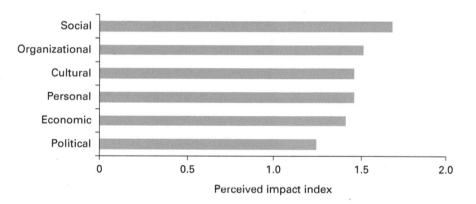

ICO participants have reached similar advanced ICT capabilities (72 percent of ICO participants have high or intermediate ICT capabilities), but significantly lower informational capabilities (only 31 percent of participants have advanced informational capabilities).

The main result from the impact survey is that there is an association between a person's informational capability and his or her human or social capabilities in all of the six dimensions of well-being investigated in this study (see figure 7.14). This is important, because expanded informational capabilities, not stronger ICT capabilities, are associated with people's perceptions about the Internet's impact on their well-being in the multiple dimensions of their lives. Moreover, an important aspect of informational capabilities is information literacy—the ability to collect, evaluate, organize, and share information with others. The ability to use ICT tools is secondary to the ability to use and place information in one's own sociocultural context.

At the same time, there are important variations in the extent to which these two variables are associated with each other. People with advanced informational capabilities perceive that the Internet has had the strongest impact on their well-being in the social and organizational dimensions of their lives. For both dimensions, the score for perceived impact is high, with PII equal to 1.68 for the social dimension and to 1.51 for the organizational dimension.

The respondents with high levels of informational capabilities indicated that the Internet has had a somewhat lesser, if still important, impact on the cultural and personal dimension of their lives, with the perceived impact in both dimensions scored as intermediate (PII = 1.46).

Furthermore, the attainment of advanced informational capabilities makes a difference in people's perception of the Internet's impact on their economic well-being. In this sphere, the perceived impact rises from a relatively low level (PII = 1.27) for respondents overall to an intermediate level (PII = 1.41) for respondents with advanced informational capabilities. While these values demonstrate that indigenous peoples with advanced informational capabilities have a more positive view of the Internet's impact on their economic well-being, respondents only ranked this dimension fifth, indicating that they believe that the Internet has had a stronger impact on other dimensions (social or organizational).

The political dimension is the area in which improved informational capabilities have had the smallest perceived positive impact on well-being, with low perceived impact only a bit higher than that of all respondents (PII = 1.24 and 1.16, respectively). In this dimension, a relatively higher percentage of respondents with advanced informational capabilities are less positive about the impact of the Internet on their well-being than in any other dimension, in spite of their ability to find, evaluate, and process information.

Informational Capabilities and Well-Being in the Social Dimension

There is a strong relationship between advanced informational capabilities and increased social well-being (figure 7.15). Approximately three-quarters of

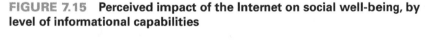

FIGURE 7.15 Perceived impact of the Internet on social well-being, by level of informational capabilities

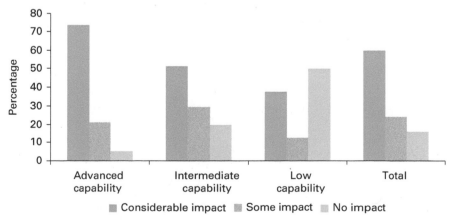

people with advanced informational capabilities (74 percent) expressed that the Internet has significantly enhanced their well-being in the social dimension. This figure is about 14 percentage points higher than that of all respondents (60 percent) and highlights the powerful effect that enhanced informational capabilities can have on people's perceived opportunities in the social dimension of their lives. The statistically strong correlation between these two variables is also confirmed by the chi-square test ($p = 0.007$).

Indigenous peoples value the positive effects of advanced informational capabilities particularly highly with regard to the delivery of social services in their community (figure 7.16). Indigenous respondents believe that the Internet has had the strongest positive effect on their community's access to education and health services (PII = 1.74 and 1.62, respectively). All of the respondents with advanced informational capabilities think that the Internet has improved indigenous communities' access to education, and 90 percent indicated that it has improved their access to health services. However, enhanced informational capabilities have a somewhat weaker effect on strengthening individual human capabilities, although the score for perceived impact on capacity at work and leadership skills is still high (PII = 1.53), indicating that respondents believe that the Internet has had a substantial impact on their well-being in these areas.

Informational Capabilities and Well-Being in the Organizational Dimension

There is a strong association between informational capabilities and the organizational dimension (figure 7.17). The majority of respondents (59 percent) with advanced informational capabilities indicated that enhanced access to and use of information provided by the Internet have strengthened indigenous peoples' organizations. By comparison, only half (50 percent) of the overall respondents believe that use of the Internet has positively influenced their organizational development. Notably, 63 percent of respondents with low informational capabilities think that the Internet has had no effect on the

FIGURE 7.16 **Perceived impact of the Internet on social well-being among respondents with an advanced level of informational capabilities**

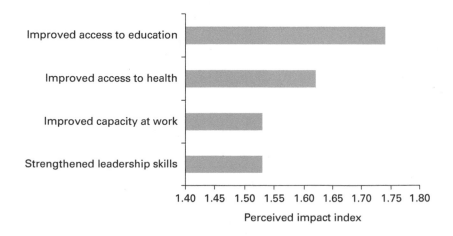

FIGURE 7.17 Perceived impact of the Internet on organizational dimension of well-being, by level of informational capabilities

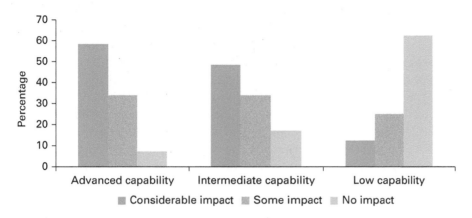

FIGURE 7.18 Perceived impact of the Internet on social capital, by level of informational capabilities

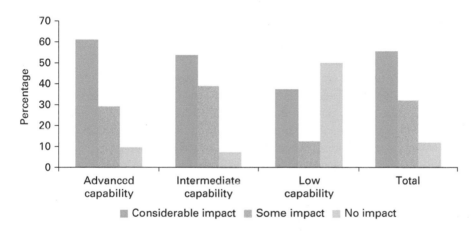

organizational strength of their communities. The chi-square test confirms the strong role that informational capabilities play in this dimension ($p = 0.004$).

To analyze the relationship between informational capabilities and social capabilities more deeply, I created two indexes—one for social capital and one for community well-being. The "social capital"[6] index is based on three questions from the impact survey: Has the use of the Internet strengthened the producer organizations in your community? Has it strengthened indigenous organizations? Has it strengthened existing networks among indigenous organizations? The statistical analysis reveals a strong correlation between informational capabilities and social capital.

In fact, a large majority (61 percent) of indigenous peoples who have advanced informational capabilities believe that the Internet has had a strong influence on strengthening their social capital (figure 7.18). Half (50 percent) of the respondents with low informational capabilities think that the Internet has had no impact at all on the formation of social capital. The chi-square test determining statistical significance confirms this relationship ($p = 0.003$).

Informational Capabilities and Economic Dimension of Well-Being

The analysis indicates important variations in indigenous peoples' views about the Internet's impact on economic well-being, as evidenced by the perceived impact index, which reaches a score of 1.27. Thus the perceived impact on economic well-being is considered intermediate. It is, however, noteworthy that 24 percent of respondents either believe that the Internet has had no effect on their well-being in this dimension or did not answer the question. These data raise the following question: What is the association between a person's informational capabilities and economic well-being? This section unpacks the variations by analyzing in more depth the association between these two variables.

First, the relationship between the two variables is statistically significant (figure 7.19). Second, the attainment of advanced informational capabilities has a somewhat weaker effect on a person's economic well-being than on other dimensions, as the number of people who stated that the Internet has had a major effect on their economic well-being is only slightly higher for persons with advanced informational capabilities (56 percent) than for all respondents (52 percent). Moreover, three-fourths (75 percent) of the respondents with low informational capabilities indicated that the Internet has had no effect at all on their economic well-being.

The chi-square test ($p = 0.008$) confirms the association between the two variables. The statistical significance is based on both the relatively weak positive effects of enhanced informational capabilities on economic development and the strong negative correlation between low levels of informational capabilities and economic development.

To understand the nature of the relationship between these two variables, I created a specific "market access" index, based on three aspects of people's capabilities to access markets: improved access to market prices, lower transaction costs through shorter travel time, and improved incomes.

The finding of no association between people's informational capabilities and market access is unexpected (figure 7.20). The chi-square statistical significance test confirms the absence of a statistically significant relationship

FIGURE 7.19 Perceived impact of the Internet on economic development, by level of informational capabilities

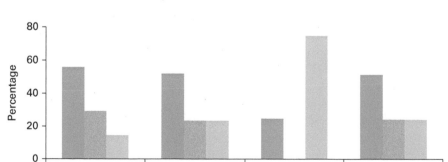

FIGURE 7.20 Perceived impact of the Internet on market access, by level of informational capabilities

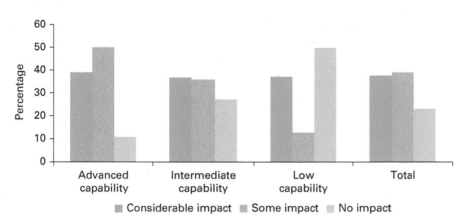

between the two variables ($p = 0.12 > 0.05$). Only a slightly higher percentage of people with advanced informational capabilities than all respondents said that the Internet has significantly enhanced their access to markets (39 and 38 percent, respectively).

This empirical evidence raises three key questions. What are the reasons for the relatively weak association between informational capabilities and access to markets? Why are there such important variations in people's views on the impact of ICTs on their economic well-being? What are some of the major barriers to a stronger association between informational capabilities and human capabilities in economic terms?

The following analysis disaggregates the data in order to examine the relationship between informational capabilities and the perceived impact of the Internet on people's economic well-being.

The relationship between people's level of informational capabilities and their perceptions about the Internet's impact on their economic well-being is relatively weak (figure 7.21). In all three areas, the respondents with advanced informational capabilities said that they have only a slightly more positive view about the Internet's impact on their economic well-being than the overall respondents (PPI = 1.44 and 1.35, respectively). Moreover, there are important variations in people's perceptions in relation to the three components of market access. A majority of people with advanced informational capabilities believe that the Internet has enhanced their access to market prices (PPI = 1.44), but they are less convinced that the Internet has had a positive impact on reducing their travel costs (PII = 1.21) and even less convinced that it has had a positive impact on their income (PPI = 0.99).

Thus, while ICTs have played an important role in improving people's access to market information, the reduction in information asymmetries has not significantly reduced their travel time or other costs or improved their income. While advanced informational capabilities have enabled people to improve their ability to access and process information related to their economic well-being (market prices), the programs have had a limited positive effect on their economic well-being. In particular, the lack of an association between advanced informational capabilities and improvements in people's incomes deserves further analysis.

I now investigate the effects of informational capabilities on each of the three dimensions of market access separately. Through this disaggregation, I aim to gain a better understanding about the Internet's impact on people's economic well-being and to identify the critical factors that can explain the absence of a strong correlation between improved informational capabilities and economic well-being. First, the data indicate a strong relationship between people's advanced informational capabilities and perceived improvements in their access to market prices. In fact, almost 59 percent of the respondents with advanced informational capabilities, in comparison to about 55 percent of the overall sample, believe that the Internet has made an important difference in improving their access to market information (figure 7.22). However, half

FIGURE 7.21   Perceived impact of the Internet on market access, by level of informational capabilities

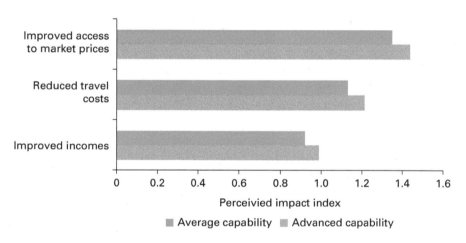

FIGURE 7.22   Perceived impact of the Internet on access to market prices, by level of informational capabilities

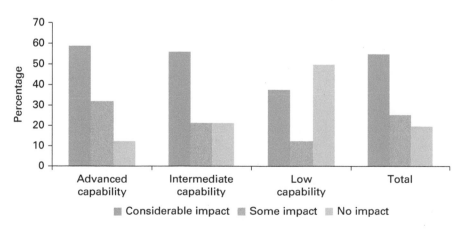

of the respondents with low informational capabilities expressed serious concerns about the positive effects of the Internet on their access to market prices. Thus informational capabilities are associated with the ability to access market prices. This finding is confirmed by the chi-square test, which shows a statistically significant relationship between these two variables ($p = 0.012 < 0.05$). Thus an important aspect of the Internet's impact on people's economic well-being is related to improved access to market prices.

Second, I investigate the lack of association between the attainment of advanced informational capabilities and the respondents' perception that the Internet has increased their income. The data analysis shows that only one-third (32 percent) of all respondents believe that they have increased their income by using the Internet; a larger percentage (40 percent) think that the Internet has not had any positive impact on their income (figure 7.23). Furthermore, only 32 percent of both respondents with advanced informational capabilities and all respondents said that the Internet has had a positive effect on their income. Thus there is no relationship between enhanced informational capabilities and improved income. The chi-square test confirms this important finding, showing no statistically significant relationship between the two variables ($p = 0.101 > 0.05$).

The third aspect of improving people's access to markets relates to the reduction of transaction costs through use of the Internet. The literature on "e-commerce" frequently mentions that the use of ICTs reduces transaction costs by improving the efficiency and effectiveness of market transactions and reducing asymmetric information in markets (Eggleston, Jensen, and Zeckhauser 2002; UNCTAD 2013; Humphrey et al. 2003). However, the survey results indicate that indigenous peoples in Bolivia believe that the Internet has had only limited positive effects on reducing their travel costs. As shown in figure 7.24, only about 40 percent of all respondents think that the Internet has had an important effect on realizing time and cost savings. Furthermore, a slightly higher percentage of respondents with advanced informational

FIGURE 7.23 **Perceived impact of the Internet on income, by level of informational capabilities**

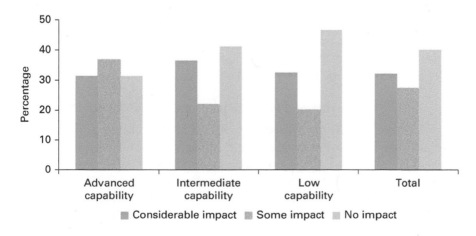

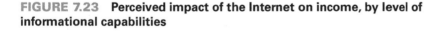

FIGURE 7.24 Perceived impact of the Internet on transaction costs, by level of informational capabilities

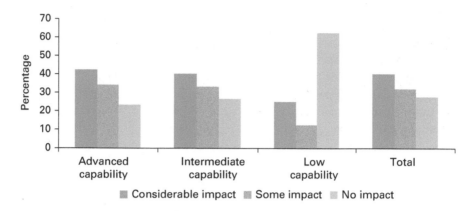

Considerable impact ■ Some impact ■ No impact

FIGURE 7.25 Perceived impact of the Internet on political well-being, by level of informational capabilities

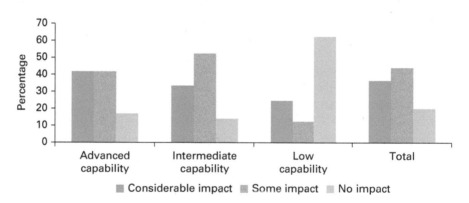

Considerable impact ■ Some impact ■ No impact

capabilities (42 percent) indicated that the Internet has helped them to save time and resources by minimizing their travel time to the closest markets. The data, however, also reveal that a significant number of respondents with advanced informational capability either believe that the Internet has had only somewhat of a positive effect (34 percent) or has not had any positive effect (24 percent) on their travel costs. This finding is confirmed by the statistical analysis, which points to the absence of an association between the two variables and by the chi-square test, with $p = 0.250 > 0.05$.

Informational Capabilities and Political Dimension of Well-Being

As mentioned, indigenous peoples perceive that the positive effects of the Internet have been lowest on the political dimension of their lives. The data analysis reveals a weak association between respondents' informational capabilities and the political dimension of their well-being (figure 7.25). In fact, the percentage of respondents who believe that the Internet has strongly influenced their well-being in this dimension is only slightly higher, reaching the relatively low value of 41.5 percent. Moreover, 41.5 percent of the respondents with high informational capabilities indicated that the Internet has had only somewhat of a positive effect on their political well-being, and 17 percent indicated that there has been no positive effect at all on this dimension of their lives. These findings are confirmed by the statistical chi-square test, which shows that there is no statistically significant association between the two variables ($p = 0.087 > 0.05$).

What are some of the underlying reasons that indigenous peoples perceive the Internet to have such a limited impact on the political dimension of their well-being? Which factors can explain the lack of an association between the two variables? Does it have something to do with differences between the central and local governments?

The limited effects that ICTs have had on people's political well-being reflect the broader political context of indigenous peoples in Bolivia. The majority of indigenous peoples are marginalized and unable to exercise their *rights to information* vis-à-vis the formal institutions of the state. This concept is disaggregated into three areas: (a) the right to seek and obtain information, (b) the right to be informed, and (c) the right to impart and publish information (Blackburn 2000).

The analysis introduces O'Donnell's concept of "vertical" and "horizontal" accountability in order to study the multiple effects that enhancing indigenous peoples' informational capabilities has on increasing their political participation in different levels of government. "Vertical" accountability refers to the right of those who are affected by the decisions of public officeholders to renew or revoke their mandates, and "horizontal" accountability refers to the obligation of public officeholders to answer for their actions to one another (O'Donnell 1999, 113). In other words, the concept of vertical accountability refers to the accountability of the state to civil society. A central aspect of this concept is the obligation of politicians and state employees to make information available to those affected by it. Blackburn describes this concept in the following way: "Vertical accountability is thereby best understood as the mirror of the right to information, whereby the one is ideally reflected in the other" (Blackburn 2000, 45).

In order to evaluate whether advanced informational capabilities can strengthen people's human capabilities to exert their right to information and thus to enhance the accountability of government, I created two vertical accountability indexes—one to describe the relationship with the central government and one to describe the relationship with local government. The vertical accountability index for the central government accounts for both aspects of accountability—first, a person's human capabilities to exert her or his rights to information, particularly improved information about national government policies and improved knowledge about citizen rights, and, second, the government's willingness to improve vertical accountability by improving the transparency of government agencies and actively disseminating information about government programs.

Only a minority of respondents believe that the Internet has had a positive effect on enhancing the central government's vertical accountability vis-à-vis its citizens (figure 7.26). About 43 percent of all respondents said that the Internet has significantly improved the vertical accountability of the central government, and 24 percent said that the Internet has not improved vertical accountability.

Moreover, less than half of the respondents with advanced informational capabilities (approximately 46 percent) said that the Internet has had a strong positive effect on improving the state's vertical accountability (figure 7.27). These findings are confirmed by the statistical chi-square test, which indicates the absence of a significant association between the two variables ($p = 0.077 > 0.05$).

These findings illustrate that the Internet has been critical in strengthening the ability of indigenous peoples to exert their right to information but has had limited effects on the behavior of the central government. The Internet

FIGURE 7.26   Perceived impact of the Internet on vertical accountability of central government, by level of informational capabilities

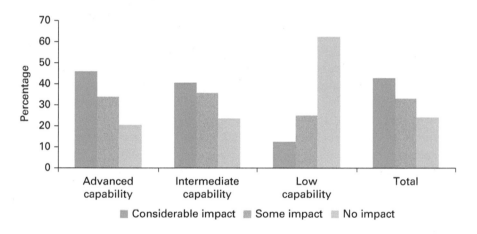

FIGURE 7.27   Perceived impact of the Internet on various components of vertical accountability

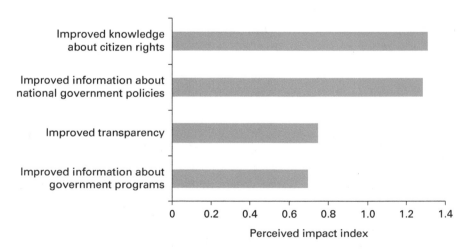

can only improve transparency and information dissemination to a limited extent. The data show that the respondents think that the Internet has had a substantial positive effect on improving their knowledge of both national government policies (PII = 1.31) and their citizen rights (PII =1.29), indicating an intermediate impact in both areas. However, they also believe that the Internet has had a weak impact on improving the transparency of the central government (PII = 0.75). The majority of respondents said that the Internet has not helped them in any significant way to improve their knowledge about government programs (PII = 0.70). Thus they are generally critical of the effects of the Internet on their political well-being. The fundamental issue seems to be limited availability of government information rather than access to ICTs.

This empirical evidence has to be seen in Bolivia's broader sociopolitical context of historically weak vertical accountability (Blackburn 2000). In particular, the relationship between Bolivia's central government and the indigenous population has been marked by exclusion and fragmentation. As described in chapter 3, indigenous peoples have historically been excluded from the political decision-making processes, in particular at the central level. Because of its tremendous geographic, cultural, and socioeconomic diversity, Bolivia is a mosaic of very different realities that continue to be separated from each other. The government has exacerbated this separation by neglecting the interests of the most vulnerable and excluded groups, such as the monolingual indigenous population. The simple fact that in 2005 between 750,000 and 2 million Bolivians were living without any sort of documentation indicates the degree to which the central state continues to neglect the rights of a large group of its citizens (Ardaya and Sierra 2002).

Thus it is not surprising that the information exchanges between the government and indigenous communities are rather limited. Only 11 percent of the respondents said that they are pleased with the information exchanges between the central government and their community, while 45 percent said that information flows are either bad or nonexistent (figure 7.28).

FIGURE 7.28 Perceived accountability of central and local government to indigenous peoples

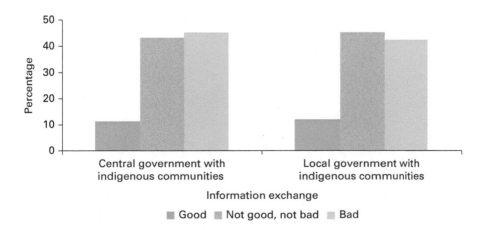

People with advanced informational capabilities are even more critical of the government's commitment to transparency; only 9 percent of this group is satisfied with the information exchanges between the central government and themselves. Notably, indigenous peoples rated the information flows with local governments only marginally better; only 12 percent of the respondents said that municipalities openly share information with their communities.

Figure 7.29 shows how people's perceptions of the central government's transparency vary depending on their ethnic identity (figure 7.29). Indigenous respondents are significantly more critical than non-indigenous respondents. In fact, a majority (53 percent) of indigenous respondents indicated that the information exchanges between the central government and rural communities are either bad or nonexistent, compared with only about 34 percent of non-indigenous respondents.

The critical views of many indigenous respondents about the lack of openness of government are also evident in the analysis of information needs described in chapter 4; improved access to government information was ranked as the second lowest priority out of nine options, with only 43 percent of respondents indicating that this information is very important to them. A majority (55 percent) of respondents said that communicating with the central government is of little or no importance to them.

What is the situation regarding information exchanges between indigenous communities and the government at the local level? The analysis applies the same conceptual framework described above and explores the differences in the effects of ICTs on the central versus local government.

As in the analysis above, I created a specific vertical accountability index describing the relationship between indigenous peoples and local government. This index is based on three aspects of a person's political capabilities: (a) improved knowledge about local government programs, (b) enhanced participation of indigenous peoples in political decision-making processes at the local level, and (c) a stronger mechanism to control the activities of the local government through local civic committees (*comités de vigilancia*).

FIGURE 7.29 Perceived accountability of government to rural communities, by ethnicity

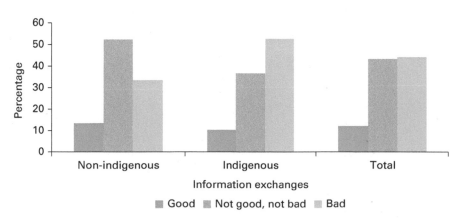

First, indigenous peoples view the impact of the Internet on local government even less positively than the effect on central government. In fact, approximately 21 percent of all respondents said that the use of ICTs has had no effect on the vertical accountability of local government, and 49 percent said that it has had a somewhat positive effect (figure 7.30).

Second, there is no significant difference of opinion between respondents with advanced informational capabilities and the overall sample with regard to the Internet's effects on improving the vertical accountability of local government. Similar to all respondents, only one-third of those with advanced capabilities (29 percent) said that enhancing their informational capabilities has expanded their political participation in local government. The statistical chi-square test confirms the absence of any association between the two variables, with $p = 0.72 > 0.05$. This is the lowest value obtained from the different indexes created for this chapter.

An important observation during my fieldwork was that, increasingly, the indigenous organizations themselves are criticized for their lack of accountability and transparency. This issue is related to the formation of a new national indigenous leadership, which has traditionally been held accountable to the organizations they lead through direct elections that take place during annual or biannual general assemblies. Traditionally indigenous communities have exerted strict social control mechanisms over their leadership in order to ensure high standards of transparency, accountability, and ethics in the conduct of indigenous organizations. However, the increasing political influence of the indigenous movement in Bolivia since the late 1990s has led to the proliferation of indigenous organizations and created new leaders at the national level, who, increasingly, are reaching positions of political power and losing their traditionally strong relationship with the grassroots.[7] Indeed, the rise of

FIGURE 7.30 **Perceived impact of the Internet on accountability of local governments to indigenous communities, by level of informational capabilities**

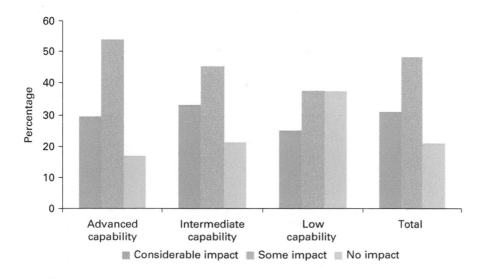

■ Considerable impact ■ Some impact ■ No impact

the indigenous movement throughout Latin America and the world has helped indigenous leaders to attain international recognition; they frequently attend high-level conferences at international institutions such as the United Nations or represent indigenous peoples' concerns worldwide. The question now is the extent to which these indigenous-led institutions are falling into the same traps of "clientelism," patronage, and corruption as the government agencies led by the non-indigenous urban elites.

Figure 7.31 illustrates that indigenous peoples are critical not only of the opacity of government agencies but also, increasingly, of the way indigenous organizations handle information. The data reveal that less than one-third (28 percent) of respondents are pleased with the way indigenous organizations share information with local communities. This figure is two times higher than the figure for central government agencies and illustrates the increasingly complex relationship between indigenous organizations and their local communities. Notably, indigenous women expressed serious concerns about the accountability of indigenous organizations, with only 19 percent of them saying that they are pleased about the degree to which the indigenous leadership shares information with them. This observation reflects the frequent, systematic exclusion of indigenous women from positions of leadership, even within the indigenous movement itself.

How much do indigenous peoples believe that the Internet can enhance the vertical accountability of central government and, more important, the accountability of their own indigenous organizations? An unexpected finding is that indigenous peoples are highly critical of the role the Internet has played in enhancing the accountability of indigenous organizations. There is no association between the two variables and, regardless of their informational capabilities, only about one-third of respondents said that the Internet has significantly improved the accountability of indigenous-led organizations. The statistical significance test confirms this finding ($p = 0.65 > 0.05$). In fact, a majority of the respondents expressed serious concerns not only about the lack of transparency and accountability of government agencies but also about the responsiveness of their own leaders (figure 7.32). This unexpected result

FIGURE 7.31 **Perceived exchange of information between government and indigenous organizations and local communities**

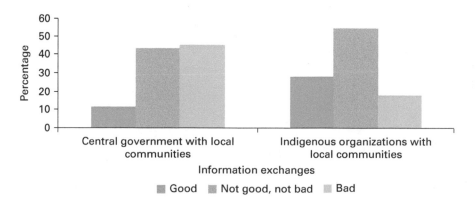
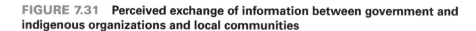

FIGURE 7.32 **Perceived impact of the Internet on the accountability of indigenous organizations to local communities, by level of informational capabilities**

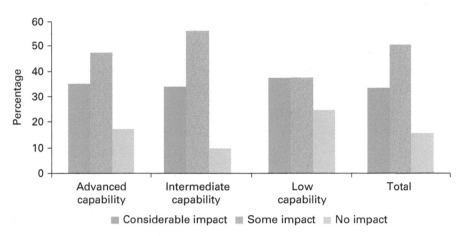

Considerable impact ■ Some impact ■ No impact

indicates that many indigenous peoples are critical of their own leadership. The survey results reveal that indigenous women in particular are critical of the indigenous leadership and believe that they continue to be excluded from gaining full access to the decision-making processes of indigenous organizations.

Informational Capabilities and Personal Dimension of Well-Being

Indigenous peoples perceive the Internet to have had a significant effect on the personal dimension of their well-being. However, in this section I argue that while there is a major positive relationship between the enhancement of an indigenous person's informational capabilities and her or his human capabilities, most indigenous peoples do not clearly separate their personal well-being from that of the broader community. In fact, for most indigenous peoples, good relationships with the overall community represent a crucial aspect of their own personal well-being, and thus the *collective* well-being generally has a much higher value among indigenous communities than among the non-indigenous population. Consequently, indigenous peoples tend to rank the effects of the Internet on their personal well-being as lower than expected.

The relationship between a person's informational capabilities and her or his personal development is weak (figure 7.33). Only a slightly larger percentage of the respondents with advanced informational capabilities than that of all respondents (53 and 51 percent, respectively) said that the Internet has significantly improved their personal development. This is the lowest perceived effect on any of the assessed dimensions. Nevertheless, the two variables are associated: among respondents with advanced informational capabilities, a high percentage (42 percent) indicated that the Internet has improved their well-being in this dimension at least somewhat and a low percentage (less than 5 percent) said that the Internet has had no impact on their personal development. The chi-square test reveals a statistically significant relationship between the two variables ($p = 0.049$).

FIGURE 7.33 Perceived impact of the Internet on personal dimension of well-being, by level of informational capabilities

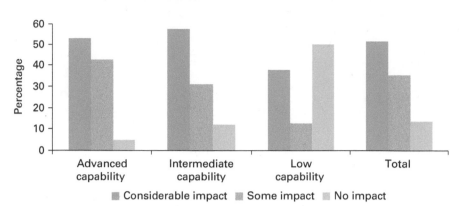

While the Internet seemingly has had a significant impact on individuals' personal development, indigenous peoples did not rate this effect very highly when asked directly about the Internet's impact on their personal development. To assess the relationship between the two variables, I created a specific "individual empowerment" index, taking into account various aspects of people's individual capabilities: (a) confidence that they have more opportunities to advance their professional development, (b) pride that they associate with the knowledge of how to use the Internet, and (c) perceived improvements in their capability to enhance their personal well-being.

In contrast to the dimension of personal development overall, a strong and statistically significant relationship exists between the enhancement of a person's informational capabilities and individual empowerment. In fact, 54 percent of respondents with advanced informational capabilities said that the Internet has had a significant positive effect on their individual capabilities to make strategic life choices, such as being better informed about educational opportunities (figure 7.33). The chi-square test confirms the strong relationship between the two variables, with $p = 0.005$.

Thus the perceived impact of the Internet on a person's individual human capabilities varies with how the question is asked. When people are asked a general question about the impact of the Internet on the personal dimension of their well-being, the ranking is lower. However, when they are asked a specific question, such as the extent to which they associate a sense of pride with the knowledge of how to use the Internet, the ranking is higher, indicating a direct relationship with individual empowerment. Indeed, more than 60 percent of respondents with enhanced informational capabilities indicated that they associate pride with the knowledge of how to use the Internet.

The significant and direct impact that the Internet has had on empowering individuals is further confirmed by qualitative, unstructured interviews, in which participants, particularly indigenous women, described the sense of pride and self-esteem they derive from knowing how to use computers and the Internet. In other words, for many indigenous peoples, technology represents the "otherness" and modernity of the mostly urban, non-indigenous Bolivian society from which they feel they have been systematically excluded. In this context, knowing how to use computers and the Internet has had a major psychological impact on indigenous individuals, particularly those in the most vulnerable groups—women, elders, and indigenous youth. This technological competence provides individuals with the sense that they have taken an important step in overcoming their marginalization and exclusion. Chapter 8 investigates the effects that use of the Internet can have on the empowerment of indigenous women.

Informational Capabilities and Social Capabilities

How does the Internet enhance not only the human capabilities of indigenous individuals but also the social and collective capabilities of indigenous communities? As the analysis has demonstrated, indigenous peoples perceive the Internet primarily as an instrument with which to improve the well-being of their communities and strengthen their organizations. In the economic dimension, the Internet is perceived to have had a relatively limited effect on personal incomes but a significant effect on the organizational capacity of indigenous-led producer associations.

To understand better the Internet's impact on indigenous communities, I created two indexes to differentiate between its effects on two aspects of social capabilities: a "community well-being" index and a "cultural identity" index. The first index takes into account four aspects of community well-being related to Internet use: (a) the belief that the Internet might enhance the social opportunities of the community, (b) the sense of overall improvements in community well-being, (c) the pride of one's community associated with Internet use, and (d) the improved sense of recognition of one's community by outsiders.

Indigenous peoples believe that the Internet has had a significant positive impact on the well-being of their communities (figure 7.34). First, a large percentage of all respondents (52 percent) indicated that the Internet has had a

FIGURE 7.34 **Perceived impact of the Internet on community well-being, by level of informational capabilities**

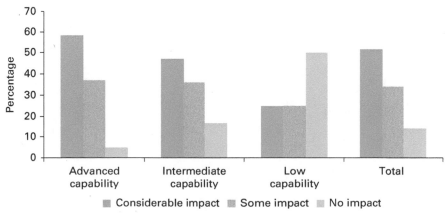

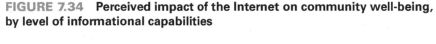

major impact on the overall well-being of their community. Second, there is a strong relationship between enhanced informational capabilities and enhanced community well-being: 58 percent of respondents with advanced informational capabilities said that the Internet has significantly enhanced the well-being of their community. That is, community well-being is the third highest-ranking dimension, after the social and organizational dimensions (74 and 59 percent, respectively). The statistical significance test also indicates a strong relationship between the two variables ($p = 000.3$).

A specific aspect of a community's social capabilities is the cultural dimension. Does the Internet strengthen indigenous peoples' cultural identity by serving as a medium for cultural expression, or is it detrimental to cultural identity? Only a small percentage of indigenous peoples (15 percent) believe that the Internet has led to the loss of their cultural identity. Remarkably, this figure is just 12 percent among respondents with advanced informational capabilities. To study the link between the expansion of people's informational capabilities and a community's social capability in the cultural dimension, I created a "cultural identity index," considering (a) strengthening the cultural identity of the community, (b) enhancing one's ability to disseminate one's own culture to outsiders, and (c) supporting the rescue and preservation of the indigenous knowledge of a community. The relationship between a person's informational capabilities and the community's social capabilities is strong and positive. Indigenous peoples said that the Internet has positively affected their cultural identity; the effects of enhanced informational capabilities, however, are more modest. Only 46 percent of all respondents indicated that the Internet has significantly enhanced their well-being in the cultural dimension (figure 7.35).

At the same time, about 54 percent of respondents with advanced informational skills think that the Internet has had a positive effect on the cultural identity of their community. This figure is almost 8 percentage points higher than for the sample of all respondents, indicating the importance of

FIGURE 7.35 Perceived impact of the Internet on cultural dimension of well-being, by level of informational capabilities

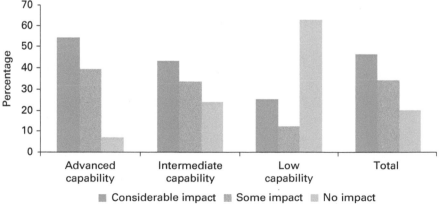

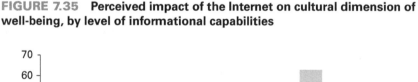

informational capabilities. This effect is confirmed by the strong statistical significance between the two variables ($p = 0.008$). Nevertheless, the absolute value of 54 percent is the third lowest of the dimensions, ranking higher than only the political and personal dimensions. So while indigenous peoples perceive the Internet to have had a significant effect on their social capabilities in the cultural dimension, they consider this effect to be less profound than in the social, organizational, or community well-being dimensions.

Conclusions

The extent to which people attain advanced informational capabilities is critical for determining whether their use of ICTs has significantly enhanced their human and social capabilities. Thus informational capabilities, and not people's ICT capabilities—their proficiency in the use of computers and the Internet—are critical for ICTs to have a significant impact on their lives.

People's informational capabilities are positively associated with their human and social capabilities in each of the six dimensions of well-being (personal, economic, political, social, organizational, and cultural). Like literacy, informational capabilities can improve people's well-being in many dimensions of their lives. However, the analysis also has shown that there are important variations in ICT impact on the different dimensions of well-being.

Respondents said that *the Internet has had the strongest impact on their social and organizational well-being.* In terms of the social dimension, the data reveal a very strong relationship between advanced informational capabilities and improved access to educational services of indigenous communities. In fact, all of the respondents with advanced informational capabilities believe that the Internet has significantly enhanced their community's access to education. Moreover, a large majority (61 percent) of respondents with advanced informational capabilities think that the Internet has been critical to strengthening their social capital—indicating that it is essential in strengthening both indigenous organizations and networks among organizations.

However, the Internet has had only a limited impact on the economic and political dimensions of indigenous peoples' well-being. In the economic dimension, although ICTs have significantly enhanced their access to market prices, they have not altered in any meaningful way existing market inequalities or significantly improved indigenous peoples' incomes. In the political dimension, the chapter has highlighted the absence of any positive association between a person's informational capabilities and her or his human capabilities. The analysis has revealed that while ICTs have helped indigenous peoples to obtain more information about their rights, they have not enhanced the transparency of both central and local government institutions.

Finally, the findings presented in this chapter have validated the hypothesis that the intermediary factor plays a central role in enhancing indigenous peoples' well-being through ICTs. It has demonstrated that intermediary organizations assume an essential role in supporting people in their use of ICTs so that this use not only becomes meaningful to them within their local cultural and sociopolitical context but also significantly enhances their informational capabilities. In fact, the central finding of the chapter is that the extent to which ICT programs support indigenous peoples to attain advanced informational capabilities is the most critical factor determining the impact of ICTs on their well-being. The principal findings from the chapter have shown that grassroots organizations have been more successful than NGO- or government-led programs in enabling indigenous peoples to transform their use of ICTs into enhanced informational capabilities and thus more advanced ICT uses, in order to improve their well-being significantly in the multiple dimensions of their lives.

Notes

1. Focus group with OMAK women leaders, April 14, 2005, and focus group with OMAK community leaders, April 19, 2005.
2. Focus group with indigenous small-scale producers and CIOEC program participants in Challapta, Oruro, July 10, 2005.
3. Interview with Claudia Camacho, coordinator of ICT project, ICO, May 23, 2005.
4. Focus group with small-scale producers, Comarapa, August 20, 2005.
5. Focus groups with grassroots leaders from the CIOEC Oruro and Potosí program in Challapata, Department of Oruro, August 27, 2005, and in Potosí, Department of Potosí, September 2, 2005.
6. James Coleman defines social capital as "variety of entities with two elements in common: they all consist of some aspect of social structure, and they facilitate certain actions of actors" (Coleman 1994). Robert Putnam defines social capital as "the collective value of all 'social networks' and the inclinations that arise from these networks to do things for each other" (Putnam 2000, 19). This study uses both definitions.
7. Interview with Mateo Martínez, director of the Fondo Indígena, January 15, 2005.

References

Ardaya, G., and G. Sierra. 2002. *Toward an Inclusive Elections Process: Final Report of Pro Citizens' Participation Consortium.* La Paz: DFID-SIDA.

Blackburn, James. 2000. *Popular Participation in a Prebendal Society: A Case Study of Participatory Municipal Planning in Sucre, Bolivia.* Ph.D. diss., University of Sussex.

Coleman, James S. 1994. *Foundations of Social Theory.* Cambridge, MA: Harvard University Press.

Eggleston, Karen, Robert Jensen, and Richard Zeckhauser. 2002. "Information and Communication Technologies, Markets, and Economic Development." In *The Global Information Technology Report 2001–2002: Readiness for the Networked World,* edited by Geoffrey S. Kirkman, Peter G. Cornelius, Jeffrey D. Sachs, and Klaus Schwab, 62–75. New York: Oxford University Press.

Grover, Varun, Kirk Fiedler, and James Teng. 1997. "Empirical Evidence on Swanson's Tri-Core Model of Information Systems Innovation." *Information Systems Research* 8 (3): 273–87.

Humphrey, John, Robin Mansell, Daniel Paré, and Hubert Schmitz. 2003. "The Reality of E-Commerce with Developing Countries." A report prepared for the Department for International Development, Globalisation and Poverty Programme by the London School of Economics, London; Institute of Development Studies, Sussex, March.

Mirani, Rajesh, and Albert L. Lederer. 1998. "An Instrument for Assessing the Organizational Benefits of IS Projects." *Decision Sciences* 29 (4): 803–38.

Molla, Alemayehu, and Richard Heeks. 2007. "Exploring E-Commerce Benefits for Businesses in a Developing Country." *Information Society Journal* 23 (2): 95–108.

O'Donnell, Guillermo. 1999. "Horizontal Accountability in New Democracies." In *The Self-Restraining State: Power and Accountability in New Democracies,* edited by Andreas Schedler, Larry Diamond, and Marc F. Plattner. London: Lynne Rienner Publishers.

Putnam, Robert. 2000. *Bowling Alone: The Collapse and Revival of American Community.* New York: Simon and Schuster.

Saarinen, Timo. 1997. "An Expanded Instrument for Evaluating Information System Success." *Information and Management* 31 (2): 103–18.

Scott, James. 1985. *Weapons of the Weak: Everyday Forms of Resistance.* New Haven, CT: Yale University Press.

UNCTAD (United Nations Conference on Trade and Development). 2013. *Information Economy Report 2013: The Cloud Economy and Developing Countries.* New York: United Nations.

Part III

The Case Studies

CHAPTER 8

Enacting and Interpreting Technology: Experiences of Aymara Women with ICTs

Overseeing the mystic archeological ruins of Tiwanaku, in the midst of the Bolivian highlands (*altiplano*) at an altitude of 3,900 meters, Aymara women are actively engaging with technology, exploring its meaning, and applying it to their daily lives. Based on their strong cultural identities as indigenous women, they are delving into the "otherness," expressing a strong curiosity and excitement toward learning to use these "machines," often perceiving computers and the Internet as a symbol of "modernity" and confiding that its use provides them with a "sense of belonging" to the rest of society.

During one of the computer training workshops organized by the Organization of Aymara Women of the Kollasuyo (OMAK), several women stood up and gathered together in a small group to discuss with passion how best to interpret the information they had just found on the Internet describing the experiences of the Urus from the small island of Taquile, located on the Peruvian side of the Lake Titicaca.[1] They had just read about the many benefits that the community is deriving from the thousands of tourists visiting the island each year. The process is managed by the local community cooperative, which is charged with overseeing the flow of tourists, including all related income-generating activities, such as housing, food, and sales from the community-owned textile store, and ensuring that the benefits are distributed equally among its members. One older woman commented,

>> How is it possible that our brothers and sisters from the other side of the Lake are able to benefit so much from the tourists? How did they manage to get all these people to buy their textiles and stay in their communities? They don't even have any ruins there, and it looks as if this is a very small island, just like the ones we have here on our side of the lake. We are here in Tiwanaku and have the most important ruins of Bolivia right here in our community, and our families and communities hardly benefit at all! We should all sit down together and think how we could learn from their experience.[2]   <<

While this story attempts to capture the excitement and enthusiasm that indigenous women have for information and communication technologies (ICTs) and the Internet, a series of questions come to mind when investigating the impact of these technologies on their well-being. Considering the dire living conditions and high levels of poverty that the Aymara endure in the extremely arid and cold plateau of the Bolivian highlands, it seems unrealistic to discuss the potential benefits of ICTs for their well-being. These women live mostly from subsistence farming and must negotiate on a daily basis pressing priorities to ensure their families' survival. What role can ICTs play in such a context? To what extent can these technologies enhance their individual and collective capabilities and thus improve their well-being? What are the perceptions of indigenous women vis-à-vis the potential impact of ICTs on their lives? What role does OMAK play in supporting indigenous women in their quest to use ICTs as an instrument to improve well-being? In which dimensions (political, social, and economic) of Aymara women's lives does the use of ICTs translate best into expanding their human and social capabilities and thus have the strongest impact?

The previous chapters have analyzed quantitative data and provided an overview of indigenous peoples' information needs (chapter 4), actual uses of ICTs (chapter 5), proficiency of ICT use (chapter 6), and informational capabilities and perceptions about the Internet's impact on their well-being (chapter 7). The focus has been on a macro-level analysis and the investigation of individual (age, gender) and socioeconomic (ethnicity, poverty) factors that influence the extent to which the uses of ICTs have a positive impact on indigenous peoples' human and social capabilities.

This chapter investigates in greater detail the local socioeconomic, political, and cultural context in which ICTs have been introduced into Tiwanaku and aims to unpack the process by which ICTs were introduced into this community. It complements the data analysis presented in the previous chapters by presenting the findings from qualitative research methods (unstructured interviews and focus groups). The case study offers a more detailed analysis of indigenous peoples' expectations of and perceptions about the impact of ICTs on their well-being and the various barriers they have encountered in relating ICTs to their daily lives.

Based on the first hypothesis presented in chapter 1—that the access to and use of ICTs have to be facilitated by an effective local intermediary organization in order to improve poor people's well-being—the case study investigates the role that an intermediary organization has played in promoting indigenous peoples' uses of ICTs. Specifically, it studies a grassroots-led program coordinated by OMAK in order to study the extent to which a high level of intermediation—the direct involvement of the intermediary organization at the local community level—has been an effective mechanism. Based on the second hypothesis of the study—that ICTs have to be appropriated by local communities in order to enhance their well-being—the case study investigates not only the extent to which indigenous peoples have acquired the knowledge and skills to use ICTs, but also the extent to which these technologies have been locally appropriated by indigenous peoples into their economic and sociocultural context.

The chapter is divided into eight sections: (a) a summary of the case study's principal findings; (b) a socioeconomic profile of the Tiwanaku municipality; (c) a brief description of indigenous peoples' views of poverty and development; (d) a description of OMAK, its main objectives, and a profile of its founder and director Andrea Flores; (e) a description of OMAK's ICT program; (f) the impact of the program on participants' informational capabilities; (g) the impact of OMAK's ICT program on the personal empowerment and social capital of indigenous women; (h) a discussion of the critical constraints on the ability of ICTs to affect the well-being of indigenous women, in particular, in the political and economic dimensions; and (i) a conclusion.

Principal Findings

The chapter argues that the lack of information in itself is a critical aspect of poverty as experienced by indigenous women—a factor shaping their systematic and deep-seated exclusion and discrimination. It illustrates that improved access to information and enhanced ICT capabilities have an intrinsic value for indigenous women's well-being. Similar to the way "literacy" has an intrinsic value for well-being by expanding a person's human capabilities in multiple areas of her or his life, the experiences of Aymara women demonstrate that ICTs can have a very similar direct effect through both (a) the expansion of their "political freedom" to access information and exert their citizenship rights and (b) the improvement of their self-confidence and self-esteem based on the acquired knowledge of how to use ICTs. In fact, the case study shows that the use of ICTs can lead to the direct "personal empowerment" of indigenous women and thus have important positive effects on the psychological dimension of their well-being.

Second, the case study demonstrates that indigenous women perceive ICTs to play a major role in enhancing the social services provided to their communities, particularly in terms of access to and quality of education. Frequently, indigenous women emphasize that the Internet could play an important equalizing role for their communities, in the sense that it has the potential to provide rural youth with the same educational opportunities as urban youth.

Third, the evidence shows that ICTs can play a critical role in strengthening existing social networks and relations among indigenous women. The case study, however, also points to some significant limitations of ICTs, in the sense that they can only expand existing networks and thus strengthen the "bonding social capital" of indigenous women. They facilitate new social networks to a much lesser degree and are, in fact, unable to catalyze the process of "bridging social capital." Thus indigenous women are not able to improve their relationship to the formal institutions of the state, market, and civil society. Their ICT use does not lead to their political or community empowerment, having only minor effects on existing structures of social exclusion, inequality, and discrimination. This major finding explains the very limited effects of ICTs on the political and economic dimensions of indigenous women's well-being.

In the political dimension of well-being, the chapter emphasizes that indigenous women are unable to use ICTs to overcome their systematic exclusion from the decision-making processes at different levels of government, since this instrument cannot alter the deep-seated structural and systemic issues responsible for their exclusion. For instance, while ICTs can play an important role in enhancing indigenous women's capabilities to exert their right to information, they cannot change the fact that although women's organizations can obtain legal status, the Law of Popular Participation (LPP) does not recognize them as legal entities and local governments do not recognize them as legitimate forms of community representation, leading to their de facto exclusion from the participatory processes of local government as established by the LPP (Clisby 2005).

Finally, the chapter illustrates the critical role that OMAK has played in assisting indigenous peoples in the process of engaging and using ICTs. A core finding in terms of the success of this program lies in indigenous women's trust in OMAK's leadership and their perception that the program is responsive to their needs. Due to the decentralized nature of the OMAK centers, the program has involved indigenous women actively in all of its activities, providing them with the necessary space to explore and interact with ICTs within their cultural context, at their own time and pace, and within their own locality. In particular, older women emphasized that OMAK's assistance, as well as that provided by other participants, has been crucial to overcoming their initial "fears" to use computers and to avoid a sense of "embarrassment" from their low educational attainment. Indigenous women emphasize that OMAK is their "own organization," where they can freely express themselves and support each other. In this sense, the project has enabled indigenous women to access a "social space" where they can meet and discuss problems and experiences openly, which has earned the respect of the broader community as a "legitimate space," freeing them from being unnecessarily questioned or harassed by their husbands or families. In this respect, the program plays a very important social role in the community, whereby the women are "empowered" to meet as a group, to share their experiences, and to look for solutions to their common problems.

Socioeconomic Profile of Tiwanaku: A Site of Cultural Richness and a Stagnant Local Economy

Tiwanaku is located in the highlands of Bolivia, just one hour (73 kilometers) from the nation's capital, La Paz. It is best known as the cultural seat of the Kollasuyo,[3] and the achievements of its early civilization across several centuries have made it one of the most important urban centers of the southern Andes. Today, Tiwanaku is the most visited archeological site in Bolivia, with thousands of tourists each day taking short day-trips from the capital to visit its ruins. Few tourists visit the small town of Tiwanaku and its surrounding communities, which have a population of 13,150 and are located next to the ruins. The large majority of tourists leave the site without interacting with the local Aymara community.

The main productive activities of the local Aymara include subsistence farming, dairy production, and fishing, with tourism-related activities playing only a minor role. In spite of Tiwanaku's historical and cultural significance, the local economy—similar to that of many towns in the Bolivian highlands—has stagnated in recent years. The information presented in table 8.1 provides an overview of Tiwanaku's main socioeconomic data, illustrating the extreme poverty rates and overall difficult living conditions of its majority Aymara population in their daily lives.

According to the 2001 census, 97 percent of the population is unable to meet their most basic needs and 74 percent of the people are living in extreme poverty. The data also reveal the extreme lack of access to social services (education and health): the population has, on average, only 4.3 years of schooling and a high illiteracy rate (28 percent). The municipality's generally low level of human development is also reflected in its human development ranking: 189 out of 314 municipalities in Bolivia.

A critical reason for the extreme poverty levels is that the agriculture sector in the Lake Titicaca region has been in crisis since the late 1970s.

TABLE 8.1　Key socioeconomic indicators for Tiwanaku

Category	Indicator
General	
Total population	13,150
Altitude (meters)	3,870
Number of rural communities	41
% rural	92
% Aymara population	95.8
Socioeconomic indicators (2001)	
Unsatisfied basic needs (% of population)	97.1
Human development index	0.537
National ranking (human development index)	189
Life expectancy at birth (years)	61.8
Literacy rate (% of population 15 years and older)	72.4
Annual consumption (US$ per capita)	644.52
Average years of schooling	4.3
Number of health centers	6
Net migration	−12,580
Extreme poverty rate (% of population)	73.5
Primary completion rate (% of population)	77.5
Ratio of female to male primary enrollment	90.8
Infant mortality rate (deaths per 1,000 live births)	65.6
% of births attended by skilled health staff	45.8
Access to potable water (% of population)	34.5
Access to sanitation services (% of population)	30.0

Sources: INE 2001; UNDP, INE, and UDAPE 2004.

This agricultural depression can be attributed largely to three factors. First, the 1952 Agrarian Reform, which resulted in the break-up of the hacienda system and provided the Aymara *campesinos* the freedom to own and cultivate their own lands, also gave rise to the *minifundista* system, which is characterized by very small plots of land. Today, the average family owns a plot no larger than 2–4 hectares (Benton 1999). Second, the green revolution, which was heavily promoted by donor-funded rural development projects in the 1970s and 1980s, aimed to modernize the agriculture sector in the highlands, but also caused severe environmental degradation, which has had an adverse effect on agricultural productivity in the majority of highlands communities (Healy 2001, 2004; World Bank 1998). Finally, many *campesino* families face intense competition in local markets and receive very low prices for their products. Most families are producing the same products (onions, potatoes, and quinoa) and lack access to credit in order to diversify production.

The increasingly difficult living conditions in Tiwanaku are confirmed by the statistical data, which demonstrate that the municipal economy stagnated throughout the 1990s. In fact, the number of people who were unable to meet their basic needs fell only slightly during this period, from 98 percent in 1992 to 96 percent in 2001 (figure 8.1). A comparison with Copacabana, the second most important tourist attraction in the Lake Titicaca region, demonstrates that, while Copacabana has made significant progress in improving living conditions, with the unsatisfied basic needs (UBN) index falling almost 7 percentage points from 1992 to 2001, Tiwanaku has stagnated. Furthermore, in the

FIGURE 8.1   **Persistently high poverty in Tiwanaku, 1992 and 2001**

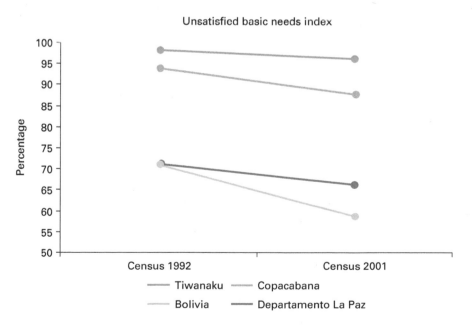

Source: UNDP and INE 2006.

department of La Paz and Bolivia as a whole, the percentage of people unable to meet their basic needs fell 5 and 12 percentage points, respectively, during the same period.

A key reason for this development is that the local tourism sector remains—in spite of its important potential—only a minor income-generating activity for the majority of the local population, as evidenced by the extremely low and stagnant figure for annual consumption per capita of US$644. The sector is dominated by either large-scale foreign-owned companies or medium-scale travel agencies from La Paz. The area's proximity to the capital city has discouraged the development of a cottage tourism industry, since the large majority of tourists return to the capital at night. The lack of infrastructure is leaving untapped demand for lodging, entertainment, and food services. As such, local communities realize only very limited benefits from tourism. According to Juana, an Aymara woman from Achaca, one of 41 Aymara communities in Tiwanaku,

>> For us it is very difficult to advance. I see the tourists come to our community every day and wonder why we can't benefit from them. They all come from La Paz and never stay here longer than a few hours. Our mayor always talks about the great future we will have thanks to the tourists and that we will be able to move ahead in our own development. I don't see it. In our community here in Achaca, we just don't feel anything good will come of it. Most families here don't have access to water or electricity, our schools don't have enough teachers, and there is no health center here at all. For us life is a struggle. I get up every morning at 5 a.m. to feed the two cows I own; since my husband has found a job in El Alto, I also have to work in the fields with the help of my sister who lives with us to help me with my three children. Sometimes, I feel it is too much for me, but I get up and know that we have to fight for a better future for our children.[4]   <<

Juana's observations are confirmed by a statistical analysis of the Millennium Development Goals at the municipal level. The following description provides a brief overview of the extent to which Tiwanaku has progressed in reaching these goals. The main finding is that the municipality has realized only modest improvements in the most critical social indicators and continues to experience difficult living conditions. In effect, the local government has been unable to provide adequate access to social services (health and education) and basic infrastructure (water and sanitation services).

For instance, the municipality has made only marginal progress in access to education. About 20 percent of the girls and boys less than 15 years of age do not complete primary school, compared with 12 percent in the department of La Paz (figure 8.2).

Tiwanaku also has a significant gap in access to health services. There are only six health centers for the 41 communities in the municipality. During a visit to two of the centers, I observed the lack of properly trained health staff and appropriate medications, making it impossible to provide adequate health services.

In spite of significant improvements since the early 1990s, infant mortality rates are still relatively high, with 66 deaths per 1,000 births in Tiwanaku in

FIGURE 8.2   Primary completion rate in Tiwanaku, 2001 and 2005

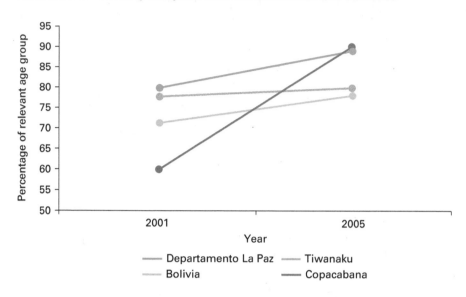

Source: UNDP and INE 2006.

FIGURE 8.3   Infant mortality rate in Tiwanaku and select areas of Bolivia, 1992 and 2001

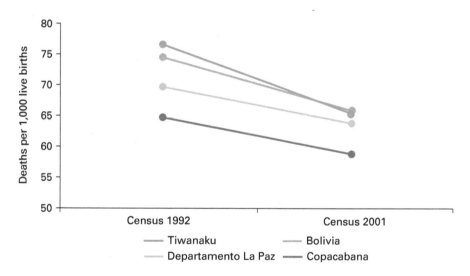

Source: UNDP and INE 2006.

2001. Copacabana has a somewhat lower infant mortality rate of almost 59 deaths per 1,000 births (figure 8.3).

Although Tiwanaku has improved the population's access to basic infra-structure services, it lags considerably behind other municipalities in Bolivia (figure 8.4). For instance, in 2001 an estimated one-third of the population (35 percent) had access to potable water, compared with three-quarters of the

FIGURE 8.4   Access to water services in Tiwanaku and select areas of Bolivia, 1992 and 2001

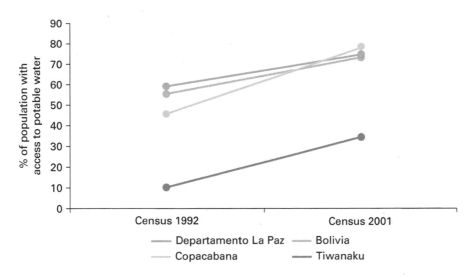

Source: UNDP and INE 2006.

FIGURE 8.5   Access to sanitation services in Tiwanaku and select areas of Bolivia, 1992 and 2001

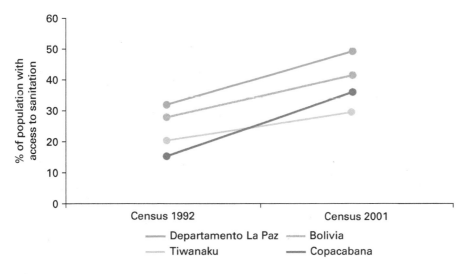

Source: UNDP and INE 2006.

population (77 percent) in Copacabana, 74 percent in the department of La Paz, and 72 percent in Bolivia overall.

The municipality faces similar challenges in improving access to sanitation services (figure 8.5). Although Tiwanaku has improved access to basic sanitation services from about 21 to 30 percent, it lags behind the department of La Paz (50 percent) and other towns in the department, such as Copacabana (36 percent).

FIGURE 8.6 **Access to electricity services in Tiwanaku and select areas of Bolivia, 1992 and 2001**

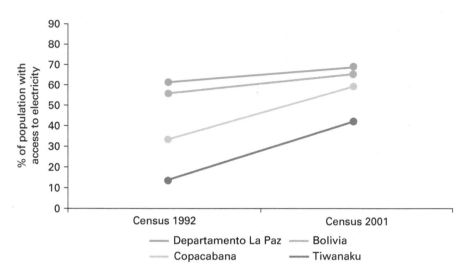

Source: UNDP and INE 2006.

Finally, Tiwanaku continues to have important deficiencies with regard to access to electricity services, with less than half of its population (42 percent) having access to electricity (figure 8.6). Tiwanaku had an extremely low rate of electricity coverage in 1992 (12 percent) and has made significant improvements in this area. Nevertheless, compared with Copacabana (59 percent) and the department of La Paz (67 percent), the population of Tiwanaku continues to have significantly less access to basic infrastructure services than other municipalities.

Indigenous Peoples' Own Views of Poverty and Development

The poverty profile of Tiwanaku is based on conventional quantitative indicators and does not take into account the multidimensional character of poverty. Indigenous peoples often contend that such conventional indicators do not describe their well-being adequately, as they fail to take into account their cultural richness and strong social networks. During a focus group with community leaders, which discussed their communities' information needs, the participants pointed out that official statistics overemphasize the individual dimension of development and do not adequately consider the collective well-being of indigenous communities.[5] Furthermore, indigenous peoples frequently assert that these conventional indicators measure people's well-being in one specific dimension, such as education or health, and fail to take a more holistic view (Haverkort, van't Hooft, and Hiemstra 2003, 258; Zoomers 1999, 89).

In light of this study's central hypothesis—that the intermediary process influences the extent to which an ICT project can or cannot enhance indigenous peoples' well-being—it is critical to understand indigenous peoples' own views about poverty, well-being, and development *before* assessing the impact of ICTs. How do the Aymara see their own development, and how do their definitions of poverty and well-being differ from those used in conventional development approaches? A common criticism voiced by many of the participants is that conventional development interventions do not understand the local sociocultural context or have profound misconceptions about indigenous communities. They do not base their activities, projects, or programs on indigenous peoples' views and aspirations and therefore are perceived to impose preconceived approaches and solutions, predisposing them to failure (Butz, Lonergan, and Smit 1991).

Community leaders frequently define development as a process by which individuals and the community reach a state of *allin qausay* or *suma qamana*— the Aymara and Quechua concept of *living well*. Central to this concept are the social relations a person maintains with his or her family and community, as well as the overall relationship between humans and nature. According to this worldview, a person reaches a situation of "well-being" when she or he lives in peace and harmony with family members, the community, and the natural surroundings (Haverkort, van't Hooft, and Hiemstra 2003, 258). For the Aymara, the concept of poverty has a different meaning than the income-centered Western concept, so that a person is considered poor when he or she no longer has social relations with his or her family and community or has lost spiritual contact with the natural environment. Considerable emphasis is placed on the interrelationships between the individual, the community, and the environment. In Aymara "to be poor" is *wajch'äta*, literally translated as "to be orphaned." This concept stresses the importance of kinship, social relations, and belonging to a community, whereby material aspects of poverty are inconsequential (Widmark 2003).

This notion of well-being emphasizes the spiritual dimension of life and is based on the principle of "dualism," a defining element in the Andean worldview (Albó et al. 1989). According to this principle, the universe, people, and all aspects of life are divided into complementary parts (man/woman, sun/moon). The harmony of life and its balance depend on a state of equilibrium that is set through the interrelationship between the halves. This complementarity or *ayni*—the interchange of strength and energy—regulates most aspects of community life, such as marriage, local political decisions, and community work. A key underlying concept is that humans and the natural environment are closely linked with each other through the spiritual world.

Given this background, the research first aimed to gain a better understanding of indigenous peoples' own conceptualization of well-being before analyzing the effects that the OMAK ICT program has had on their well-being. Thus I carried out several focus groups with indigenous leaders in order to solicit their views about what development and poverty mean to them and their communities. Many participants in the focus groups emphasized that they do not perceive themselves to be poor. A frequently voiced statement was, "*Tampoco somos tan pobre*" ("After all, we are not that poor").[6] They highlighted that conventional development programs give too much importance to the material aspects of well-being and portray their communities as being deprived, weak, and marginalized, while they view their communities as being rich in social relations, kinship ties, indigenous knowledge, and culture. One participant expressed her feelings about poverty in the following way:

>> All of us are poor. We face a lot of struggles and have to make a lot of sacrifices to make a living and to be able to have enough money to send our children to school and to pay for the most basic things we need to live. But just because we have no money, does this really mean that we are completely poor and need to be ashamed of ourselves in front of others? ... I think many young people see it like this, but we have to also recognize that we have a lot of things we can be proud of: our community is strong, we don't have a lot of conflict with other families, and we all help each other, if one person falls ill or has lost everything. We can be proud of our culture and our traditions. We do not have to be ashamed only because we don't have all these material things like the people in the city.[7]   <<

At the same time, many participants recognized their difficult living conditions, particularly due to the lack of access to land and social services. The existing plots are usually too small to cultivate enough food to ensure a family's subsistence. Thus almost every family needs to engage in nonagricultural activities—mostly temporary or permanent work in the nearby city of El Alto or La Paz or even abroad—in order to make a living. One participant expressed this concern,

>> I am quite worried about our communities, since all the young people want to leave their homes and search for a better life in the city. It wasn't like this several years ago, when we were still able to produce enough food on our fields, so we could eat well. The main problem is that the land will never grow, while our families do. We have depended for generations on our lands, but now it is not enough anymore. I feel that we have to produce today four times as many onions as we did a couple of years ago, in order to get the same money we need to live. To work in the fields does not give us enough to feed our families anymore. I just wish that our children would not have to leave our communities to search for work somewhere else and that we were able to improve our production on our fields so it is enough for all of us.[8]   <<

In addition to the problems stemming from the *minifundio* system, many farmers mentioned frequent droughts and extended periods of severe frost. Furthermore, agricultural production throughout the region suffers from acute soil erosion. Despite these problems, most participants attributed their poor agricultural production to the lack of competitive markets in which to sell or exchange their products:

>> In Tiwanaku, we don't have as much access to our own indigenous markets, as our brothers have in Jesus de Machaca.[9] I have seen how our brothers from Peru go to their local markets, and they carry textiles, cheese, and *charque* and exchange it for onions, potatoes, and corn. Everybody benefits from this *trueque*[10] there, while we are forced to sell our products to the traders who come to our markets, and we receive very little money for our products. I think our main problem is that we are missing the right markets for our products.[11]   <<

The majority of men tended to emphasize the importance of land, agricultural production, access to credit, and the need for more productive investment in their communities, while the majority of women stressed the vital need to improve access to social services, in particular, education and health care. One older Aymara woman from Tiwanaku pointed to the critical importance of education for their development, as follows:

>> When I was a girl, my parents wouldn't let me go to school. Still today, I suffer from this, since I hardly know how to read or write. I am so glad that this has changed today and that all of our children, also the girls, can go to school and advance in their lives. I think about how many more things we could have done in my generation if we had had the chance to go to school. For me, to improve the education in our communities is the most important issue, so that our children can advance and not have the same problems we have experienced.[12]   <<

Participants frequently mentioned four educational issues: (a) inadequate training of rural teachers, (b) lack of access to education beyond the third or fourth grade, (c) failure to implement bilingual and intercultural education programs, and (d) lack of adequate infrastructure in schools:

>> The teachers are not very well qualified here in our communities. Frequently, they don't even speak or understand Aymara and don't know anything about our culture. I also feel that they do not like teaching since they frequently either come too late or leave early. We have to do something to change this; otherwise our children will not learn anything.[13]   <<

In relation to health services, participants emphasized that they rarely visit the health posts in the municipality, preferring instead to use traditional medicine, which is readily accessible in their communities. The majority of the Aymara go to health posts only for severe illnesses. These health posts have inadequate resources, including improperly trained health professionals, poor equipment, and undersupply of medications, which reinforces the distrust of their services. Thus, while the participants stressed that health is the most important dimension of their well-being, they also expressed strong distrust of Western medicine and denounced the municipality's failure to improve health services.

Finally, a commonly held perception is that the central government has abandoned or forgotten them and their communities. Participants frequently described their situation as having been "marginalized" and isolated from the rest of society and their voices left unheard by key decision makers. As a result of this state failure, they continue to lack access to the most basic services (electricity, water, and sanitation), while the quality of social services (education and health) they do receive is very low. Rosa Morales, OMAK president, encapsulated the views voiced by many:

>> Our main problem is not that we are poor … . We are used to living under difficult conditions and to working hard in order to make a living. No, our biggest problem is that the rest of society has impoverished us … . The terms under

which we have to sell our goods, for instance, are not equal; the traders from the city always dictate the price ... . There is just no equality. For us, Aymara *campesinos*, everything just seems to be much more difficult, even to go to the court and to obtain a birth certificate takes several days for us. The government doesn't seem to listen to our needs or demands or doesn't seem to care about us. In the meantime, everybody else continues to treat us as second-class citizens. We have to come together and organize ourselves to improve our living conditions. Nobody else will do it for us![14]   <<

OMAK: The Grassroots Aymara Women's Organization of the Kollasuyo

OMAK is a grassroots women's organization that works to improve the well-being of Aymara women based on their own terms—that is, through their worldview, culture, and values. A core goal of the organization is to promote the dignity and well-being of women, while seeking to strengthen mutual respect and harmony between women and men. The organization stresses the need for self-determination and has based all of its activities on a model of autonomous indigenous development. Holding true to its approach, the organization is managed exclusively by indigenous professionals, with indigenous women holding key leadership positions. As such, OMAK works to empower indigenous women at the grassroots level, so that they may (a) define their own development needs and priorities, (b) make their own decisions about which activities to engage in, and (c) take on leadership positions in the organization and lead project implementation. OMAK was founded in 1983 and obtained its legal status as a grassroots organization in 1991.

Andrea Flores, director and founder of the organization, is a charismatic, professional Aymara woman *de pollera*.[15] She was born in Tiwanaku and migrated to El Alto to pursue her studies in communications at the Universidad Mayor de San Andres in La Paz.[16] As the majority of Aymara women or *cholas*[17] in El Alto, she has maintained her strong links with her rural community of origin, and her identity is very much anchored by traditional Aymara values. Her own life story encapsulates the multiple barriers and forms of discrimination that Aymara women experience when confronted with the mostly *mestizo*-dominated urban society of Bolivia. She said,

>> For me as an Aymara woman, I know what it means to struggle for one's own rights and to be confronted with the daily forms of neglect and rejection from the rest of society. Most people—although they say that Bolivia is a multicultural society—deep inside have strong prejudices against us. Wherever you go, you can find these barriers. It sometimes feels that you are running against a wall, and if you are alone, you have great difficulties to overcome these barriers.[18]   <<

Thus OMAK's main objective is to strengthen the capacity of Aymara women to organize themselves and to form small autonomous women's groups at the grassroots level. OMAK has organized over many years a network of women's centers in eight municipalities—Achacaci, Tiwanaku, Palca, Luribay, Inquisivi, Corocoro, Chonchocoro (Viacha), and Curahuara de Carangas—coordinating their activities from the central office in El Alto. This network expands its reach throughout seven provinces, mostly in the department of La Paz and one in Oruro, including approximately 400 members. Each center has its own legal status, organizational chart, and managerial independence; women are in charge of their own projects, and the central office plays a facilitating role. All key decisions are made locally either by the independent centers or by the network as a whole. Andrea Flores explains the rationale behind this approach:

>> Early on during my university years, I started to understand that it is much harder for us as indigenous women to achieve the things we would like to do— so I decided to first become active in the student movements, and I went on the streets to ask for radical change, ... the one we so urgently need. I almost became very radical and thought that we needed a socialist revolution ... . Only much later I understood that what we really need is to organize ourselves as women and to have a space for ourselves to express our frustrations, anger, but also our dreams and aspirations, to help each other in our daily struggle to feed our children and to overcome all these different forms of discrimination in our society, which we all share. Working together as Aymara women, I thought, would help us to learn from our common experiences and provide us with more strength to face our daily struggles for a better life.[19]   <<

OMAK's most significant asset is its strong roots in the Aymara culture and its anchoring and constant reaffirmation of core principles and values from the *ayllu* system.[20] This organization espouses the importance of mutual respect between men and women and upholds the Andean concept of dualism with the aim of reaching *chachawarmi*—a holistic state of harmony between men and women within communities and nature.[21] OMAK advocates the strengthening of the Aymara culture by recovering the lost balance within families and communities attributed to the multiple dependencies and forms of oppression since Spanish colonization.

Based on these cultural values, the mission of OMAK is to support Aymara women in their struggle to overcome all forms of ethnic discrimination, racism, and social exclusion, with the ultimate aim to establish a pluricultural, multiethnic society with equal opportunities for all. OMAK aims to reach these objectives by (a) establishing a network of autonomous women's centers, (b) strengthening the capacity of indigenous women to exert their right to information and to become more knowledgeable with regard to the political and social issues affecting their lives, (c) providing capacity building to Aymara women on topics related to leadership, human rights, community organization, and productive activities, and (d) supporting the women's centers in the design and implementation of social and productive projects. Andrea Flores described the challenges in reaching these objectives:

>> Several years ago, we had an important restructuring of our organization, since too many of the decisions were taken at the central level. After we found out that we had serious problems in implementing our projects in several communities, our board of directors came together, and we had a long and intensive discussion about what went wrong with our projects and how the women of the centers

would have to take on more responsibilities for the projects that they were involved in. Well, it was not easy for all of us, but I think we learned our lesson, and now all the centers have their own president and decision-making processes, without the center interfering in this process. It is, of course, not always easy to accept the priorities of the local base. However, we realized that this was the only way to achieve our mission and to enable Aymara women to have their own places and spaces, to take their lives into their own hands, and to overcome all the different forms of oppression and dependencies.[22]   <<

This quote illustrates the challenges in leading an authentically grassroots organization. The leadership has a responsibility to maintain integrity in the balance of power and decision-making processes, to ensure that these remain in the hands of its members, and to see that the center does not assume a dominant role. This balance is particularly vulnerable during an organization's period of growth, when the center's coordinating role may become usurped by the need to disburse funding quickly and to realign and expand its activities. Such rapid evolutions do not provide the necessary platform for the often lengthy processes of consultation that need to take place in order to uphold the grassroots nature of the organization—a trial of sustainability.

OMAK's continuity since 1983 and its broad-based and growing membership are a strong indication that the organization has been able to balance the decision-making processes between the center and the local branches and to navigate the complex local sociocultural landscape. Furthermore, the fact that the average OMAK member has been an active network volunteer for more than five years validates OMAK's objectives and vision and affirms the strong commitment of members to the organization. One woman expressed her solidarity with the network, sharing the following:

>> OMAK is different than other organizations. We women feel that this is our own organization. It has helped us to find a space of our own, where every woman, no matter her age or education, can freely express herself and does not have to be embarrassed or ashamed … . When I first joined the meeting, I was so afraid to speak that I only listened to the conversation. Somehow, all my life I was told to listen and felt that I had nothing to say … . Every decision in my life was taken by either my parents or my husband … . Only now, after having participated in our women's group for two years, I have seen that other women share the same experience … . I have the feeling that I can talk as well and share with them my problems and dreams for a better future.[23]   <<

Thus OMAK has maintained its grassroots character and responsiveness to a deep-seated need, providing indigenous women with a space of their own in which to express and share their experiences, problems, and aspirations. A central aspect of the organization is the decentralized network of autonomous women's centers at the community level. Based on the strong demand for more exchanges of information and experiences, as expressed by many women in the centers, creating an information network through the use of ICTs has become an important priority for OMAK.

OMAK's ICT Program

Since 2000, OMAK has focused on the innovative use of ICTs as a conduit for empowering indigenous women and supporting them in their daily struggle to overcome multiple forms of racial discrimination and social exclusion. The key objective of its overall program is to strengthen indigenous women's human and social capabilities in the areas of human rights, indigenous leadership, and the management of small-scale productive activities. The role of the ICT capacity-building program was complementary in nature and meant to reinforce the other training modules, rather than focus exclusively on the field of ICTs themselves.

At the inception of the ICT program in early 2000, the large majority of participants had never before used a computer, and some had only seen others using them. In the first phase of the program, OMAK organized a "computer course for indigenous leaders" in its El Alto headquarters. In this first course, 15 female leaders from the network of women's centers were trained in basic computer and Internet skills. This pilot unleashed such strong demand from network members that OMAK decided to expand the program in order to reach a much larger number of women, particularly at the grassroots level. This much more ambitious education and ICT program aimed to (a) develop an information network among all of its centers, (b) strengthen its institutional capacity to use ICTs, and (c) expand the reach of its ICT capacity workshops to the women's centers at the grassroots level (Bustillos Rodríguez 2003). Based on this concept, OMAK obtained funding from a Dutch foundation to construct its first women's center in Tiwanaku, which would serve as a common meeting place for indigenous women to organize educational, informational, and productive activities.

Furthermore, the organization secured six months of volunteer support from an intern sponsored by the Organization of American States Trust of the Americas Program, who worked on the program's activities during 2003. Based on the strong individual agency and the determination of the women in Tiwanaku, the construction of this women's center was completed in 2004—with most of the labor supplied by the women and their families, minimally supported by paid construction workers. Soon after the inauguration of the Tiwanaku center, OMAK initiated a series of training workshops, with the curriculum including ICTs.

Impact of the Program on the Well-Being of Indigenous Peoples

This assessment of the program draws on several years of experience with indigenous women's ICT use. I first provide an overview of the program's impact and then outline indigenous peoples' own perceptions of its effects on their individual and collective well-being. I chose the Tiwanaku center because it has the most extensive experience with implementing an ICT project with indigenous women. Several findings are also related to the effects of the ICT program on OMAK's overall network.

The empirical results are based on a subsample of the impact user survey (n = 23), which focused exclusively on OMAK participants. The outcomes from the empirical data illustrate that the OMAK respondents perceive the Internet to have had a positive impact across all the dimensions of their lives (figure 8.7).

First, OMAK participants perceive the Internet to have had a substantial impact on their social (PII = 1.70), personal (PII = 1.65), and economic (PII = 1.52) well-being. In the social dimension, they believe that the Internet has had a high impact on their communities' access to education and health services (PII = 1.70). They also said that the Internet has had a high-intermediate impact on their organizational (PII = 1.48) and political (PII = 1.39) well-being. Thus they see the Internet as supporting indigenous peoples in both their personal and their social (organizational, political) development—both core objectives of OMAK.

An unexpected finding lies in the fact that OMAK participants are considerably more positive about the Internet's impact than the overall sample in five of the six dimensions of well-being. However, the difference is particularly marked in the social, economic, and political dimensions. For instance, in the social dimension, the perceived impact index for the OMAK respondents (PII = 1.70) reaches a value that is 0.26 higher than the one of the overall sample, where it reaches an intermediate level (PII =1.44). In both the economic and political dimensions, OMAK participants view the impact of the Internet more positively than the respondents from the overall sample, with the perceived impact index reaching a value that is 0.25 higher for the economic and 0.23 higher for the social dimension, respectively. The cultural dimension is the only area in which the perceptions of OMAK participants coincide with those of the overall sample, with an intermediate score for perceived impact of 1.26 for both groups.

FIGURE 8.7   **Perceived impact of the Internet on various dimensions of well-being among OMAK participants**

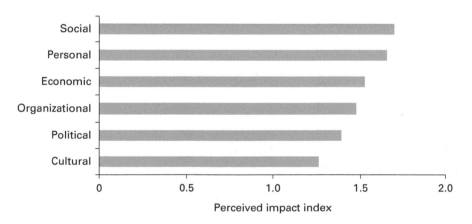

Perceived impact index

A second major finding is that, different from the overall results, OMAK participants have a more positive view of the Internet's positive effects on both their individual and social capabilities.

In order to examine the effects of the Internet on various dimensions of people's lives, I use the same set of indexes as developed in chapter 7. Respondents have a particularly positive view about the Internet's impact on (a) the strengthening of social capital within their communities, (b) their personal empowerment, and (c) the overall well-being of their communities (figure 8.8). With respect to social capital, the score for perceived impact is high (PII = 1.65), indicating that the respondents believe that the Internet has had a substantial impact on this area of their well-being. Since OMAK is organized as a network of grassroots women's centers, the strengthening of social capital, for both its individual participants and their communities, is a central objective. Thus OMAK respondents hold a more positive view than the overall sample of respondents (PII = 1.43).

Furthermore, OMAK participants believe that the Internet has had a higher impact on their personal empowerment (PII = 1.57) than the overall respondents (PII = 1.38). The same is true with regard to the impact on community well-being: the majority of OMAK respondents said that the Internet has had a high impact on their communities' well-being (PII = 1.52), while the overall sample rated the impact as intermediate (PII = 1.37).

FIGURE 8.8 **Perceived impact of the Internet on specific dimensions of well-being among OMAK respondents**

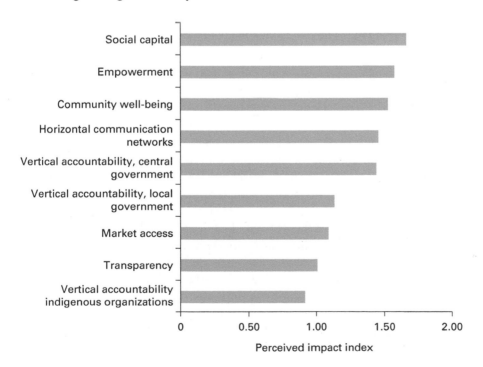

Perceived impact index

Respondents are less positive with regard to the Internet's impact on (a) improved access to markets, (b) improved accountability of indigenous organizations, and (c) enhanced government transparency. Their view that the Internet has had a small impact on their ability to access markets (PII = 1.09) may reflect the barriers, such as lack of access to capital, that make it difficult for indigenous peoples, particularly women, to commercialize their products.[24] It is important to note that the program did not seek to use the Internet for improving participants' economic opportunities. The relatively negative view about the Internet's impact on the accountability of indigenous organizations (PII = 0.91) is a major unexpected finding that seems to point to the increasingly complicated nature of the relationship between the indigenous leadership and communities at the grassroots level, whereby women continue to be, by and large, excluded from leadership positions within most of the national and regional indigenous organizations.[25] This finding seems to indicate that indigenous women not only face major obstacles in gaining the right to participate in the decision-making processes within the formal state institutions (local governments), but also are largely excluded from exercising their voice and choice within the traditional indigenous power structures.

Finally, the data reveal that the OMAK respondents have a less positive view about the Internet's ability to improve the transparency of government institutions, as evidenced by the low perceived impact index of 1.00. These figures indicate that indigenous peoples perceive the Internet to have had a limited impact on their ability to access government information.

An Integrated Approach to ICTs

How was the project designed and implemented, and to what extent did ICTs form an integral part of OMAK's overall program activities in Tiwanaku? The analysis seeks to distill whether indigenous women were able to explore the meaning and application of technology on their own terms and to appropriate it to their locality's sociocultural context and relevant needs.

The program initiated its activities through a detailed information needs assessment, which provided participants with a space in which to reflect about the meaning and significance of information and communication for their individual and collective well-being *before* engaging in the use of ICTs. The two principal objectives of the assessment were to (a) provide indigenous women, *mallkus* (community leaders, who are all men), and youth with separate spaces in which to define their own information needs and priorities and (b) facilitate a process in which participants were first engaged in "reflecting" on the role that information plays in their lives and communities. Thus the program significantly contributed to raising the "critical awareness" of indigenous peoples with regard to both information itself and the extent to which the right to information constitutes an important aspect of political freedom.

During my fieldwork, I had the opportunity to attend four of these workshops on two occasions. The focus groups were facilitated by local indigenous professionals and conducted in indigenous languages and Spanish. I was impressed by the fact that the large majority of participants were actively engaged in the discussions and demonstrated a high level of energy and enthusiasm, particularly the women and youth. At various times during the discussion, participants expressed their sense of "isolation" from the rest of society and deliberated on the fact that an important aspect of social exclusion can be attributed to the lack of information and the absence of communication channels. One of the participants summarized this feeling as follows:

>> For us in our communities, it is quite hard to know what is happening in the rest of Bolivia. We are quite isolated. We don't have access to newspapers or television. The only way we can be informed is by listening to our radios, and usually we only listen to them early in the morning before we go out to work in our fields. Also we cannot always trust what they say. Sometimes they try to confuse you. It is quite hard for us to be informed about anything or to form our opinions. Now, for example, there is a lot of discussion of the Constitutional Assembly, but how should we know whether we support it, when we have so little information about it? I wish we could learn more about these issues and be more active.[26]   <<

The needs assessment included a mapping exercise to discern the existing information and communication channels and to identify some of the key barriers to improving access to information. An intriguing finding is that most indigenous women feel that indigenous men have much easier and better access to information than they do. In particular, the local *mallkus*, who are all men, were identified as the communities' key recipients and keepers of information, with most women of the view that they do not share their information with other community members.

>> Look at our husbands. They always come together in order to chat with each other. After they come back from the fields, they can just leave the home and go to the plaza in order to meet with their friends. There they talk about all the things that are happening in our community. The *mallkus* even have their own office in the municipality, and they always receive a lot of information from the mayor on different projects he is currently planning for the communities. For us women, it is much more difficult, since we spend most of our time at home working in the kitchen or taking care of our children. My husband also selects the programs we can listen to on the radio at home. Sometimes, he also just takes it away, and then I have nothing. I often really feel isolated from the rest of our community and the world.[27]   <<

Another key finding from the needs assessment is that indigenous women, in particular, have very high expectations about the role that the Internet can play in enhancing their access to information and in improving their well-being. Specifically, in the areas of education and health, 85 percent of the women surveyed indicated that they believe that the Internet can provide them with relevant information.

Finally, participants expressed a high need for improved communications with family members and friends who have migrated either to the urban centers of La Paz, Cochabama, or Santa Cruz or abroad to Argentina,

Spain, or the United States. The youth group, in particular, described the difficulty involved in communicating with their family members and friends who have left Tiwanaku, citing the lack of reliable, affordable telephone services.

A critical aspect of the information needs assessment was the process itself, which enabled indigenous peoples to reflect through small group discussions on the role that information plays in their lives. During the discussions, many participants gained a much better understanding of the existing information ecology and the barriers to the flow of information in their communities. In particular, indigenous women described in great detail their lack of access to information and their very limited ability to communicate with the outside world. The methodology used by OMAK is based on Paolo Freire's concept of "critical conscientization," whereby through a process of reflection and action (praxis) the oppressed are first made aware of the economic, social, political, and cultural reasons for their oppression and then identify concrete steps to take action against the oppressive elements in their reality (Freire 1972). The process of defining their own information needs and priorities thus constituted an important aspect of the overall program, significantly contributing to the empowerment of indigenous women and meeting one of the program's key objectives.

A second critical finding is that the women's groups and OMAK's leadership perceive the ICT project to be an integral part of their overall program and not an isolated activity. In this way, the project was based on the organization's overall rights-based approach, which aims to empower indigenous women to gain greater control over their lives and to support them in acquiring new skills and improving their individual and collective well-being. Therefore, the ICT project has to be viewed in the broader context of OMAK's human rights and advocacy work, described above. This is a substantial advantage of the program, since it allowed ICTs to be embedded in the overall objectives of the organization, providing a clear rationale and a sense of meaning for its uses. This integration helped the organization to avoid a common mistake—overemphasizing the technology itself—and provided a much broader platform for discussing and reflecting on issues related to the right to information and inequalities in information flows within communities. For instance, one of OMAK's activities relates to indigenous women's health and nutritional education and their right to improved health care. This subject was integrated into the ICT project, and participants were taught how to search for and access information related to women's health, child nutrition, and pregnancy nutrition.

Impact on Informational Capabilities

What impact did the program have on indigenous women's informational capabilities? The project did not include just women; it also included men and youth, with separate workshops to provide women and youth with a secure and inviting space in which to express their opinions and viewpoints freely and to give men a chance to participate and gain their support. The program focused on the catalytic role of information in indigenous peoples' development, so that the training courses in ICT skills formed part of a much broader curriculum, including information and well-being.

The first finding is that the program participants not only attained advanced ICT capabilities but also achieved a certain level of information literacy, communication capabilities, and ability to share their own local knowledge and experiences with others, attaining advanced informational capabilities (figure 8.9). The data indicate that OMAK participants have reached relatively high overall informational capabilities (ICI = 1.48) with communications capabilities receiving the highest value (1.63).

Moreover, as shown in figure 8.10, OMAK participants have obtained a more advanced level of informational capabilities (ICC = 1.48) than the participants of the overall sample (ICC = 1.36). Only participants in the two CIOEC (Coordinadora de Integración de Organizaciones Económicas Campesinas

FIGURE 8.9 **Four dimensions of informational capabilities among OMAK respondents**

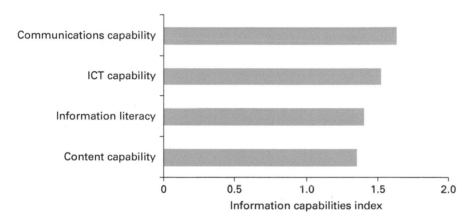

FIGURE 8.10 **Perceived impact on informational capabilities, by program**

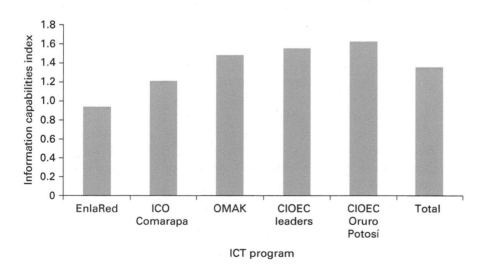

Indígenas y Originarias) programs said that they have attained more advanced informational capabilities (for CIOEC leaders program, ICC = 1.56; for CIOEC Oruro and Potosí program, PII = 1.63).

This finding is noteworthy given OMAK's challenging local context of high rates of poverty and illiteracy. In spite of these difficulties, the program has demonstrated that even the poorest and most excluded groups can attain advanced informational capabilities, enabling them to make meaningful use of ICTs. In order to validate this finding, the rest of this section provides a more detailed account of indigenous women's experiences when engaging with technologies. It is based on participatory observations made during several ICT training workshops, as well as on selected excerpts taken from 14 semi-structured interviews with project participants. The subsequent description examines the extent to which its participants have attained advanced levels in any of the four dimensions of informational capabilities, as defined in the theoretical framework: (a) ICT capabilities, (b) communications capabilities, (c) information literacy, and (d) capability to produce and share local knowledge and experiences with others.

ICT Capabilities

In terms of ICT capabilities, many participants voiced their initial fear of using technology and their concern with "breaking" or "damaging' the computer. One participant expressed her anxieties in the following way:

>> The computer was at first something completely new for us. Most of us had only seen computers from far away, and we had never seen it close up. I remember that I was at first very afraid even to touch it. Of course, I was curious since I have seen the people in El Alto use it in the computer centers. But this was different. Suddenly I was sitting in front of this "machine," and I was just really afraid to break something or to make a mistake. I remember that I was so embarrassed, when I made a mistake and suddenly the computer turned off.[28]   <<

Many described how they slowly overcame their fears through the encouragement of other women in the group. A critical factor in reaching the comfort level needed to learn how to use the computer was the establishment of a "secure" and "nonthreatening" environment—the women in this group had a long-established relationship with one another as OMAK members, and they all had known each other for several years, having worked together on other projects. It is particularly important for indigenous women, who frequently suffer from low self-esteem due to the multiple forms of oppression suffered in their lives, to have the support of a peer group. The inviting setting provided by the workshops was central to the success of OMAK's training programs. Another aspect of women's insecurities and fear of technology is related to their low formal educational attainment.

In one training workshop, I observed how an Aymara woman around 56 years of age learned, with the help of her daughter, how to type her name and the name of her community on the keyboard. It was humbling to see how this woman, in spite of her limited formal education, persevered and managed to complete the task. Thus a major unexpected and positive outcome of the project has been its support of several "functionally illiterate" women by teaching them the basics of reading and writing. In this sense, ICTs have played an important catalytic role in enhancing their human capabilities. This experience was described by most women as an extremely empowering one, as it enabled them to use the reading and writing skills they rarely had an opportunity to use. In these contexts, computer courses can be extremely motivating for people of all ages.

Several women expressed how much they value learning how to perform basic word-processing skills and to use the spreadsheet application Excel, which they think is easily applicable in their daily routines. One woman spoke about the usefulness of the computer:

>> Once we had overcome our original fears to use computers, we realized in our group that it was actually not so complicated to use them. I found it most useful to first learn how to use basic programs, such as Excel, which allows me to make all my calculations and the budget for the center. One thing we all found extremely helpful was to learn how to write our own documents and to print them. We also learned how to write our own CV [curriculum vitae] in the computer and to save it on a disk. This was one of the most useful things I have learned.[29]   <<

This quote indicates that indigenous women value basic ICT skills more than learning how to navigate the Internet, as they can apply these tools directly to their daily lives (for example, using Excel to perform financial calculations for the women's group budget).

As shown in figure 8.9, participants have obtained advanced levels of ICT capabilities in general, with the ICT capability index reaching the relatively high value of 1.52. This finding is noteworthy given the low levels of educational attainment. Participants' ICT capabilities and education levels, however, are not associated with each other. This finding is confirmed by the Kruskal-Wallis test, which shows that there is no statistically significant relationship between the participants' ICT capability and their education level ($\chi^2 = 2.828$; $p = 0.243 > 0.05$).

Communication Capabilities

Participants mostly use the Internet for communicating and, to a lesser degree, for conducting searches or research. OMAK participants said that the Internet has improved their communication capabilities, with the communications capability index reaching a value of 1.63—the highest value of all the components of informational capabilities. Furthermore, even participants with relatively low ICT capabilities said that they have attained advanced communications capabilities. In fact, the Kruskal-Wallis statistical test confirms that there is no statistically significant relationship between the participants' score for impact on ICT capability and their score for impact on communications capabilities ($\chi^2 = 1.183$; $p = 0.554 > 0.05$).

As confirmed in the following quote, the program has had a very positive effect on the ability of participants to communicate with family members and friends who live outside the community:

>> For me the most important skill I have learned in the course is how to read and write e-mails. Now, I can write frequently to my son who lives in Argentina, and our communication has really improved. The main problem with the telephone is that it is too expensive and not very reliable, since the operator does not always pass on the messages to my son. Today, I use the Internet once a week in El Alto, after I have been to the local market. This works quite well and is not so expensive.[30]   <<

Information Literacy Skills

Many women were interested in finding information on Aymara culture and history to use as a reference point in their children's education. They were pleased to find a site called Aymaranet, which contained rich material on Aymara culture, history, customs, and poetry. Remarkably, this site is produced by an Aymara professional residing in Washington, DC, who uses this space to maintain a strong social network with other Aymara migrants within the United States. Although it originates in a location far from Tiwanaku, the site generated the most interest among the women and is perceived as being most useful.

This raises the issue of the extent to which formal education is a prerequisite for indigenous peoples to make meaningful use of the Internet. One of the key lessons from the OMAK experience is that indigenous peoples benefit most from Internet use when it is accompanied by a capacity-building program that simultaneously strengthens their information literacy skills. In fact, most of the difficulties that indigenous peoples encounter in using the Internet are related to the analysis and interpretation of information rather than to use of the technology itself. As shown in figure 8.9, participants have reached intermediate capabilities with regard to information literacy—with an information literacy index of 1.40. Although this value indicates lower impact than on both communications capabilities (1.63) and ICT capabilities (1.52), it is still significantly higher than the ranking among the overall participants (1.28).

Participants have improved their information literacy skills by discussing the information and services found online with other participants in small groups. Thus a key finding from the OMAK experience is that for vulnerable groups such as indigenous women, the assistance of an intermediary organization is essential to ensuring that users acquire the necessary information literacy skills—a prerequisite to maximizing Internet use. This finding is contrary to the frequent assertion in the ICT literature that the Internet leads to a process of "disintermediation," as it puts people directly in contact with each other and eliminates the need for a facilitator or mediating agent. This assertion does not hold true for indigenous women, who require the hands-on support and guidance of an intermediary. They frequently lack the formal education, information literacy skills, and the self-confidence to interact with technologies in a meaningful way on their own. An inhibiting factor in their use can be attributed to the Internet's design, which is based on "Western" norms and values from the "developed" countries and is therefore rather "alien" to indigenous women. The encumbered first experience with the "otherness" inherent in this technology may hamper future interactions and further explorations to applying its potential meaning to their own lives. Indigenous peoples who do not receive assistance often become quickly disillusioned about the Internet's usefulness and, as a consequence, stop using it after a short period of time.

In spite of these barriers, the program has shown that indigenous women are able to derive benefits from Internet use if they receive support. The program also has shown that indigenous women can learn quickly how to use the Internet in a meaningful way once they have acquired the basic information literacy skills necessary to find, interpret, process, and share information. According to one OMAK participant,

>> First, I was very confused about the large amount of information available on the Internet. It is very difficult to understand how to find the things you are looking for. But, then, with the help of our instructor and by doing it together in small groups, we learned how to use it. In a way the Internet is like a newspaper or a TV [television]. We cannot believe everything we read or see. Many people outside of our communities don't know anything about our culture, customs, and beliefs. So it is not a surprise to me that the information we find on the Internet is also like this. We just have to be aware that not everything is true we will find there.[31]   <<

Content Capabilities

The fourth dimension of informational capabilities is the ability not only to consume but also to produce original content and share it with others. A key contribution of an intermediary organization is to support indigenous peoples in their production of information based on indigenous knowledge and local experience.

This issue is particularly important for indigenous peoples due to (a) their strong demand for local information and (b) the fact that, for most rural communities, no local Internet content is available. As such, the lack of local and relevant content signals another critical role in which intermediary organizations can play a support function, with the "content gap" representing a substantial barrier limiting indigenous peoples' benefits from Internet use. OMAK has taken on this role by pointing participants to the few websites and networks with content developed and populated by indigenous professionals. Participants identified this aspect of the workshops as being particularly valuable to them. Furthermore, OMAK initiated the construction of its own website through Aymaranet, aiming to provide a space for members and the organization to create their own content and share experiences with other indigenous peoples and the general public. Finally, the program has developed a partnership with the Red Indígena, an initiative of the Fondo Indígena that specifically supports indigenous peoples in (a) producing and sharing local content and (b) establishing a strong virtual network that enables them to exchange their experiences and knowledge more easily with one another.

Nevertheless, the statistical analysis reveals that OMAK participants have reached the lowest capabilities in this dimension of informational capabilities, with the content capability index reaching a relatively low value of 1.35 (as shown in figure 8.9). This finding is also confirmed by the qualitative data. In fact, one of the most important barriers identified by women is the Internet's lack of relevant local content. Once they had the opportunity to surf the net, most of the participants, irrespective of gender, were interested in finding information on their community. During several training workshops, I observed the same phenomenon time and again. The first piece of information that the indigenous participants would look for was of a local nature and related to their community and culture. Frequently, they expressed disbelief or a sense of frustration when they were unable to find information about their locality. In the case of Tiwanaku, the women were frustrated about the lack of information applicable to their productive activities and livelihoods:

>> I was first very excited when I learned how to search for information. It was really fascinating to see how we could have access to so much information in such a short time. At the same time, my group and I were quite disappointed once we saw that there was almost no information available about Tiwanaku. For example, we wanted to look up the prices for yogurt and cheese in the local markets, and even with the assistance of our instructor we were not able to find anything. Then, we had the idea to get more information on how to improve our dairy production. We thought we could learn something on how to improve our cheese and yogurt production, but this was also very hard to get. There was so much information, and we needed a lot of assistance to find it or understand it. It is quite difficult to find information that is directly relevant to our needs.[32]   <<

This quote encapsulates the multiple challenges facing indigenous peoples in making meaningful and applicable use of the Internet. As such, a major restriction to Internet use is its urban-biased content, including its propagation of Western-centered values and norms, as held by non-indigenous societies. In fact, only a minor percentage of Internet content has been created by indigenous peoples, and the biases of non-indigenous authors become apparent once indigenous peoples surf the web. Most of the information found cannot be applied directly to their culturally distinct realities. Another major limitation is that the majority of the web content is presented in a rather academic language and is too abstract to be useful for solving the practical issues encountered by indigenous peoples in rural communities (as in the search for information on yogurt or cheese production shared by Fernanda Villegas). Finally, little content is available in indigenous languages. Most of the Tiwanaku participants had difficulty understanding why they could not find information about their municipality, including the municipal budget and local government projects. Another area of great interest for Tiwanaku's indigenous women was related to local markets and prices for dairy products—an area in which they specialize. While they were able to find a complex market price index of 78 products across Bolivia's main urban centers, they were unable to find any directly applicable information on local markets.

This section has demonstrated that indigenous women have strengthened their individual and collective agency to use ICTs in a meaningful manner. A central aspect of the program has been its focus on informational capabilities instead of just ICT skills training. A critical enabling factor for the local appropriation of technology can be attributed to the OMAK peer-group relationship, which provided participants with the necessary sense of security and confidence to learn how best to use ICTs, according to their needs, aspirations, and cultural context.

Impact on Empowerment

The empirical results reveal that the OMAK participants perceive the Internet to have had a substantial impact on their personal development. With a perceived impact value of 1.65, this dimension received the second highest score of all dimensions. Based on this unexpected result, I created an empowerment index to unpack the link between the use of ICTs and the advancement of people's personal development. The index is based on the results from three questions posed in the impact survey: (a) To what extent has your knowledge to use the Internet made you proud? (b) To what extent has Internet use opened up new opportunities to improve your well-being? (c) To what extent do you feel that the Internet has provided you with more knowledge to make decisions?

The theoretical basis for this index is Naila Kabeer's concept of personal empowerment, which encompasses options, choice, and control as critical factors in the process of women's empowerment. Kabeer argues that empowerment is "the expansion in people's ability to make strategic life choices in a context where this ability was previously denied to them" (Kabeer 1999, 473). This notion of empowerment draws strongly on Amartya Sen's "capability approach"—the main theoretical framework of this study. In fact, Sen conceives development as "a process of expanding the real freedom that people enjoy" and stresses the critical importance of personal empowerment by arguing that "greater freedom enhances the ability of people to help themselves and to influence the world, and these matters are central to the process of development" (Sen 1999, 18). This particular definition of personal empowerment was chosen because it stresses the central duality on which the process is based—first the need to undo a previous situation of "powerlessness" and the subsequent influence exercised by these individuals over their reality, as shaped by their sociopolitical context.

OMAK respondents have a significantly more positive view of all aspects of personal empowerment than the respondents from any of the other programs (figure 8.11).

OMAK members indicated that they feel proud of knowing how to use the Internet, with a mean perceived impact value of 1.68, compared with 1.43 for the overall sample. They expressed that the Internet has had a stronger positive effect on their lives than the participants of the overall sample. In fact, the overall empowerment index reaches a value of 1.57 for the OMAK participants in comparison to a value of 1.38 for the overall participants.

These data do not reveal the process by which the program has led to women's empowerment. This section thus provides a detailed account of the way in

FIGURE 8.11   **Perceived impact on personal empowerment among OMAK respondents**

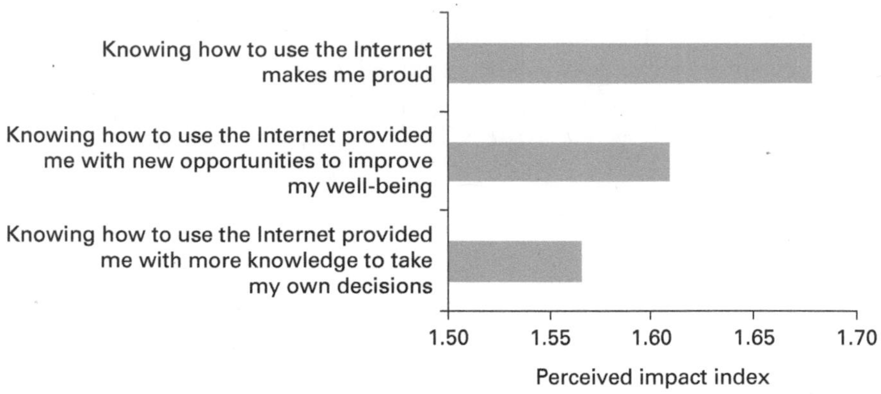

Perceived impact index

which the program has contributed to indigenous women's personal empowerment. The voices of indigenous women themselves, as captured during the interviews, allow for a greater understanding of this process.

Teresa Nacho, 34 years old, is an Aymara woman from the community of Tiwanaku. Her previous occupation was as a domestic servant, and she is now a nursing student at the Universidad Católica in Tiwanaku and an active OMAK member. In her words,

>> When I was nine years old, I had to leave my family and work as a domestic servant in La Paz. My family did not have any money. We are many—five brothers and sisters—so there was just not enough money. I had to work, just like my two sisters. The hardest for me was that I was not allowed to play or go to school anymore, like the other children I lived with. I had to do all the housework for the family I worked for.[33]   <<

Forced child labor is common among indigenous women. Due to high poverty levels, girls, in particular, either never enter or drop out of school at an early age and often become house servants in order to unburden their families of an extra mouth to feed. This form of discrimination is also evident in the official statistics, which indicate that indigenous women's educational attainment averages a mere 3.5 years (Jiménez Pozo, Landa Casazola, and Yáñez Aguilar 2006). As startling as this number is, it is not as revealing as the individual life stories of Aymara women and the hardships most have to endure. In fact, it is estimated that approximately 200,000 Aymara women work as domestic servants in La Paz and are subject to multiple forms of "oppression" by the mostly *mestizo* families who rely on their services (Albó 1994). During the year that my family and I lived in La Paz, we observed the broad-based acceptance of this form of oppression within Bolivian society. The most shocking aspect of a domestic servant's life experience is her lack of power to make life choices. Most indigenous women internalize their powerlessness and accept their role of servant as a given. Many suffer from deep-seated psychological effects, including low self-respect and self-esteem.[34] Xavier Albó provides a detailed account of this process:

>> Throughout this process there is no formal discrimination based on explicit rules of ethnic exclusion of the apartment type. No group is formally rejected. However, society is organized as if everybody belonged to the dominant groups. Every aspect of society is thought from the perspective of the dominant ethnic minority. These considered "lower" ethnic groups become the invisible others. Yet these groups are needed by those on the top, as domestic servants, rural workers, peddlers, and other forms of cheap labor ... . Then oppressive and/ or paternalistic relationships develop, in which the subordinated position of non-dominant groups is made and kept very clear ... . The overall result of this set of relationships is that people belonging to the lower-status ethnic groups regard themselves as "second-class citizens" or even "foreigners in their own land." In daily life, this is only a feeling, but is one of the more frequently stated complaints of peasant unions and indigenous political organizations. (Albó 1994, 129)   <<

This arrangement places many indigenous women in a situation of isolation since they are rarely allowed to spend a significant amount of time with their families or friends. Teresa commented on the effects of her separation from her family:

>> As a domestic servant, you basically have to do whatever the *señora* tells you to do. As a child, I was still lucky because I was allowed to go to school in the evenings and on weekends. Actually, the first *señora* I worked for was very encouraging. But later I had to change families, and as an adult I found it much harder to be an *empleada*.[35] The most difficult part then was that I was not allowed to see my daughter, who had to live with my parents in El Alto—not even on her birthday ... . Then one year, the family I worked for decided to move to Santiago de Chile. I had no options but to go with them. During this year, I was not able to see my daughter or family even once. I felt very alone there.[36] <<

Finally, most domestic servants suffer from racial discrimination, public humiliation, and violence. For instance, the doorman in the building where we lived in La Paz would not allow Teresa to enter, although we had invited her for lunch. The concepts of powerlessness and the highly unequal power relationship between the oppressor and the oppressed, as described by Freire, typify Aymara women's experience as domestic servants. How did Teresa overcome her situation of powerlessness, and what role did OMAK and the ICT project play in supporting her personal empowerment? She tells us,

>> I have participated now for several years in the OMAK program, mostly through the women's group in El Alto and recently in Tiwanaku. The last family I worked for were foreigners, and they were much more open than the Bolivians. I had more freedom and was allowed to leave the house more frequently. Through a friend from El Alto, I met the other OMAK women. Our discussions in the group helped me a lot. I became quite active and took on more responsibilities. One year I was even the treasurer. Although it was a lot of work, it helped me a lot. I felt that this was something I could do, without anybody else telling me how to do it.[37] <<

The literature on community psychology emphasizes the role that active involvement in community organizations can play in overcoming a sense of powerlessness and reestablishing a sense of self-control and confidence (Berger and Neuhaus 1977; Zimmerman 2000). Teresa describes how she overcame her sense of powerlessness and decided to leave her job as a domestic servant:

>> Three years ago, I decided to stop working as a domestic servant. My dream to be a nurse was always with me. The other women from the center helped me to remember this. They encouraged me to think more about what it is that I could actually do. Also, I needed to spend much more time with my daughter, since she started school and my parents were not able to take care of her anymore. I started to take on more flexible work, doing laundry for families and cleaning houses on daily arrangements, but no longer as a live-in. This allowed me to spend more time with my daughter. We then had to move back into my parents' house for a while.[38]   <<

With regard to the role that the program played in supporting her personal development, Teresa said,

>> As soon as I heard about the plans to organize a program in OMAK that would teach us how to use computers, I was very interested in participating. I always saw the young people in El Alto use the Internet, and I was curious about how to use it. First, it was, of course, a challenge, particularly for the older women. For me it was actually not so difficult, since I was one of the few who completed school when I was a child. I was even able to help other women to use the computers. The nice thing was that we were able to do this together as a group. Most of us were very proud once we received our course certificates.[39]   <<

Her description illustrates that an important aspect of the program is that the women—in spite of their very limited formal education—have been able to acquire a new skill set, which has had a very positive effect on their self-esteem. During the course, I observed that indigenous women, particularly those who have experienced numerous obstacles in their lives, including being denied access to education, have a strong desire to learn new things, to advance in their personal development, and to strengthen the capacities of the group. For Teresa, the training had a very positive effect, since it was one of the factors that enabled her to attain her dream and to enter the nursing program:

>> During the computer course, we learned many useful things. I was so excited that I was able to look for information on which universities offer nursing courses. I had gotten some basic information before, but with the help of the Internet I found the Catholic University's program, which had more affordable options in smaller towns away from El Alto and La Paz. I was very excited when I learned about this program. Also, I learned how to type my CV in the computer and—the most useful thing—how to send e-mails. So, about two months after our course, I applied to the university, and several weeks later I was accepted. It was like a dream to get the letter. First, I could not believe that I, who worked for most of my life as an *empleada*, would be allowed to study. I was really very happy.[40]   <<

Teresa's life story leads me back to my main research question: To what extent have ICTs and the OMAK program contributed to her personal empowerment and enhanced her well-being? The first aspect of Teresa's impressive personal development can be attributed to OMAK's intermediating role as a responsive and effective grassroots women's organization, providing her with a strong support network through the women's group. As evident from Teresa's account, OMAK has played a pivotal role in providing indigenous women with a social space of their own, where they can come together, express themselves freely, and give and receive support. The support of the group has played a critical role in her personal empowerment. Furthermore, OMAK has provided her, through its decentralized structure, the opportunity to take on an important leadership role (as treasurer). The ICT program has also enabled her to support other women in learning how to use computers and the Internet. This aspect of the program, which provides women with the opportunity to take on leadership positions in the implementation of its activities, is one of the most important reasons for its success. OMAK's decentralized grassroots co-management model has empowered participants in all of its stages—design, planning, and implementation—by strengthening their sense of control and self-esteem. The program's activities have enabled women like Teresa to feel more in control of their own lives and supported them in gaining a stronger sense of competence and self-esteem. This seems to be the primary positive effect of the program, with their newly acquired ICT skills being a secondary benefit.

At the same time, the accounts also point to the positive direct effects of ICT use on women's empowerment. Most participants stressed that their newly acquired knowledge of computers and the Internet has provided them with a strong sense of achievement and the feeling that they can overcome at least some of the barriers they face in their daily struggle. As figure 8.7 has shown, about two-thirds (65 percent) of female participants in the ICT training program indicated that their knowledge of Internet use makes them very proud. This figure is larger than the proportion in any of the other ICT programs and is significantly higher than among male OMAK participants.

Teresa's account also shows that her knowledge of Internet use has increased her critical awareness—a key dimension of empowerment. She learned about the Catholic University's nursing program in rural areas through the Internet. She learned to take on a leadership role and to help less-educated women to use ICTs. According to her account, this has increased her sense of being able to exert "control" over her own actions. A crucial factor of the ICT program is its inviting and encouraging nature, which has provided all of the participants with a sense of achievement, with the granting of the final certificates enabling the women to share a sense of pride among themselves and with others.

Furthermore, during the training workshops, I observed the empowering effects that technology use can have on indigenous women. During one workshop in Tiwanaku, I observed three women leaders from a remote community type a one-page description of their hometown and upload the text with an accompanying picture to a website designed for the course. They struggled for more than an hour, eventually completing the task. Once they had completed the process—writing, saving, and uploading the file onto a website—they expressed a strong sense of pride and accomplishment. One woman summarized the experience as follows:

>> For most people in the cities we do not even exist. Our small community has only 12 families, and we felt that people have forgotten about us. The government doesn't seem to care that we have lived all these years without water, light, or even a stove in our homes. But, don't we also have a right to live like humans? For me today, I have learned how to share the story of my community with others, and now it is there in the computer. Now our community must exist. You can even see a picture of the community on the Internet. Nobody can say anymore that we don't exist.[41]   <<

This woman expressed her strong gratitude to the instructor and her sense of pride in knowing that her community and culture were no longer invisible— other participants and the world could now learn about them. This feeling of being forgotten or abandoned by the state or society is commonly voiced by indigenous peoples living in Bolivia's more isolated and remote communities. In fact, geography plays a substantial role in determining the importance attributed to technology. Particularly for people living in remote communities, the knowledge of ICT use has the greatest potential for empowerment, as they strongly associate technology with modernity and ascribe it with attributes that make closing the gap with the urban world possible. In this sense, sheer knowledge of ICT use provides users with a sense of belonging that fortifies their self-worth and esteem by enabling them to interact with the rest of society on more equal terms, and to overcome their isolation and the perception of backwardness.[42] The extent to which this psychological effect is immediate and short term or long term is not known. It is, however, notable that the most direct effect of ICT use I was able to observe was the direct positive effect on people's psychological and personal development.

Impact on Social Capital

The empirical results reveal that the OMAK program has had a considerable positive effect on the organizational dimension of participants' well-being. As shown in figure 8.7, many OMAK participants believe that the Internet has had significant positive effects on the organizational dimension of their community life, as evidenced through a relatively high value for perceived impact (PII = 1.48). They believe that ICTs contribute to their collective well-being by strengthening their organizations and social networks. In order to gain a more in-depth understanding of the effects of ICTs on this dimension, I created a social capital index based on the theoretical concept of social capital (Coleman 1994; Putnam 2000). The index captures the effect of the Internet on social relationships and includes the following three statements on the potential impact of ICTs: (a) the Internet has strengthened social networks between indigenous organizations, (b) the Internet has strengthened indigenous organizations, and (c) the Internet has strengthened the producer organizations (cooperative) in my community. The data analysis finds that OMAK respondents have a relatively positive view of the Internet's impact on these aspects of social capital.

FIGURE 8.12 **Perceived impact on existing social capital among OMAK respondents**

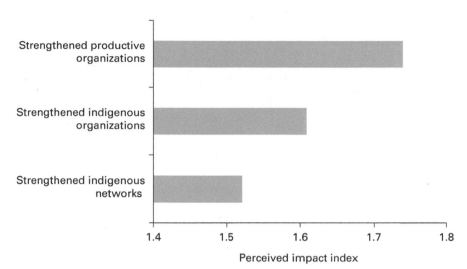

Figure 8.12 shows that the indigenous respondents have a positive view about the Internet's impact on all three areas of the social capital index. The index reaches its highest value for the question concerning the impact that the Internet has had on strengthening productive organizations (PII = 1.74). Second, the OMAK respondents indicated that the Internet has made a substantial contribution toward strengthening indigenous organizations (PII = 1.61). Furthermore, the results in figure 8.12 indicate that the majority of the participants also have a positive view of the Internet's impact on the strengthening of indigenous networks, with a relatively high value for perceived impact (PII = 1.52).

In order to understand the specific aspects of the program's contribution toward strengthening social networks and to unpack the process by which ICTs play a role in supporting this process, I now present a view from within the organization.

Jorge Flores, 53 years old, is a Quechua male from a community close to Cochabamba and spouse of Andrea Flores, director of OMAK. He is a co-manager of OMAK and responsible for the logistical and organizational aspects of OMAK programs, which include liaising and managing communications with the network and providing background support for OMAK program activities. Although Jorge cannot speak on behalf of Aymara women, nor does he pretend to do so, he plays a key role in supporting the organizational aspects of OMAK's work and is committed to improving the living conditions of indigenous women.

In response to a question related to the Internet's organizational impact on OMAK, he commented,

>> For me the Internet is a vital tool that allows us to be able to better communicate and coordinate our activities with other grassroots organizations or NGOs [nongovernmental organizations] that work in a similar area as OMAK. We are part

of several national networks, which work on human rights and women's issues. For instance, we are part of a national network of 25 organizations that work on solving the problem of the many undocumented indigenous people in Bolivia. All of the communications with these organizations are now by e-mail, and I help Andrea to keep up with our e-mails almost every evening.[43]   <<

Answering a question concerning the extent to which e-mail has provided OMAK with new opportunities to expand its work or to support the independent women's groups, Jorge explained,

>> For me, the Internet has just made some things easier. Communications have really improved, in particular with like-minded organizations and also to some degree with our major donors. But I would not say that it has fundamentally changed our work. Before we relied much more on the telephone, and the costs were much higher. We also relied much more on "word of mouth" to make OMAK more known to many actors, in particular at the national and international level. Now most people know our organization. We are even part of the UN [United Nations] Working Group on Indigenous Populations since 1990, and Andrea has been invited several times to participate in the UN meetings in Geneva and New York. The main problem I continue to see is that this does not help us at all to improve our communications with the grassroots. Sometimes it can even distract us from our main objective, which is to do everything possible to support the Aymara women in the women's groups.[44]   <<

The Internet seems to have improved communications with international donors, like-minded organizations, and actors in the national and international arenas, but it has failed to strengthen social relations among women's groups or to improve their communications with OMAK's central offices in El Alto. Jorge described the many organizational challenges encountered in the program:

>> For us, since we are an organization made up of a strong grassroots network of eight independent women's centers, our main goal is to improve the communications and information exchanges between the centers and then also between the centers and us in the central office. This has been really challenging, since there is no connectivity yet in most of the rural communities, and even the telephone is highly unreliable. Within our ICT program, we have, for instance, taken computers from La Paz to Luribay or Curahuara de Carangas in order to install them in the women's centers. First, this worked well since there are many things the women can learn without having access to the Internet. However, then we had many problems with the dust. We had dust everywhere, and our machines broke down. It is really quite different to do this work in the more remote and rural areas. We had also no support at all from any of the mayors' offices. Most of the work was done by the women themselves.[45]   <<

The lack of connectivity and access to ICTs in rural areas is a major barrier for many indigenous peoples to use these technologies. As Jorge pointed out, there is very limited government support to install and maintain ICT services in rural communities. Nevertheless, many indigenous women have found creative ways to overcome these barriers and continue to explore how best to use these technologies according to their own needs:

>> There was a time, when we were quite concerned about the program's continuity, since in most rural communities there is no Internet access. We also had the Luribay problem with the dust, so we were afraid that most of the women would stop using the technology. Another problem was that we did not have the support of our Canadian volunteer anymore, as he could only stay with us for about six months during the beginning of the program. The other issue we faced was that once the project funding dried up, we had to find a way to continue the program on our own. All these things were not easy, and the women were the ones who showed the most strength and determination to continue the project. Without their insistence, I think the program would have died a long time ago. What was amazing to see was that once the women, in particular, the younger ones, had received the most basic training, they looked for creative ways to continue using the technology. In many remote places they are continuing to use it once a week, when they go to the markets in the closest towns. They continue to communicate with each other, and in this way it has really strengthened our network. But I have to say, it is based a lot on the strong initiative of the women themselves.[46]   <<

The main challenges facing many ICT programs are continuity and maintenance. Most programs are funded by international donors and typically limited to a two- to three-year financing window. Jorge's account describes how OMAK addressed this challenge by creating a strong network among indigenous women from a wide range of communities. This strategy formed the basis for the long-term sustainability of the program, even after the program financing had been discontinued. Furthermore, OMAK's program has been implemented in highly diverse local contexts, as the eight women's centers are located in areas that diverge significantly in terms of poverty and remoteness. While the center in Tiwanaku is located in an area with extreme poverty and widespread lack of access to basic infrastructure services, the centers in Curahuara de Carangas and Corocoro are located in even more remote municipalities with even higher rates of poverty. Thus OMAK's major contribution is its ability to work in extremely challenging environments, with the aim to reach one of the most disadvantaged groups in Bolivia—poor rural Aymara women. Jorge's account emphasizes the individual agency of Aymara women in continuing to use ICTs on their own, as the program has had difficulty maintaining its activities. He also explained the positive effects that the program has had on women's social relationships:

>> In several cases, the women have come together in their groups and discussed how to find local solutions to their problems. For instance, in Curahuara de Carangas, they started lobbying the mayor for his support. The important advantage there is that the mayor is one of us—he is an indigenous leader who had worked for many years in the indigenous movement before becoming mayor. So, they were successful, and today they can access once again the Internet directly in their communities. Now there is a lot of communication between the different groups again, although they continue to come together in the events we facilitate as well.[47]   <<

Most of the women in the program use ICTs as a means to communicate better within their existing networks. They use ICTs to complement other more traditional forms of communication, such as community meetings. ICTs thus primarily strengthen the "bonding social capital" of indigenous peoples and support the "bridging social capital" to a much lesser extent. The critical difference between these two concepts is that "bonding ties" refers to people who are similar to oneself—for example, in terms of ethnicity, culture, or demographic factors—while "bridging ties" refer to people who are different from oneself (Woolcock and Narayan 2000). Networks based on bonding social capital are homogeneous, based on similarities, while networks based on bridging social capital are inclusive across different groups.

In other words, ICTs can reinforce indigenous peoples' existing horizontal networks, but they do not improve their vertical communications with others from government agencies, small businesses, NGOs, or civil society who are not already part of their original social networks. Bonding networks tend to be used extensively among the most vulnerable and marginalized groups, since their primary function is to protect and secure a minimal livelihood—for instance, in emergencies such as a natural disaster or economic shock. In contrast, bridging networks tend to create innovations and to enhance social mobility (Barr 1998). Putnam defines them as follows: "Bonding networks are good for 'getting by,' whereas bridging networks are crucial for 'getting ahead'" (Putnam 2000, 23).

This finding has important implications. It indicates the limitations of ICTs, which cannot readily provide indigenous peoples with new economic and social opportunities due to their inability to catalyze their communications and networks with the rest of Bolivian society. This might be particularly true for indigenous women at the grassroots level, who have limited contacts and experiences in networking with the non-indigenous population and formal state institutions, the market, and civil society. In this sense, ICTs do not alter existing social networks; rather, they serve as a mere vehicle for strengthening them.

Despite these limitations, the indigenous leadership is increasingly using ICTs to engage in a dialogue with the non-indigenous population at the national level and to form alliances with other indigenous organizations and international donors. However, the high level of attention recently being given to "indigenous issues" in international forums and networks has created tensions within the indigenous movement. Jorge described the tensions as follows:

>> One of the issues I see is that, while many people at the grassroots still have serious difficulties in accessing these technologies, the indigenous leaders are increasingly using the Internet to "represent" indigenous peoples' issues—an important and new responsibility for many leaders. For instance, two years ago, we had an incident where one of our partners was using e-mails to contact all sorts of international organizations, and then she was invited to several conferences. The problem was that this person claimed to represent OMAK's grassroots network of Aymara women, while nobody [in the organization] knew about it.

Well, what I want to say is that the internationals cannot know, since they are not very familiar with Bolivia. However, I believe that all of the leaders have an increased responsibility to use the technology for the good.[48]   <<

The tensions and conflicts of interest that can result from the technology's "misuse" could have important unintended consequences for many indigenous organizations and the movement as a whole. During my fieldwork, I noticed increased factionalism within Bolivia's indigenous organizations, which is finding a platform for expression on the Internet. For example, CIDOB (Confederación de Pueblos Indígenas de Bolivia), the main organization representing the indigenous peoples of the eastern lowlands, was split into two distinct and independent organizations, with each faction claiming on its website to represent the official CIDOB. This is certainly a larger and more complex issue that goes beyond the topic of this research, but it is related to Jorge's call for the increased responsibility of indigenous leaders to use these powerful instruments wisely.

At the grassroots level there is great concern about this very issue, reflected in the results of the impact survey, where a majority of respondents expressed skepticism about the Internet's ability to improve communications between organizations and local communities. As shown in figure 8.13, 70 percent of OMAK respondents said that they either do not believe or are only mildly convinced that the Internet can enhance this type of communication.

In effect, while the Internet plays a critical role in strengthening indigenous organizations and existing social networks, it cannot significantly improve communications between the grassroots and the newly emergent indigenous leadership.

To the contrary, as Jorge noted, the strong appropriation of the technology by relatively few leaders could create new tensions within indigenous

FIGURE 8.13 **Perceived limited impact of the Internet on accountability of indigenous organizations, by program**

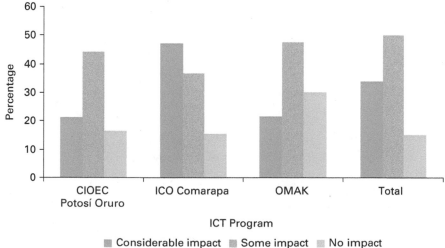

organizations and the grassroots. At the same time, the improved ability to communicate with like-minded counterparts and international support networks—NGOs and donor agencies—offers important opportunities to strengthen organizational capacity and identify new resources and funding opportunities, which could ultimately be of great benefit to local communities.

In sum, a key finding from the analysis is that the Internet seems to have the potential to strengthen horizontal social networks at the grassroots level, but not to enhance the flow of communications and information between the grassroots and the formal institutions of the state, the market, and civil society. Many indigenous peoples at the grassroots level expressed a deep frustration about their inability to use ICTs to strengthen their bridging social capital, especially with the central and local governments, and to improve their economic well-being. As such, it remains highly unlikely that a local indigenous women's group could succeed in establishing a new business relationship with a dairy producer, catalyzed solely by the use of ICTs and through it be able to improve its ability to market its dairy products. The analysis also has shown that ICTs only complement but do not replace more traditional forms of communication. Thus community meetings are essential to maintaining strong social networks among indigenous peoples.

Constraints Faced by Aymara Women in Enhancing Their Well-Being through the Use of ICTs

However, promoting indigenous women's Internet uses is a complex process, and the program also encountered considerable barriers in this regard. As this section will demonstrate, while the Internet is the ICT tool offering the greatest potential to strengthen indigenous peoples' well-being, it also presents the greatest complications and challenges. It presents excerpts from interviews with women who are OMAK participants and with a local priest who is a leading figure in Tiwanaku. It draws on 3 of the 14 interviews conducted (see appendix J). The analysis of these three interviews is supplemented by a broader analysis of Tiwanaku's main actors and institutions and their relationship to OMAK and the ICT program.

Interview with Rosa Morales

Rosa Morales is a 54-year-old Aymara woman from the community of Achaca. She has a primary education and is president of OMAK's women's group in Tiwanaku. She said,

>> When I came in, we had been through a lot of presidents, and many women did not see the value of OMAK for their lives. We did not have our official papers in order [legal status of the women's organization]. The women trusted in me for two years. We have done a lot of paperwork. We have suffered through a long legal process. We have fought against the bureaucracy. As Aymara women, I have to defend our right because we need to live like people. We need a better education for our children. We had this dream of our own center where we

could meet, produce our cheese and yogurt, and learn from each other. We have had this dream for a long time. Before we had nothing—no space for ourselves to meet, organize, and learn. After my second term and four years of hard work, we finally managed to build our center and are now even learning how to use the Internet. We were just last year accepted as an independent organization with our own legal status. Well, I think we have come a long way.[49]   <<

To be president of the local OMAK women's center in Tiwanaku is difficult, since this position requires a lot of time and dedication in the midst of a commonplace plight for survival, where most families struggle to produce enough or to find temporary work just to make a living. For Rosa, her toughest constraint is the extreme poverty her own family faces and the barriers she has encountered as an indigenous woman in dealing with the formal institutions of the state and traditional indigenous forms of government—both dominated by men.

>> The *mallkus* [traditional authorities] when they first heard about our plans to build a women's center did not give it any importance. They thought that we would never manage to achieve our dreams. But then later, once we started with the construction of the center, they came to me and complained. They don't like it that women form their own organization. They asked, why do you need your own organization? Do we represent the entire community? They even said, how is it possible that this woman, this ignorant woman, claims to be the president of her own organization?[50]   <<

In spite of the resistance from the traditional authorities, Rosa has been president of OMAK Tiwanaku for six years. She is highly respected for her courage, strong will, and ability to get things done in spite of the many constraints she is facing in the different spheres of her life—attributes that inspire her OMAK counterparts.

>> For me it was important that we organize ourselves as women and that we do not continue to live like that in a completely "uncivilized" way. I just wanted to do good for my community, and I wanted things to work better for our families. For us, being able to use the Internet means that we can begin to belong to the modern world and be part of the rest of society.[51]   <<

The creation of the women's center and the launching of the first ICT training courses created significant tensions within the community:

>> Once we were successful with our center, the men started to become jealous about our activities and threatened not to allow us to go to our meetings and computer courses anymore. Some men threatened to hit their wives if they went to the center without their approval. They really got very jealous about the computers and wanted to be included in this important program.[52]   <<

During our conversations, the women often spoke of physical violence. Bolivia has a high rate of violence against women, and OMAK has recognized this as a major issue and has incorporated strategies in its programs to address it. However, most projects do not reflect this in their project design. In this sense, OMAK's approach is quite different, since it openly acknowledges the problem and takes proactive steps to reduce the violence against women:

>> Unfortunately, it is quite common that men are regularly beating their wives for all sorts of reasons. Many times they are jealous because the women spend some time away from home, or they just believe that it is their "right" to beat their wives. One of the first things we discuss in every new group is that we women also have rights and that we do not have to accept being beaten by our husbands or anyone else. Frequently, women help each other in small groups to deal with this big problem. We also organized workshops with men in which we talked openly about this topic. Another important lesson we have learned, particularly with respect to our computer and Internet workshops, is that we have to include the men as well. Otherwise they get very jealous. Based on these concerns we had to react and organize separate workshops for men, women, and the youth. I am happy that through this approach we seem to have managed to avoid serious problems of violence. It, however, always remains a problem, no matter in what activity women are engaged in.[53]   <<

Another critical challenge that the program has encountered is its difficult relationship with the Tiwanaku municipality, which is often overshadowed by deep-seated racism and machismo. Although the Law of Popular Participation mentions gender equality as one of its objectives, it does not specifically recognize women's organizations as territorial-based organizations and does not imbue them with legal status or statutory protection as participatory bodies, as provided for the *ayllu*, unions, and other indigenous and grassroots organizations. The logic for omitting women's organizations from the legal framework is that they do not represent the "entire" community and thus are not entitled to participate in the political decision-making processes of local governments (Clisby 2005). Rosa Morales described the culture of machismo as being entrenched in the municipality and the principal barrier in developing a constructive relationship with the local government authorities:

>> In the municipality, I think that the men could be much more considerate towards women leaders. The men are in charge of their time and can return at whatever hour. If they have a meeting, they can return after midnight or in early morning. But a woman cannot do that because a woman has more responsibilities and she is a mother. So they should give us some priority, a bit more space, and this does not happen in the municipality. For example, I went to look for the mayor because I wanted to talk to him, as I represent many women in his municipality. He should give us some attention, but he seemed to be more interested in talking to his friends while I waited. So I asked him, May I come in? Yes, but wait a moment. Then one more hour went by … . I work, and I also have to spend time with my two daughters. I have responsibilities at home. All the time I can I dedicate to my community, but this is how it is because I have taken on this responsibility. If I was a man, he would say, come in. But because I am a woman I cannot go and help myself and share a drink with him. I am just not able to talk to him in the same way men can do. This is what I am seeing a bit of. And it should be the opposite: that because we are women, we should be given *priority*.[54]   <<

The mayor's lack of interest in the project and the difficulty of engaging the municipality constructively in it were confirmed in a meeting with him in La Paz. I accompanied Andrea Flores and the local president of OMAK to this meeting. It had taken weeks to get an appointment, which had been confirmed. When we arrived at his office, we were informed that the mayor had had to travel and that instead a councilwoman (the only woman in a position of power in the municipality) would listen to our concerns. While this meeting was cordial and the councilwoman listened politely to OMAK's concerns and even accepted OMAK's proposal for joint collaboration with the municipality, there was an unspoken consensus of its futility. It was well-known that the mayor was the decision maker on any activity undertaken by the municipality and meeting with him was an unavoidable prerequisite for any possible collaboration between his office and OMAK. Several weeks passed before the mayor again agreed to meet with us, but this time his office proposed meeting in the Café Central, a popular coffee shop, where he seemed at ease and was said to spend much of his time. He was a relatively short man, with outsized cowboy boots and leather jacket. Before sitting down, he briefly greeted Andrea and shook hands with her husband and me. During the entire meeting he conveyed to us that he was very busy and that while he appreciated the work OMAK was doing in Tiwanaku, he had many more pressing issues to attend to, such as the construction of several new rural roads and a planned rural electricity project. In reaction to OMAK's proposal for a joint project to improve Tiwanaku's access to Internet services, he responded in a politely arrogant manner:

>> I have known you, Andrea, for many years, and I am very grateful for all the good work you are doing in our communities, but we have just finalized our Annual Operating Plan, and we have the urgent need to improve basic infrastructure in our municipality. We are planning to finish a new road in the next couple of weeks, and I receive so many demands to improve the electricity and water services. In fact we have already planned all of our investments for this year. The Internet project you have been talking about sounds really interesting, and I would be very interested in having this service in our municipality. So please can we sit down when it is quieter, in a couple of weeks, when I have more time to discuss this further?[55]   <<

The mayor's attitude was not surprising and matched his reputation; he was astute and gifted at exploiting local politics and power relationships and putting his own interests above the needs of the communities. In fact, an analysis of the use of funds channeled from the Law of Popular Participation to local governments confirms that there is a customary mismatch between community demands for improved social services (better health and education services) and municipal projects (improvements to the city's public square and building of rural roads). Particularly in the highlands, the population has lost confidence in municipal government as a result of frequent irregularities and inefficiencies in the administration of funds.

The results from the impact survey regarding the Internet's impact on the political dimension complement the findings from the qualitative analysis. As shown in figure 8.14, OMAK participants perceive that while the

FIGURE 8.14   Perceived impact of the Internet on vertical accountability among OMAK respondents

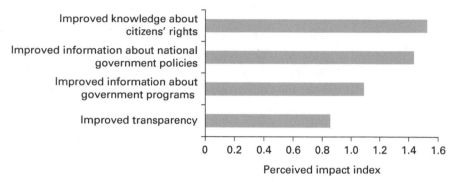

Perceived impact index

Internet has improved their access to information about their citizenship rights (PII = 1.52) and national government policies (PII = 1.43), it has had only a very limited effect on enhancing their access to information about government programs (PII = 1.09) and improving the transparency of government (PII = 0.86). In particular, the very low perceived impact on government transparency seems to indicate that the use of the Internet has not enabled rural communities to overcome the constraints on improved political participation.

An important insight from the meeting and frequent conversations with Tiwanaku's citizens is that OMAK plays a crucial role in filling the gap left by the state's failure to respond effectively to local demands and to provide services to its citizens, particularly in rural indigenous communities. According to Rosa Morales,

>> We currently do not even have access to electricity in our homes. I have many times requested this from the authorities. There are so many marginal areas in our municipality, and they should help us. They should not betray us. But sometimes they just look for the pretty faces, and since they see that we are poor they try to harm us. So we have come together with the help of OMAK and have managed to get electricity services installed in our center. We have asked for this to happen so many times, until they finally gave in and provided us with this service.[56]   <<

Rosa is occupying a political space within the municipality; in spite of the constraints she faces in her daily work, she has been quite successful in asserting her authority and in claiming spaces for the improvement of indigenous women's well-being. The analysis confirms that while indigenous women continue to be confronted with deep-rooted structural barriers of "ethnic exclusion" and machismo, their solidarity, leadership, and organization are the elements that pave the way for success in exerting their rights and in "claiming" spaces of their own. In this context, a critical positive impact of the program has been not so much the direct benefits that indigenous women obtain from using ICTs as the strengthening of their leadership skills and individual agency in the process of implementing the project in such a complex institutional setting as is Tiwanaku.

Interview with María Morales

María Morales is a 26-year-old Aymara woman from the Achaca community. She is a student of tourism management at the Universidad Católica, Tiwanaku, and an active member of OMAK.

>> I think that OMAK's most important aspect is that we can form a network among Aymara women and use our centers to improve our yogurt and cheese production. Together we have managed to find a small market for our products, and we frequently attend together the local markets to sell our products. What we should do next is to strengthen our relationship among the different centers and exchange our goods. For instance, in Luribay, they produce a lot of fruits and marmalade, in Curahuara de Carangas, the women are great artisans, and they produce a lot of handicrafts. And then we can use the Internet to better market our products. I heard of an example in Ecuador, where indigenous women are selling their hammocks through the Internet. We should do the same.[57]   <<

Of the persons I met, María was the most interested in economic issues and had the most innovative ideas on how to use the Internet to improve the economic well-being of Aymara women and their families. Her experience is unique in many aspects. She was the youngest member of the group and one of the very few Aymara women who had been accepted at Tiwanaku's Catholic University. Only 5 percent of the students attending the university in Tiwanaku are from one of the municipality's 41 communities. Her studies in tourism management reinforced her interest in using the Internet to improve economic well-being:

>> For me, it has been very important to know how to use computers and the Internet. When I first joined the university, I had never used a computer before and was instead looking at the other students in the university to see how they used the computers without any problems. I am so glad that I finally have learned myself how to use them ... . I could not imagine how I would do my homework without one. Now since I know more, I am happy to show my mother and the other women in the group how to use computers. But it is much more difficult for them to use this technology, since they only went to school for a few years.[58] <<

In terms of the economic aspects of the program, OMAK has strengthened the capacity of its members in dairy production (cheese and yogurt) and in local marketing and commercialization. The livestock itself plays a critical role in building women's assets, increasing their incomes and improving nutrition—in particular, for children and pregnant women—in a region where food shortages and extreme poverty frequently lead to chronic malnutrition. María was enthusiastic about the potential of the women's center to improve the economic well-being of its members and their families:

>> An advantage of our women's center is that we not only come together to discuss our problems, but we also help each other in improving our agricultural production. As a group, we have learned how to produce much better cheese and yogurt than before, and we now also help each other to commercialize our products in local markets. Two weeks ago, for instance, we participated in a

local fair for small-scale dairy producers in La Paz's Plaza Avaroa, and we were very happy with the results—we sold more cheese than ever before. All of this is only possible because we are helping each other. Each one of us alone would not be able to produce enough cheese or yogurt in order to sell large amounts, since we only own two or three cows each; however, as a group we can make a real difference.[59]   <<

She also spoke of the many challenges they face in trying to interface with market demands:

>> Things, however, have been not been so easy for us. We had, for example, serious problems producing enough of the same-quality cheese. A month ago, we wanted to participate in a fair, but before we left for the market we realized that we just did not have enough of the same products we could offer. In the end we decided not to go. Also, it is very hard for us to transport our products to the different markets. We can only take as much as we can carry on our backs. Another problem is that not enough people know us or our products. Now we at least have a name for the cheese and yogurt, but most people don't know us. It is hard to compete with the big companies. Here I believe that the Internet could really help us. We could use it to market our products much better and to send out messages to previous customers that we will have a stand at the next fair, when it is close to their communities.[60]   <<

The more I engaged María in a discussion on how the Internet could support productive activities, the more apparent it became that indigenous women are confronted with a set of structural impediments and deep-seated market inequalities that cannot be altered by simply enhancing their ICT use. As María mentioned, the transport sector is dominated by *mestizos*, and indigenous women lack the capacity to produce standardized products in large quantities. Nevertheless, it seems that OMAK's important contribution is to have organized the women into small producer groups, helping them to improve their dairy production and to sell their products in local markets.

It became clear that OMAK's ICT program has not yet had any significant positive impact on their economic well-being. First, its two activities—the ICT program and the productive project—were not directly linked with each other, in spite of the OMAK leadership's best intentions to exploit these synergies, and second, significant inequalities were evident in the agricultural markets between the *mestizo* businessmen and the small-scale indigenous producers, reaffirming that the structural barriers in the market were too great for the mere use of ICTs to change.

Chapter 3 describes in more detail the structural barriers of high poverty rates and social exclusion facing indigenous peoples in Bolivia. Indigenous women face many challenges that go beyond the challenges related to new technologies when using the Internet to improve their access to markets. Specifically, they also face barriers such as (a) lack of access to capital, (b) limited capacity to produce enough of a specific product to serve national markets, (c) lack of technical know-how and technical equipment to produce large quantities, (d) dependence on middlemen to transport products to markets, (e) concentration of agricultural markets, particularly in the dairy sector, and (f) their overall high poverty levels and low educational attainment. It has been argued that these barriers are structural in the sense that they are based on the inequality in agricultural markets between large-scale and small-scale producers in Bolivia. The analysis seems to indicate that information asymmetries (lack of information about local market prices) are only one of many factors preventing indigenous women from improving their access to markets.

Interview with Padre Claudio

Padre Claudio is a 58-year-old Catholic priest and director of the local university campus of the Catholic University. He is a highly respected community figure, who is committed to Tiwanaku's advancement out of poverty. I sought his views on many occasions during my fieldwork.

We met during my first visit to the OMAK center, at Andrea's insistence, because he was the first person to have accessed the Internet and to have provided Internet services to Tiwanaku and the surrounding provinces. During this initial visit, he gave me a tour of the campus and spoke with great conviction about the Internet's value for his work and for the university:

>> The Internet is essential. It opens the door to the world for us. Imagine, I lived for more than 25 years in different rural areas across Bolivia and never before did I have the opportunity to have the same information as the people in the city. You have to know that here we do not have access to national newspapers and only very limited access to TV. That's the reason why I asked for this satellite to be installed here two years ago. Unfortunately, we had many problems with the system, and once it broke down, the company that installed it never come back to fix it. Now we have been without a connection for several months, and it feels like we have been put back into the "darkness" again.[61]   <<

While Padre Claudio emphasized the critical importance of the Internet for himself and his students, I was quite surprised to hear that he was deeply skeptical about OMAK's program. He showed me his local community radio station and shared his thoughts about the OMAK program:

>> Yes, of course, I have heard about the computer courses OMAK has organized for the Aymara women. But you know most women don't even know how to read or write, since most of them did not attend any school. I know that this is changing, and the younger ones have a better education. However, in general I think that illiteracy is a huge issue here. Also, people are not used to reading anything. That's the reason why we have no newspapers. It is not part of their culture. It is basically an oral tradition, and that's why I have built the community radio station here on campus. Also, most people don't have any time to read or learn how to use the Internet. Very different with the radio, since they can listen to it while they work in the fields, or many women also listen to it when they work in their kitchen. We have now even developed a program to teach the *campesinos* English, since this is really important in today's world.[62]   <<

As part of the tour, he also showed me with great enthusiasm two greenhouses he had managed to build on the campus and proudly demonstrated the great variety of vegetables several students were in the process of harvesting. Based on his many years of work in the *altiplano* and his professional expertise in agricultural extension, he provided me with a detailed account of the many barriers indigenous peoples face in improving their agricultural production and their relationship with the markets.

>> One of the key issues of the local economy is the lack of adequate technology and know-how by most of the subsistence farmers. Agricultural production has suffered for years from the increasing subdivision of the lands, the lack of technical assistance, and the increasing problems of soil erosion. Most of the *campesinos* just produce enough to feed their children, and this has become increasingly difficult. While most communities have maintained a strong sense of cultural identity and their traditional social organizations have survived first the Spanish colonization and then the Republic, they have lost to a large extent their traditional and communal agricultural practices. Basically, since the land reform, the land has been divided in individual plots, and most farmers just don't have the means nor do they have access to credit, in order to improve their agricultural productivity.[63]   <<

Regarding the potential role that the Internet and ICTs could play in enhancing the economic well-being of Aymara farmers, he spoke about the structural inequalities that exist in local agricultural products and labor markets.

>> To be quite honest with you, the Internet cannot play any meaningful role in improving the economic conditions of the *campesinos*. They are just stuck in between two worlds: their own traditions, in which they highly value collective work (*minka*) and where they exchange all kind of goods with each other (*trueque*), and the market, where they have to sell their products, such as potatoes, onions, or quinoa for very low prices to the *mestizo* middlemen, who control the transportation and the commercialization of all those goods. The conditions of this trade are highly unequal, and the ones who benefit the most are the businessmen in the urban centers, who hire the middlemen to approach the *campesinos* to sell their products. The Internet won't be able to change anything. The only benefit I can see is that the people can get more information on how to improve their production. However, this will not change their dependency on the middlemen nor improve their limited access to credit. I see the benefits to be minimal at least for now.[64]   <<

The one area where Padre Claudio thought the Internet could play an important role was in forming a new generation of indigenous professionals and raising their awareness about their rights and the situation of the Aymara communities more broadly in Bolivian society:

>> For me the Internet can really be of great help in our work here at the university. I am very proud of my students and the way they are able to combine their knowledge of the rural communities and the *altiplano* with the "modern agricultural practices" from around the world. The Internet is vital for the formation of these young leaders, and it can play an important equalizing role, in the sense that my students for the first time have access to the same knowledge as the students in La Paz or Buenos Aires. A second role that I can see for this technology

is as a tool to raise the awareness of indigenous youth. It can help us to make these young people "think" and "reflect" about their own identities and the way they could become important agents for change. The Internet enables them to explore the world, and I hope this will enable them to put their own studies into a broader context.[65]   <<

The discussions with Padre Claudio helped me to place the OMAK program better into the broader social, economic, and cultural context of Tiwanaku. He shared with me his very broad experience in working for more than 25 years with indigenous peoples, particularly indigenous youth, in the Bolivian highlands. Based on his experience, he drew a somewhat critical conclusion about the Internet's impact on indigenous peoples' well-being. He pointed out that many indigenous peoples face critical barriers, such as high illiteracy rates or the sheer lack of time to learn how to use the Internet, that impede them from reaping the benefits of its use. Nevertheless, he also left me with a rather optimistic outlook, when he said that the Internet can open a "new world" to his students and indigenous youth by enabling them to place their own communities into the context of the world, but that they need to learn English first for this process to take place.

Conclusions

The case study has shown that a critical success factor of the OMAK program is its adoption of an autonomous approach to development based on the women's own cultural values, norms, and aspirations. In comparison, most NGO or government programs are perceived as external to the communities, and their mostly non-indigenous professionals are seen as trying to "impose" their own vision of development. Conventional approaches frequently lack the cultural affinity or openness to comprehend indigenous peoples' needs and aspirations and instead focus on the more technocratic process of implementing the planned activities. Furthermore, these programs are usually based on a preconceived notion of communities' needs and bring ready-made interventions. Indigenous communities often reject these programs, which they see as paternalistic in nature, leading to their lack of sustainability.

Another critical aspect distinguishing OMAK's approach from the majority of other interventions is its in-depth understanding of the existing social organizations, cultural diversity, and power relationships in indigenous communities. Its approach is highly flexible and varies depending on differences in each locality's political, social, and cultural context. To address the important cultural diversity among the indigenous peoples of the highlands, the organization works exclusively with indigenous professionals who have a good understanding and appreciation of the local sociocultural environment.

Furthermore, the analysis has shown that OMAK's leadership has many years of experience in dealing with internal conflicts within indigenous communities and takes proactive steps to negotiate the objectives of the program with the different local actors before initiating its activities. For instance, in Tiwanaku, the local *mallkus* threatened to boycott the project, saying that it would create tensions between men and women. OMAK responded by adjusting its project and including men in the training workshops. At the same time, it maintained its original objective of empowering indigenous women by organizing several additional workshops specifically targeted to them. This example illustrates that OMAK is very conscious of the existing power relationships at the local level and recognizes the importance of consulting the main actors (municipal and traditional authorities from the *ayllu*, local school, and church) and obtaining their consent before initiating its activities. In particular, its strategy of engaging with both the traditional indigenous authorities and the local government representatives demonstrates its deep understanding of the local dynamics within indigenous communities.

Moreover, the chapter has demonstrated that OMAK fully integrated its ICT project into its overall organizational objectives and vision. The project was not presented nor perceived as a stand-alone intervention. In this manner, the ICT project addressed and encompassed the organization's overall rights-based approach and, in doing so, attained its ultimate goal—to empower indigenous women so that they may gain greater control over their lives and strengthen their capabilities to improve their individual and collective well-being. Therefore, the ICT workshops formed an integral part of OMAK's broader human rights, leadership, and advocacy work. This is a critical advantage, since the use of ICTs was embedded in the main objectives of the organization, which provided them with a clear rationale and sense of meaning. A fundamental positive outcome of this approach is that the participants not only enhanced their ICT capabilities, but also significantly strengthened their informational capabilities, so that ICT uses become catalysts for improving indigenous peoples' well-being across multiple dimensions of their lives.

Finally, the chapter has provided a detailed account of some of the structural constraints that indigenous women confront in their daily lives, which can significantly restrict the impact of ICT use on their human and social capabilities. The analysis has illustrated how multiple forms of discrimination entrenched within the existing social structures act as considerable impediments, diminishing the potential positive effects that ICTs might otherwise have on the political and economic dimensions of their lives. Nevertheless, the analysis has demonstrated the important contribution of ICTs for strengthening the individual and collective agency of indigenous women, enabling them to improve their well-being in both realms. Indigenous women's ICT use plays a critical catalytic role for social change in their communities and in Bolivia as a whole.

Notes

1. The Urus are an indigenous group living in the wetlands and on the shores of Lake Titicaca; they are ancestors of both the Quechua and the Aymara and are believed to be the oldest indigenous group in the Lake Titicaca region (Healy and Zorn 1994).
2. Elizabeth Corneo, ICT training workshop, Tiwanaku, July 19, 2005.

3. Kollasuyo refers to the Bolivian highlands, one of the four administrative regions of the Inca Empire, located south of Lake Titicaca (Healy 2001).

4. Juana González, Aymara leader, Achaca, Tiwanaku, April 15, 2005.

5. Focus group with community leaders, Tiwanaku, April 19, 2005.

6. Focus group with community leaders, Tiwanaku, April 19, 2005.

7. Focus group with women leaders, Tiwanaku, April 14, 2005.

8. Focus group with women leaders, Tiwanaku, April 14, 2005.

9. Jesus de Machaca is a neighboring municipality that has maintained the *ayllu* as the central form of its traditional government and whose communities continue to rely more on the subsistence economy and indigenous markets and less on their relationship with the market economy.

10. *Trueque* is the traditional exchange of goods by the indigenous communities. This practice was the basis of the economic richness of the Inca Empire and is still prevalent in several communities.

11. Focus group with community leaders, Tiwanaku, April 19, 2005.

12. Focus group including older Aymara women, Tiwanaku, April 14, 2005.

13. Ibid.

14. Rosa Morales, focus group with indigenous women leaders, Tiwanaku, April 14, 2005.

15. *Mujeres de pollera* are urban indigenous women who wear the traditional gathered skirt of the Bolivian Andes (Albó 1994, 439).

16. The Universidad Mayor de San Andres, the main public university in La Paz, has recently provided an increasing number of indigenous peoples the opportunity to pursue a university education.

17. *Chola* is a derogatory term for indigenous women who have migrated to the urban center but maintain many aspects of their traditional way of life and culture (Albó 1994).

18. Interview with Andrea Flores, director of OMAK, June 5, 2005.

19. Ibid.

20. *Ayllu* is the traditional social community organization of indigenous peoples in the highlands. The *ayllu's* key values of reciprocity and solidarity stress the collective aspects of well-being within indigenous communities and are based on a combination of kinship, territorial ties, and symbolism (Albó 1975).

21. *Chachawarmi* stresses the complementarities between men and women in the family and the broader community, with mutual respect between men and women being central for reaching a state of individual and collective well-being.

22. Interview with Andrea Flores, June 5, 2005.

23. Interview with Elvina Montenegro, member of the women's group from Luribay, July 19, 2005.

24. Focus group with OMAK grassroots women's leaders, Tiwanaku, July 19, 2005.

25. Interview with Andrea Flores, president and director, OMAK, June 5, 2005.

26. Focus group with community leaders, Tiwanaku, May 28, 2005.

27. Ibid.
28. Ibid.
29. Angelica Morales, treasurer of the women's group in Tiwanaku, July 20, 2005.
30. Eulalia Ruiz, OMAK program participant, Tiwanaku, July 20, 2005.
31. Interview with Eulalia Ruiz, OMAK program participant, Tiwanaku, July 20, 2005.
32. Interview with Fernanda Villegas, OMAK program participant, Tiwanaku, July 20, 2005.
33. Teresa Nacho, OMAK program participant, Tiwanaku, July 21, 2005.
34. Andrea Flores also described the negative effects of being a domestic servant. She stressed that one of the reasons she founded OMAK was to enable women to overcome some of the negative effects they have suffered from working as domestic servants as children and young adults. She stressed that she suffered under the severe lack of self-esteem when she worked for many years as a domestic servant for a *mestizo* family in La Paz herself. Interview with Andrea Flores, director of OMAK, June 5, 2005.
35. *Empleada* is the Spanish expression for domestic servant.
36. Teresa Nacho, OMAK program participant, Tiwanaku, July 21, 2005.
37. Ibid.
38. Ibid.
39. Ibid.
40. Ibid.
41. Participant in ICT training workshop, Tiwanaku, July 19, 2005.
42. Focus group with grassroots women leaders, Tiwanaku, July 19, 2005. Felix Gutiérrez, who has worked for many years on promoting the use of ICTs by indigenous peoples at the grassroots level in the highlands of Bolivia, also made this observation. Interview with Felix Gutiérrez, director, Indigenous Communications Initiative, Red Wayra, La Paz, July 7, 2005.
43. Interview with Jorge Flores, OMAK co-manager, May 10, 2005.
44. Ibid.
45. Ibid.
46. Ibid.
47. Ibid.
48. Ibid.
49. Interview with Rosa Morales, president of OMAK women's group in Tiwanaku, July 21, 2005.
50. Ibid.
51. Ibid.
52. Ibid.
53. Interview with Andrea Flores, June 5, 2005.
54. Interview with Rosa Morales, president of OMAK women's group in Tiwanaku, July 21, 2005.
55. Interview with Lino Condori, mayor of Tiwanaku, La Paz, August 15, 2005.

56. Interview with Rosa Morales, president of OMAK women's group in Tiwanaku, July 21, 2005.
57. Interview with María Morales, Aymara youth leader, Achaca, Tiwanaku, July 22, 2005.
58. Ibid.
59. Ibid.
60. Ibid.
61. Interview with Padre Claudio, Catholic priest and director of the local campus of the Catholic University, Tiwanaku, July 22, 2005.
62. Ibid.
63. Ibid.
64. Ibid.
65. Ibid.

References

Albó, Xavier. 1975. "La paradoja Aymara: Solidaridad y faccionalismo." Cuadernos de Investigación 8, Centro de Investigación y Promoción del Campesinado, La Paz.

———. 1994. "Ethnic Violence: The Case of Bolivia." In *The Culture of Violence*, edited by Marcial Rubio C. and Kumar Rupesinghe. New York: United Nations University Press.

Albó, Xavier, Kitula Libermann, Armando Godinez, and Francisco Pifarre. 1989. *Para comprender las culturas rurales en Bolivia*. La Paz: MEC, CIPA, and UNICEF.

Barr, Abigail. 1998. "Enterprise Performance and the Functional Diversity of Social Capital." Working Paper 98-1, University of Oxford, Institute of Economics and Statistics, Oxford.

Benton, Jane. 1999. *Agrarian Reform in Theory and Practice: A Study of the Lake Titicaca Region of Bolivia*. Aldershot: Ashgate.

Berger, Peter L., and Richard John Neuhaus. 1977. *To Empower People: The Role of Mediating Structures in Public Policy*. Washington, DC: American Enterprise Institute for Public Policy Research.

Bustillos Rodríguez, Nidia. 2003. "Weaving Tapestries of Solidarity with Virtual Thread: Information and Communication Technologies at the Service of Grassroots indigenous Women in Bolivia." *Indigenous affairs* 2 (2003): 26–31, IWGIA, Copenhagen, Denmark.

Butz, David, Steven Lonergan, and Barry Smit. 1991. "Why International Development Neglects Indigenous Social Reality." *Canadian Journal of Development Studies* 12 (1): 143–57.

Clisby, Suzanne. 2005. "Gender Mainstreaming or Just More Male-Streaming? Experiences of Popular Participation in Bolivia." *Gender and Development* 13 (2): 23–36.

Coleman, James. 1994. *Foundations of Social Theory*. Cambridge, MA: Harvard University Press.

Freire, Paolo. 1972. *Pedagogy of the Oppressed.* London: Penguin Books.

Haverkort, Betrus, Katrien van't Hooft, and Wim Hiemstra, eds. 2003. "Antiguas raíses, nuevos retonos, El desarrollo endógeno en la práctica." COMPAS-AGRUCO, Plural, La Paz.

Healy, Kevin. 2001. *Llamas, Weavings, and Organic Chocolate: Multicultural Grassroots Development in the Andes and Amazon of Bolivia.* South Bend, IN: University of Notre Dame Press.

———. 2004. "Towards an Andean Rural Development Paradigm?" *NACLA Report on the Americas* 38 (3): 28–33.

Healy, Kevin, and Elayne Zorn. 1994. "Taquile's Homespun Tourism." In *Cultural Expression and Grassroots Development: Cases from Latin America and the Caribbean,* edited by C. D. Kleymeyer. Boulder, CO: Lynne Rienner.

INE (National Institute of Statistics). 2001. *National Census of Housing and Population 2001.* Ministry of Planning and Coordination, National Institute of Statistics, La Paz.

Jiménez Pozo, Fernando, Wilson Landa Casazola, and Ernesto Yáñez Aguilar. 2006. "Bolivia." In *Indigenous Peoples, Poverty, and Human Development in Latin America,* edited by Gillette Hall and Harry Patrinos. London: Palgrave Macmillan.

Kabeer, Naila. 1999. "Resources, Agency, Achievement: Reflections on the Measurement of Women's Empowerment." *Development and Change* 30 (3): 435–64.

Putnam, Robert D. 2000. *Bowling Alone: The Collapse and Revival of American Community.* New York: Simon Schuster.

Sen, Aymarta. 1999. *Development as Freedom.* New York: Knopf Press.

UNDP (United Nations Development Programme) and INE (National Institute of Statistics). 2006. "Bolivia atlas estadístico municipal." UNDP, Social and Economic Policy Analysis Unit, and INE, La Paz.

UNDP (United Nations Development Programme), INE (National Institute of Statistics), and UDAPE (Social and Economic Policy Analysis Unit). 2004. "Indice de desarrollo humano en los municipios de Bolivia." United Nations Development Program, National Institute of Statistics, and Social and Economic Policy Analysis Unit, La Paz.

Widmark, Charlotte. 2003. *To Make Do in the City: Social Identities and Cultural Transformations among Aymara Speakers in La Paz.* Uppsala University, Department of Cultural Anthropology and Ethnology.

Woolcock, Michael, and Deepa Narayan. 2000. "Social Capital: Implications for Development Theory, Research, and Policy." *World Bank Research Observer* 15 (2): 225–49.

World Bank. 1998. "Bolivia: Rural Productivity Study." World Bank, Washington, DC.

Zimmerman, Marc A. 2000. "Empowerment Theory." In *Handbook of Community Psychology*, edited by Julian Rappaport and Edward Seidman, 43–63. New York: Springer US.

Zoomers, Annelies. 1999. *Linking Livelihood Strategies to Development: Experiences from the Bolivian Andes*. Amsterdam: Royal Tropical Institute.

CHAPTER 9

Exploring the Links between ICTs, Decentralization, and Well-Being

One early morning more than 30 local government officials from all nine departments of Bolivia gathered for a one-week training course at the Central University of La Paz to discuss how information and communication technologies (ICTs) might help local governments to provide better services to communities, enhance their transparency, and improve their communications with the central government. During the training workshop, which was organized by the government-led EnlaRed Municipal Program, participants frequently expressed their high expectations about the contribution that ICTs could make to transforming how government institutions work. One of the participants from Batallas, a rural municipality in the Bolivian highlands, expressed his hope that ICTs could improve the delivery of services to its citizens:

>> The most important issue for our municipality is to improve the basic services to our population; most families in Batallas are very poor and do not have access to water, electricity, and sanitation. We have several important projects in our municipality to improve these services, but most people don't know about them. I am convinced that with the help of ICTs we can better inform our communities about our ongoing activities. This is very important for us to improve our relationship with the communities.[1]   <<

They also expected ICTs to enhance government efficiency and increase institutional capacity. Marcelo Carbone, the representative from Comarapa, a municipality located in the Santa Cruz Valley in eastern Bolivia, emphasized his municipality's need for modern communications technology:

>> For our municipality, ICTs can play a key role in strengthening our institutional capacity—many people in the municipality currently are still using typewriters or even hand-written notes to communicate with other government offices or

citizens. I am convinced that the introduction of computers and ICTs can help us
to become more efficient and thus to improve our services.[2]   <<

The government officials who participated in EnlaRed started with high
expectations. Nevertheless, Bolivia's long history of political clientelism and
corruption, the opacity of its public sector, and the paternalistic relationship
between the state and the indigenous population raise questions about the
ability of ICTs to enhance the performance of local government (Blackburn
2000; Goudsmit and Blackburn 2001).

The program took place in the context of the broader decentralization
agenda in Bolivia, which was guided by the assumption that local governments are more efficient and less corrupt than the national government
because they are "closer" to the people, better able to understand the
dynamics of local economic and social development, and more accountable to local communities (Samoff 1979; Oyungi 2000; Faguet 2003;
World Bank 1999). The 1994 Law of Popular Participation (LPP) signaled a
dramatic shift in national policy from a centralized, top-down approach to
a decentralized, participatory, locally empowering approach (Booth, Clisby,
and Widmark 1996). The LPP transferred authority and resources from
the central state to local municipalities in order to make government more
efficient and improve the implementation of poverty alleviation policies
(World Bank 1999). The decentralization program delegated new responsibilities to municipal governments and territorial grassroots organizations
and allocated 20 percent of national tax revenue to them on approval of
their annual operating plans for the social sectors—mainly health and education. It also established mechanisms for social control, called vigilance
committees, which were made responsible for overseeing the expenditure of
funds (Kohl 2003; Blair 2000). According to Blackburn (2000), implementation of the program faced severe obstacles, since powerful regional elites
dominated municipal planning processes and captured the resources provided by the central government. Furthermore, the results of the bottom-up
participatory process had to fit with national and departmental plans and
respond to the spending priorities of the national government (Goudsmit
and Blackburn 2001, 590). Despite its shortcomings, the LPP recognized
traditional indigenous forms of organization and transferred significant
resources to local governments. It was a formidable attempt to shift power
relations within Bolivia's *mestizo*-dominated society, so that indigenous
peoples and other marginalized groups would have the space to voice their
concerns, interests, and needs.

This chapter investigates the role that ICTs can play in supporting decentralization and in helping to improve the well-being of indigenous peoples,
particularly in the political dimension. It uses a case study approach to investigate the process by which a government-led program introduced ICTs into
indigenous communities and the extent to which indigenous peoples' perceptions about the impact of the use of ICTs on their well-being are influenced
by the manner in which the program is implemented within a specific local
socioeconomic context. Thus emphasis is placed on the process of project

implementation and the role of the intermediary organization in facilitating the use of ICTs in indigenous communities.

The case study presents qualitative evidence from the government-led EnlaRed Municipal Program—the principal ICT program in support of local governments in Bolivia, which was coordinated by the Federation of Municipal Associations of Bolivia (FAM). The program was chosen in order to study the Internet's impact on indigenous communities within the context of a government-led ICT program with a low level of intermediation, which is characterized by the limited direct involvement of the intermediary at the local level. Based on the central role that the theoretical framework assigns to the intermediation process, the case study forms a counterpart to the previous chapter, which focuses on a grassroots-led program.

The case study also complements the quantitative results presented in chapter 7, which show that indigenous peoples perceive that the Internet has had a somewhat limited impact on the political dimension of their lives. By investigating some of the factors that have led to this perception, the chapter addresses the central research question of the study: Can ICTs enhance indigenous peoples' human and social capabilities and under what conditions? To answer this question, it is critical not only to study the social dimensions in which the majority of indigenous peoples believe that the Internet can make an important difference in their lives, but also to analyze an area, such as the political dimension, in which the perceived impact is much more limited.

In particular, the case study addresses the following questions: Did the use of ICTs lead to any institutional changes in local governments, alter the relationship between local government officials and indigenous communities, and thus strengthen indigenous peoples' human and social capabilities in the political dimension? Has the use of ICTs enhanced the institutional capacity of local governments to improve the delivery of social services to local communities? Has the use of ICTs enhanced the transparency of local governments and made them more responsible and accountable to the needs of local communities? In which dimensions (political, organizational, and social) have ICTs had the biggest influence on local governments and to what extent have they enhanced the human capabilities and social well-being of the program's participants? Finally, did EnlaRed achieve its main objectives, and what constraints and limitations did it face during its implementation?

The program sought to improve the transparency and accountability of local governments, to enhance local governance structures and the relationship between local government and indigenous communities, and to strengthen the institutional capacity and thus the performance of local governments, in particular, in relation to the delivery of social services. Thus, while the program focused on the political dimension, it also sought to improve local democratic processes and to strengthen the popular participation of indigenous peoples in local governments. The program also sought to address institutional and social aspects of local development.

The program conducted the following activities: (a) implemented an ICT capacity-building program for local government officials; (b) developed and promoted the use by local, regional, and national government officials of a national portal on issues related to municipal development (www.Enlared .org.bo); (c) assisted municipalities in developing and implementing their own websites at the local level; and (d) helped to formulate a national ICT strategy within the framework of the ongoing Estrategia Nacional de las Tecnologías de Información y Comunicación (ETIC) initiative sponsored by the United Nations Development Programme.[3]

The chapter presents the qualitative results from unstructured interviews and focus groups, complemented by quantitative results from the user impact survey. Specifically, it presents the results of 16 unstructured interviews with local and central government officials, program participants, ICT specialists, representatives of nongovernmental organizations (NGOs), indigenous community leaders, and two focus groups with program participants and participatory observations of two national ICT training workshops organized by EnlaRed.

The quantitative research is from two user surveys and a subsample of the impact survey ($n = 22$), which was conducted as part of my research in July 2005. The first user survey, conducted by the project team in October 2004, presents the results of an information needs assessment carried out in the 15 participating municipalities. The second user survey, conducted in September 2005, focused on the participants' proficiency in the use of ICTs. Finally, the chapter uses statistical data from the Bolivian National Institute of Statistics to provide an overview of the principal socioeconomic indicators in the municipalities.

The case study applies the alternative evaluation framework presented in chapter 1, focusing particularly on (a) the role that the intermediary organization EnlaRed played in promoting the use of ICTs in indigenous communities and (b) the extent to which local government officials and indigenous peoples appropriated ICTs into their local socioeconomic and cultural context. It draws on the studies of Ciborra (2002), Ciborra and Navarra (2005), Madon (1993, 2004), Walsham (1993), and Heeks (1999, 2003) and the methodology of Pettigrew (1985, 1998), which emphasizes the importance of the economic, social, political, and cultural context into which the technologies are introduced. These factors thus are essential for assessing the effects of EnlaRed on local governments and the well-being of indigenous peoples.

The chapter is organized into the following main sections: (a) an overview of the case study's principal findings; (b) a socioeconomic profile of the Batallas and Comarapa municipalities; (c) a description of the EnlaRed Municipal Program and its objectives; (d) an evaluation of the impact of the program, including its impact overall and its impact on it participants' informational capabilities, political well-being, organizational well-being, and social development; and (e) a short conclusion.

Principal Findings

The chapter argues that the EnlaRed program failed to reach its core objectives of enhancing the transparency, governance, and performance of local governments and thus did little to enhance indigenous peoples' well-being.

The case study illustrates that the program did not enhance participants' human and social capabilities because it did not enhance their informational capabilities. The chapter uses two case studies—one from the highlands (the municipality of Batallas) and one from the eastern lowlands (the municipality of Comarapa)—to show that the program did not promote the meaningful use of ICTs within the specific social, political, and cultural contexts of the 15 participating municipalities. In fact, the majority of the participants said that the Internet has had only a minor impact on all of the principal dimensions of well-being. The empirical findings also reveal that participants in EnlaRed have significantly more negative views on the overall impact of the Internet on their lives than participants in all of the other programs assessed in this study.

Based on the theoretical framework presented in chapter 1, the case study illustrates that a critical reason for the low overall impact of EnlaRed is that its participants did not attain advanced *informational capabilities*. Furthermore, the program failed to analyze the existing informational ecologies of the communities in which it worked. Furthermore, it did not provide participants with a space for broader critical reflection on the role of information and ICTs in improving local governance and human capabilities. Instead, it focused on a narrowly technology-driven view of ICTs and provided only practical training in the use of ICT applications, such as e-mail and website design. Such an overemphasis on technology itself resulted in the program's inability to consider the broader institutional processes related to introducing ICTs into communities and led to the lack of appropriation of ICTs within the local political and socioeconomic context.

In the *political dimension* of well-being, the case study demonstrates that the program failed to enhance indigenous peoples' human and social capabilities. Specifically, it did not meaningfully improve the transparency and vertical accountability of local governments nor enhance indigenous peoples' popular participation in the decision-making processes of local government. These findings are supported by the empirical data, with the value for perceived impact for the political dimension reaching a low value of 0.94, which indicates that the majority of participants believe that the Internet has had only a minor positive impact on the political sphere of their lives. This finding is particularly noteworthy since the program's main objectives were political in nature. A critical reason for its failure to enhance local governance is its overly abstract national agenda and its centralized approach, which prevented it from bringing about any meaningful institutional changes at the community level.

In the *organizational dimension*, the evidence shows that while the program succeeded in strengthening horizontal networks and relationships among local government officials, it failed to invoke any organizational changes or strengthen the institutional capabilities of local bureaucracies.

Moreover, the chapter provides evidence that the program failed to achieve its third major objective: improving indigenous communities' collective capabilities in the *social dimension* of their lives. According to the empirical data, EnlaRed participants perceive that the Internet has had only a minor positive impact on access to education and health services in their communities, as shown by a low value for perceived impact of 0.88. This finding is notable since the social dimension is the one dimension in which participants of the overall survey have the most positive views, with a relatively high value for perceived impact of 1.44.

Why did the program fail to achieve its major objectives? The case study makes an important contribution to the overall study by analyzing—based on the theoretical framework—some of the underlying factors for why this program has had such a low impact on indigenous peoples' human and social capabilities and thus on their well-being.

Based on the first hypothesis of the study, the case study investigates the intermediary process by which the project was introduced into indigenous communities. The program's fundamental problem was the institutional weakness of its intermediary, the Federation of Municipal Associations of Bolivia. A central government agency that suffered from internal factionalism and political instability, FAM was largely ineffective in promoting ICT use at the local level and far removed from the local sociopolitical, economic, and cultural context of its network of members, largely small-scale rural municipalities. These institutional weaknesses had serious consequences for EnlaRed, which was closely associated with FAM and thus lacked the necessary credibility and trust with local governments that are essential for successful implementation of e-government programs. An important consequence of the program's political context was that FAM selected the participating municipalities based on purely political considerations and not on the demands of local authorities or on the demonstrated technical preparedness of the municipalities chosen.[4] The project team was forced to engage with several local governments that had neither the political will nor the technical capacity (basic Internet access) to participate in the project.

Furthermore, based on the second hypothesis of the study, which highlights the need of ICTs to be locally appropriated by indigenous communities, the case study investigates the extent to which the program was effective in adopting ICTs to the specific local socioeconomic, political, and cultural context of the municipalities in which it was implemented. A principal finding of the case study is that the program suffered as a result of its conceptual design, which overemphasized the role of technology itself and failed to place ICTs into the much broader sociopolitical context of the struggle for enhanced transparency, more accountability, and improved local governance. In other words, there was a substantial gap between the project design and the reality on the ground; this "design-reality gap" is common among e-government programs in developing countries (Heeks 2002, 2003). The overly centralized approach made it impossible for the project team to respond adequately to the specific needs and demands of local realities. Excessive centralization also prevented the program from acknowledging the rich cultural diversity of different indigenous groups and from adapting the ICT services provided to the sociocultural context of local areas. Since the project worked in 15 communities in all nine departments of Bolivia and with a number of indigenous groups, it was unrealistic to assume that a single approach would be applicable everywhere.

Finally, the case study shows that the program was limited by the broader socioeconomic context in which it was implemented. In particular, it suffered from various external constraints, including the financial commitment and political expectations of its donors, the inadequacy of Bolivia's physical and regulatory infrastructure, and the cultural, socioeconomic, and geographic conditions of the participating municipalities.

The complete dependence on donor funding was another core constraint on the project. The program suffered from both the lack of donor continuity and strong donor influence and control. In fact, the U.S. Agency for International Development (USAID), the bilateral donor agency of the U.S. government, exerted a significant amount of control not only over project funding but also over daily operation of the program. Dependence on donor financing resulted in an overly technocratic approach in which the program provided specific services (like training in the use of ICTs) to local governments for a certain time period, but could not respond to the range of local demands or significantly improve the long-term well-being of local communities.

Finally, the program lacked the necessary enabling environment for its activities, in that Bolivia's central government had no coherent national strategy for expanding telecommunications and ICT services into rural areas. The absence of a much-needed regulatory framework and the shortage of public investments in rural telecommunications infrastructure meant that a large majority of rural municipalities consistently lacked access to the necessary telecommunications infrastructure and connectivity that would enable them to use the ICTs effectively. A consequence of the central government's failure to lead was that the project could not count on the necessary Internet connectivity in the majority of the participating municipalities; most participants had only very scattered access to the Internet through public access points in nearby intermediary cities.

Socioeconomic Profile of Batallas and Comarapa: Two Diverse Municipalities

Batallas and Comarapa are two municipalities with significant differences in their geographic location, social indicators, and cultural and political context. Batallas was chosen because its main demographic characteristics, socioeconomic data, and local cultural context are similar to those of Tiwanaku, described in the case study of the OMAK (Organización de Mujeres Aymaras del Kollasuyo) program, which is the subject of chapter 8. The choice of Batallas was intended to minimize the differences in external variables and to isolate the effects of the ICT intervention itself. Comarapa, a municipality in the eastern Bolivian lowlands, was chosen because its socioeconomic conditions are different from those of Batallas and were expected to be conducive to implementation of the program. Thus the choice of two municipalities that are quite different in their geographic location, socioeconomic situation, poverty levels, and cultural context was intended to shed light on whether variations in the program's outcome were the result of external factors.

Socioeconomic Conditions in Batallas and a Comparison with Tiwanaku

Batallas is located in the Bolivian highlands in the province of Los Andes, approximately 60 kilometers from La Paz. It is connected to the capital city by the main highway that leads from La Paz along the eastern shores of Lake Titicaca to Peru via the major tourism center of Copacabana. As shown on the map in appendix B, Batallas and Tiwanaku are located adjacent to Lake Titicaca at an altitude of approximately 3,800 meters. The statistical data provided in this section demonstrate the similarity of both municipalities' socioeconomic indicators, in particular the persistently high levels of extreme poverty and the very limited access to basic infrastructure and social services.

According to the 2001 census, living conditions in Batallas are very similar to those in Tiwanaku, with 96 and 97 percent of the population, respectively, unable to meet their most basic needs and 67 and 74 percent, respectively, living in extreme poverty (table 9.1). The municipalities also share common cultural and social characteristics; in both locations, more than 90 percent of the population is Aymara, and agricultural unions are the main form of social organization. Both municipalities lack access to education and health services. Batallas has slightly higher social indicators; its population completes, on average, 4.9 years of schooling compared with 4.3 years in Tiwanaku and has a slightly lower illiteracy rate (22 and 28 percent, respectively). Both municipalities have a similarly low level of human development, with a score on the human development index (HDI) of 0.557 for Batallas and 0.537 for Tiwanaku.

A critical reason for the persistently high level of poverty experienced in both municipalities is the agricultural crisis brought on by creation of the *minifundista* system, which led to the break-up of land into small plots, making it impossible for the Aymara farmers to grow sufficient agricultural products—including potato, quinoa, or cabbage—to ensure the livelihoods of their families (Kay and Urioste 2007). As chapter 3 has shown, the distribution of land between *minifundistas* and large landholders remains highly unequal in Bolivia. In fact, according to official data from the National Council for Agrarian Reform, 87 percent of usable land (28 million hectares) is owned by only 7 percent of landowners. The remaining 93 percent of small-scale landowners own only 4 million hectares. Whereby most of the large landholdings are located in Bolivia's eastern lowlands, the large majority of small-scale subsistence farmers live in Bolivia's highlands. Moreover, only 5 percent of the land belonging to the big landowners is being put to productive use (Urioste 1992, 184–85). Kevin Healy provides a detailed account of the negative consequences of this agricultural crisis for small subsistence farmers in the highlands of Bolivia (Healy 2004).

A strong indication of this ongoing economic crisis in the Andean highlands is the high rate of seasonal and permanent emigration in both

TABLE 9.1   Key socioeconomic indicators for Batallas

Category	Indicator
General	
Total population	18,693
Altitude (miles)	3,860
Number of rural communities	41
% rural	93
% Aymara population	95.7
Socioeconomic indicators (2001)	
Unsatisfied basic needs (% of population)	0.557
Human development index	95.7
National ranking (human development index)	156
Life expectancy at birth (years)	60.4
Literacy rate (% of population 15 years and older)	77.7
Annual consumption (US$ per capita)	$691
Average years of schooling	4.9
Number of health centers	5
Net migration (per 1,000 persons)	−15.22
Millennium Development Goals (2001)	
MDG 1: Eradicate extreme poverty and hunger	
Extreme poverty rate (% of population)	67.3
MDG 2: Universal primary education	
Primary completion rate (% of population)	87.5
MDG 3: Gender equality	
Ratio of female to male primary enrollment	94.16
MDG 4: Reduce child mortality	
Infant mortality rate (deaths per 1,000 live births)	71.3
MDG 5: Improve maternal health	
% of births attended by skilled health staff	30.9
MDG 7: Ensure environmental sustainability	
Access to potable water (% of population)	32.8
Access to sanitation services (% of population)	36.2

Sources: INE 2001; UNDP and INE 2006; UNDP, INE, and UDAPE 2004.

municipalities, reflected in the negative net migration rates of 15.2 (out of 1,000) in Batallas and 12.5 in Tiwanaku in 2001, as reported by the census data of the National Institute of Statistics (figure 9.1). The temporary emigration to urban centers has become a key livelihood strategy for many rural indigenous peoples in the Bolivian highlands; consequently, nonagricultural incomes have increasingly become a key source of income for indigenous families (Zoomers 1997).

FIGURE 9.1 **Poverty rates in Batallas and Tiwanaku, 1992–2001**

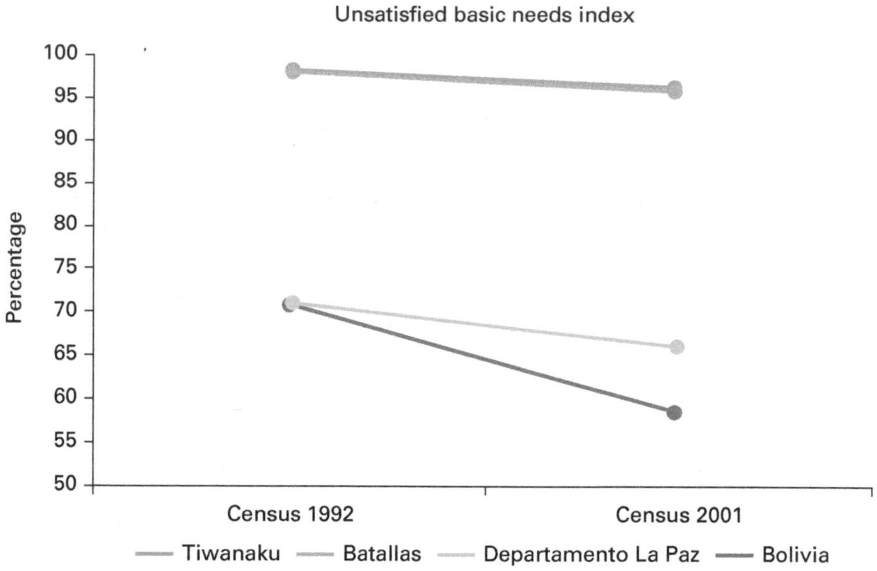

Unsatisfied basic needs index

Source: UNDP and INE 2006.

The difficult living conditions in both municipalities are also confirmed by the analysis of core socioeconomic indicators over time. In both municipalities, the living conditions of the large majority of the population improved somewhat during the 1990s, with the share of people unable to meet their basic needs falling slightly from 98 to 96 percent in Batallas and from 98 to 96 percent in Tiwanaku (figure 9.1). However, the poverty gap between the two municipalities and the rest of Bolivia widened, as poverty rates in all of Bolivia fell more quickly—from 71 percent in 1992 to 58 percent in 2001.

Access to basic health services is limited in both municipalities (figure 9.2). In Batallas, infant mortality rates worsened slightly, rising from 60 deaths per 1,000 live births in 1992 to 71 in 2001. In Tiwanaku, infant mortality rates fell from 77 deaths per 1,000 live births in 1992 to 66 in 2001. In both municipalities, maternal mortality rates—as measured by the percentage of births attended by skilled health workers—also declined. In 2005, only 34 percent of births in Batallas and 50 percent of births in Tiwanaku were attended by a professional health worker (figure 9.3).

Access to education has improved in both municipalities. Batallas has achieved almost universal levels of primary education, with enrollment rates of almost 98 percent in 2005, while Tiwanaku lags behind, with only 78 percent of boys and girls completing a basic level of education (figure 9.4).

Another key challenge for both municipalities is the lack of access to basic infrastructure services, particularly potable water and sanitation services. As shown in figure 9.5, Batallas and Tiwanaku continue to lag behind both the La Paz Department and the national average with regard to their population's

FIGURE 9.2   Infant mortality rates in Batallas and Tiwanaku, 1992 and 2001

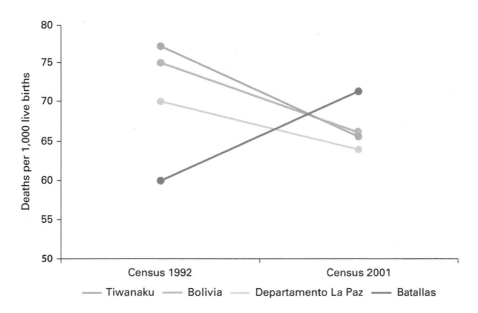

Source: UNDP and INE 2006.

FIGURE 9.3   Share of births attended by a skilled health worker in Batallas and Tiwanaku, 2002–06

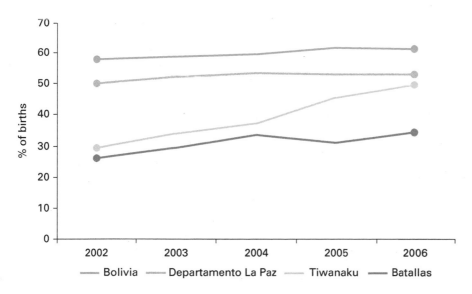

Source: UNDP and INE 2006.

access to potable water. In 2001, only about one-third of the population in both municipalities had access to potable water, compared with more than two-thirds of the population in La Paz and Bolivia as a whole.

Furthermore, both Batallas and Tiwanaku have very limited access to basic sanitation services. Despite improvements since 1992, only 33 and

FIGURE 9.4   Access to education in Batallas and Tiwanaku, 2001 and 2005

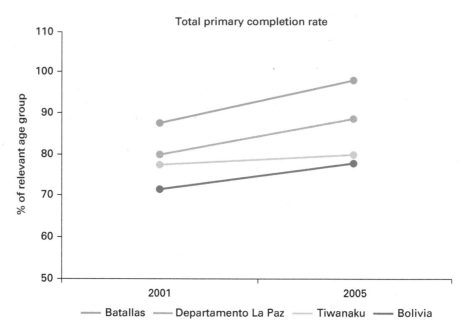

Source: UNDP and INE 2006.

FIGURE 9.5   Limited access to potable water in Batallas and Tiwanaku, 1992 and 2001

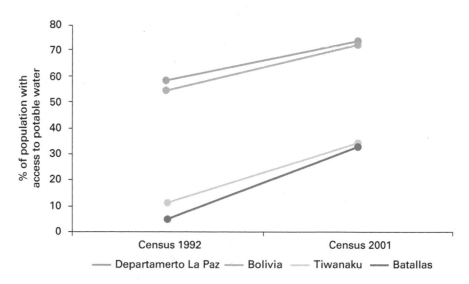

Source: UNDP and INE 2006.

FIGURE 9.6   Limited access to sanitation services in Batallas and Tiwanaku, 1992 and 2001

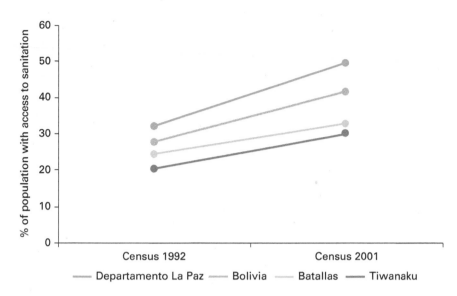

Source: UNDP and INE 2005.

30 percent of the population, respectively, had access to sanitation services in 2001 (figure 9.6).

Finally, annual consumption per capita is low in both municipalities—US$655 in Batallas and US$691 in Tiwanaku—which reflects their relatively low level of socioeconomic development.

Socioeconomic Conditions in Comarapa

Comarapa is located in the center of the Santa Cruz Valley, approximately six hours from the city of Santa Cruz in the eastern lowlands of Bolivia. Its low altitude, warm temperatures, and stable climate provide highly favorable conditions for agriculture. Indeed, the majority of its 14,600 inhabitants engage in medium-scale farming. The municipality is ethnically mixed; slightly more than half of the population (53 percent) is *mestizo*, and the rest (47 percent) are Quechua. Most Quechua families own and live on small-scale plots in the higher altitudes of the Andean slopes surrounding the valley; the *mestizo* population dominates the economic and agricultural activity in the center of the municipality, which is located in the valley.

Comarapa's overall socioeconomic development is somewhat more advanced than that of Batallas and Tiwanaku (table 9.2). A smaller percentage of the population (76 percent) is unable to meet its most basic needs, compared with more than 90 percent of the population in Batallas or Tiwanaku. The poverty rate in Comarapa is also slightly lower, with 58 percent of the population living in extreme poverty, compared with 67 percent in Batallas and 74 percent in Tiwanaku.

TABLE 9.2   Key socioeconomic indicators for Comarapa

Category	Indicator
General	
Total population	14,660
Altitude (miles)	1,870
Number of rural communities	27
% rural	72.1
% indigenous	47.1
Socioeconomic indicators (2001)	
Unsatisfied basic needs (% of population)	75.8
Human development index	0.59
National ranking (human development index)	145
Life expectancy at birth (years)	61.8
Literacy rate (% of population 15 years and older)	85
Annual consumption (US$ per capita)	941
Average years of schooling	5.6
Number of health centers	
Net migration (per 1,000 persons)	−8.58
Millennium Development Goals (2001)	
MDG 1: Eradicate extreme poverty and hunger	
Extreme poverty rate (% of population)	58.0
MDG 2: Universal primary education	
Primary completion rate (% of population)	77.8
MDG 3: Gender equality	
Ratio of female to male primary enrollment	97.61
MDG 4: Reduce child mortality	
Infant mortality rate (deaths per 1,000 live births)	78.3
MDG 5: Improve maternal health	
% of births attended by skilled health staff	92.4
MDG 7: Ensure environmental sustainability	
Access to potable water (% of population)	60.5
Access to sanitation services (% of population)	39.0

Sources: INE 2001; UNDP and INE 2006; UNDP, INE, and UDAPE 2004.

Moreover, according to basic social indicators, living conditions are better in Comarapa than in Batallas. Average years of schooling are 5.6 years, and the illiteracy rate is 15 percent. The level of human development is intermediate, with an overall HDI of 0.59.

With regard to overall poverty levels, the statistical data confirm the improvements that Comarapa realized during the 1990s. In fact, the percentage of people who are unable to meet their basic needs decreased from 85 percent in 1992 to about 76 percent in 2001 (figure 9.7). While the population of Comarapa experienced gains in their living conditions overall during this time period, the population in Batallas and Tiwanaku did not, and poverty levels continue to be higher in the highlands than in many lowland communities.

As shown in figure 9.7, the eastern lowlands region of Bolivia has made significant progress in poverty reduction. In the department of Santa Cruz, the level of extreme poverty declined from around 60 percent of the population in 1992 to only 38 percent in 2001; in the department of La Paz, poverty rates declined only from 71 to 66 percent in the same time period.

Comarapa also has relatively low rates of maternal mortality, and about 90 percent of women received professional support during childbirth in 2002 (figure 9.8). This is far above the national average of 61 percent and significantly higher than in Batallas and Tiwanaku (49 and 34 percent, respectively).

Infant mortality, however, remains high in Comarapa, with 78 deaths per 1,000 live births in 2001 (figure 9.9). These rates are even higher than in Batallas and Tiwanaku (71 and 65 deaths, respectively, per 1,000 live births). Although they have declined since 1992 (85 deaths per 1,000 live births), infant mortality rates in Comarapa are still far above the average for Bolivia as a whole and for the department of Santa Cruz.

Finally, access to basic infrastructure services is better in Comarapa than in either of the other municipalities in the Bolivian highlands. Access to potable water has improved considerably since the early 1990s, increasing from 24 percent in 1992 to almost 61 percent in 2001 (figure 9.10).

However, only about one-third (39 percent) of the population has access to basic sanitation services (figure 9.11).

Therefore, living conditions are somewhat better overall in Comarapa than in Batallas or Tiwanaku, suggesting that the conditions for implementing

FIGURE 9.7 Poverty rates in Comarapa, 1992 and 2001

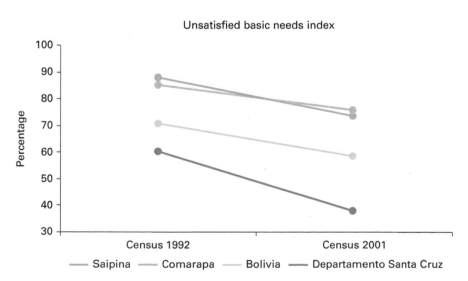

Source: UNDP, INE, and UDAPE 2004.

FIGURE 9.8   Maternal mortality rates in Comarapa, 2002–06

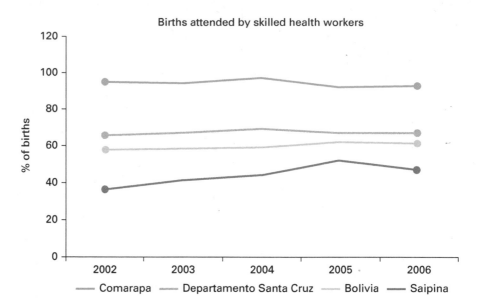

Source: UNDP, INE, and UDAPE 2004.

FIGURE 9.9   Infant mortality rates in Comarapa, 1992 and 2001

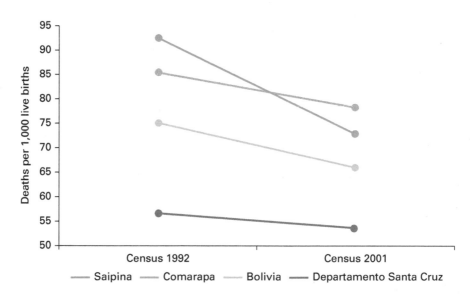

Source: UNDEP, INE, and UDAPE 2004.

an ICT program are more favorable. The empirical evidence in this chapter seeks to determine the extent to which this correlation can be confirmed and whether EnlaRed was indeed more successful in Comarapa than in Batallas or whether it had only limited impact in both municipalities, regardless of the overall socioeconomic situation of the municipality.

FIGURE 9.10 **Access to potable water in Comarapa, 1992 and 2001**

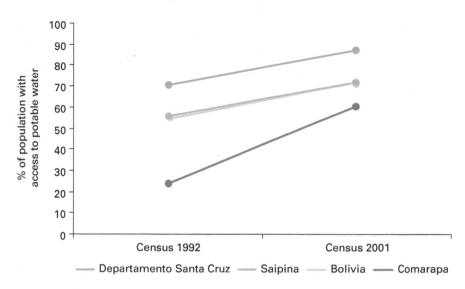

— Departamento Santa Cruz — Saipina — Bolivia — Comarapa

Source: UNDP and INE 2005.

FIGURE 9.11 **Access to sanitation services in Comarapa, 1992 and 2001**

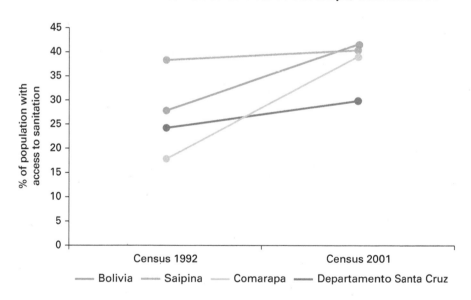

— Bolivia — Saipina — Comarapa — Departamento Santa Cruz

Source: UNDP and INE 2005.

The EnlaRed Municipal Program

The EnlaRed Municipal Program was implemented in two phases. The first phase, from March 2001 to June 2004, was implemented by the International City/County Management Association (ICMA) and financed by the USAID. The ICMA is a U.S.-based international professional organization for managers, administrators, and assistants in cities, towns, and counties that provides technical and managerial assistance, training, and information resources to its members and the local government community. During its first phase, the program sought (a) to create a central Internet portal to assist affiliated municipalities to respond to the needs and demands of their citizens (www.Enlared .org.bo), (b) to develop an information system for the Mancomunidad of the Chiquitania,[5] and (c) to gather lessons learned and offer recommendations on how to encourage local municipalities and communities to use ICT applications (Ayo 2003).

However, the program's impact on local governments was limited by its ambitious and poorly focused objectives; its concentrated, top-down organization; and the perception among participants that the ICMA was a foreign and distant organization unconnected with the particular realities of the Bolivian municipalities. To analyze the program's impact, I conducted a series of unstructured interviews during field visits to the Chiquitania region in the eastern lowlands in April 20–25 and August 2–7, 2005. I interviewed a local indigenous leader, the mayor of Concepción, the director of Apoyo Para el Campesino-Indígena del Oriente Boliviano (APCOB)—a local NGO promoting indigenous peoples' development in the eastern lowlands—and an ICT specialist of APCOB (see appendix J for a detailed list of interviews). In the interviews, I asked about the extent to which local government officials and community leaders have used the Internet for their daily work and the role that EnlaRed has played in promoting use of the Internet in their communities. I also visited four indigenous community centers and two municipalities.

During the site visits and interviews, it became clear that the municipalities and communities were not aware of the information management system and were not even connected to the Internet. Moreover, people did not use the Internet in a local telecenter or some other public space. A tremendous "design-reality" gap was evident between project descriptions, consultant reports, and realities on the ground. Most of the individuals interviewed were not aware that the project existed, even though its first phase had been completed less than a year before the fieldwork was carried out. In the municipality of Concepción, located in the Chiquitania, approximately five hours from the city of Santa Cruz, the newly elected mayor expressed his frustration with the planning and execution of the program:

>> Frequently the needs of local communities are not considered. The projects are in fact designed on the desks of the ministries or of the international donors. A consequence of this approach is that many projects are not sustainable at the local level and end once the financing dries up. This program must be one of those unsustainable projects—at least for my municipality there are currently no concrete, visible benefits from this project.[6]   <<

Because the first phase did not have any lasting impact on local communities, I focus here on the second phase, when USAID officials decided to refocus the program and strengthened its partnership with FAM, Bolivia's national association of local governments.[7] During this phase—from the beginning of September 2004 to the end of November 2005—FAM assumed responsibility for project coordination. It organized a competitive bidding process in mid-2004, which was won by the Swiss-funded Programa de Apoyo a la Democracia Municipal (PADEM).[8] The program focused on increasing the capacity, accessibility, and accountability of municipal governments in order to strengthen the participation of community-based organizations and citizens in local governance. This program was selected, in part, because of its seven-year experience in working with local governments on issues related to good governance, participation, and empowerment.[9]

The new project team aimed to make the program more responsive to specific local needs and to integrate ICTs into the local and regional processes of policy making and citizen participation. Emphasis was placed on finding new ways to achieve the development objectives of the project.

In order to set realistic goals, the team focused on providing technical assistance to 15 municipalities and 9 regional municipal associations, covering all of Bolivia's 9 states.[10] A major challenge was that the 15 municipalities are highly diverse, including rural, poor indigenous communities such as Batallas in La Paz or El Puente in Tarija, *mestizo*-dominated rural municipalities such as Comarapa or Montero in Santa Cruz, and large urban municipalities such as Trinidad or Cobija—the capitals of Beni and Pando, respectively. These municipalities differ not only in their access to electricity and telecommunications services but also in their socioeconomic, political, and cultural context. To understand the local context, the program team initiated its activities in October 2004 by conducting a detailed ICT needs assessment at the local level.

The main objective of the assessment was to determine the differences in awareness, ICT use, and proficiency of use among local government officials. As part of this baseline study, the team held community meetings and conducted interviews and structured surveys with mayors, councilmembers, and technical staff of the municipalities; representatives of local territorial grassroots organizations; members of the vigilance committees, and other community leaders. In total, 634 people—424 men and 210 women—provided information on their information and communications needs and their current use of ICTs (Calla 2004).

The assessment found that there was little awareness or use of the Internet in any of the rural municipalities and that, even in the urban municipalities like Tupiza or Uyuni, local governments were not using the Internet for work. Out of 127 local government officials, only a quarter (24 percent) used the Internet for work; the overwhelming majority (two-thirds) did not use the Internet at all, and one-tenth did not even use computers. The needs assessment revealed profound differences in ICT readiness among the municipalities. The details of the program are discussed later in the case study; what is important to note here is that the program was initiated even in rural municipalities with little or no demonstrated technical preparedness.

Overall Impact of the Program

The empirical analysis of the program's impact is based on data from two user surveys conducted by the project team and a subsample of the impact survey ($n = 22$) conducted as part of my research in July 2005. The first user survey (*n* = 190), administered as part of the ICT needs assessment in October 2004, was designed to generate a baseline of information about the awareness and proficiency of ICT use. The second user survey (*n* = 186) was conducted in September 2005 in the same 15 municipalities as part of the project's monitoring and evaluation activities. The main difference between these two surveys and my impact survey is that the project surveys did not include any questions about people's perceptions of the project's impact on their personal well-being and the well-being of their communities, but instead assessed their proficiency in ICT use and evaluated the ICT training component of the program.

Before analyzing participants' perceptions of the project's impact on their human and social capabilities, it is important to note several basic statistical indicators of people's ICT use. The first critical finding from the data analysis is that the large majority of local government officials from the 15 municipalities that participated in the project were unaware of the project's existence. Only a small percentage of respondents to the baseline survey conducted in 2004 and the user survey conducted in 2005 indicated that they had heard of the project, and an even smaller percentage said they had used the services provided by the project. Out of 127 respondents, only 16 (13 percent) had heard about the project and only 14 (11 percent) were using the portal developed by the ICMA. This figure is astonishing, since in October 2004 the project had been operating at a national level for three years. These data confirm that, particularly in its first phase, the project failed to make any long-term positive impact on local governments or on the participating communities.

Awareness of the program increased significantly during the second phase, but the project continued to suffer from relatively low awareness at the local level (figure 9.12). In 2005 only 70 of 190 respondents (about 40 percent) indicated that they had heard of the project—a major improvement from the

FIGURE 9.12 Awareness of the EnlaRed program

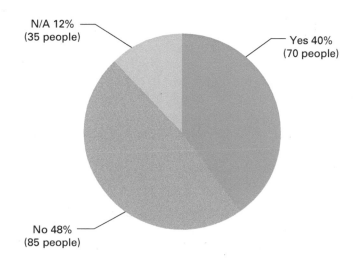

N/A 12%
(35 people)

Yes 40%
(70 people)

No 48%
(85 people)

Note: n = 190.

extremely low figure of 13 percent a year earlier. However, the project continued to be underused: only 31 respondents (17.5 percent) indicated that they frequently used the services provided by the project.

In spite of the significant improvements during the second phase of the project, the continuously low levels of awareness and use among local government officials raise questions about the program's overall effectiveness and are a good initial indicator for the limited impact of the project.

The principal finding of this section—that the project had a very low overall impact—is confirmed by the empirical results from the impact survey I carried out during my fieldwork. Results from this survey indicate that—based on the participants' own assessments—the program did not enhance in any significant manner the human and social capabilities of its participants. Participants said that the Internet has had a relatively minor impact on their political, social, and economic well-being and an intermediate impact on their organizational well-being (figure 9.13).

With the exception of the organizational dimension—where the perceived impact index reaches an intermediate level of 1.35—the EnlaRed participants believe that the Internet has had a relatively low impact on their well-being in all of the other dimensions (political, social, and economic).

Another noteworthy finding is that the EnlaRed participants also have considerably less positive views on the Internet's impact on their lives than the overall sample. In the social dimension, the score for perceived impact is only 0.88, compared with 1.44 for the overall sample. In the personal dimension, the score for perceived impact is 1.06, compared with 1.41 for the overall sample. In fact, as the data in figure 9.13 show, EnlaRed respondents have a less positive view of the Internet's impact on their lives in five out of the six dimensions. The only exception is the organizational dimension, in which the EnlaRed respondents have a more positive view about the Internet's impact on their lives than the overall respondents.

At the same time, the difference in people's perceptions is particularly notable in the political and social dimensions, since the core objective of the project

FIGURE 9.13   Perceived impact of the Internet on various components of well-being among EnlaRed participants

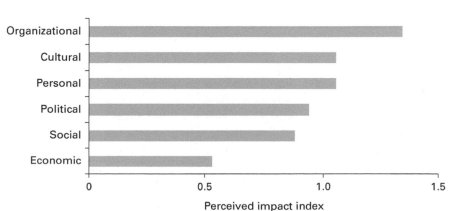

Perceived impact index

was to improve indigenous peoples' well-being in both of these dimensions. In fact, in the political sphere, most of the program's participants said that the Internet has had only a minor impact (PII = 0.94).

Compared with participants in the OMAK program, the subject of chapter 8, EnlaRed participants said that the Internet has had a low impact on their social well-being. For instance, participants in the OMAK program said that the Internet has had a very positive impact on their social well-being, with a high score for perceived impact of 1.70, which is almost twice the score for EnlaRed participants (PII = 0.88). With respect to the economic dimension, the EnlaRed respondents have a particularly negative view, with the perceived impact index reaching a very low value of 0.53.

Finally, both EnlaRed respondents and the overall sample said that the Internet has had an intermediate impact on their cultural well-being, with a score for perceived impact of 1.06 and 1.16, respectively.

Only in the organizational dimension do EnlaRed project participants have a more favorable view than the overall sample, with a score of 1.35 and 1.33, respectively.

The empirical evidence from the impact survey thus finds that the participants from EnlaRed have more negative perceptions of the Internet's effects on their well-being overall than the participants from other programs. In contrast with the overall results, EnlaRed participants are less positive in their assessment of the extent to which the Internet has had an impact on their individual well-being and on the social capabilities of local governments and their communities.

I now turn to the effects of the Internet on various components of well-being. The empirical analysis shows that respondents believe that the Internet has had the most significant impact on strengthening their social capital, establishing horizontal communication networks among local government officials, and enhancing the vertical accountability of local governments (figure 9.14). Remarkably, these three areas were not among the core objectives of the program and instead seem to be unintended consequences.

FIGURE 9.14 Perception of ICT impact on specific areas of well-being among EnlaRed respondents

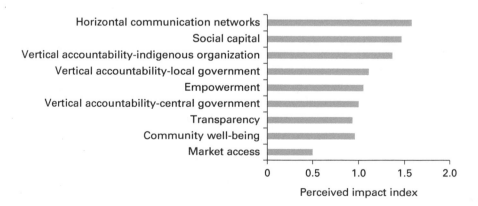

Moreover, with respect to the strengthening of social capital, the results indicate that the majority of respondents deem the Internet to have had an important positive impact on their lives. The perceived impact of 1.47 is slightly more positive than that of the overall sample (1.43), but less positive than that of OMAK participants.

They also said that the program has had a very positive impact on the establishment of horizontal communication networks. The majority of respondents said that the Internet has had a substantial effect on this area, with a high score of 1.58. Moreover, the EnlaRed respondents have a more positive view than either the OMAK respondents (PII = 1.45) or respondents overall (PII = 1.33).

Participants said that the Internet has had a low-intermediate impact on the vertical accountability of local governments (PII = 1.12). This area was a key priority of the program, as vertical accountability is critical for improving the relationship between local governments and citizens. Nevertheless, respondents from both the overall sample and the OMAK program said that the impact has been slightly higher, with a score of 1.15 and 1.13, respectively. Even in this area, which represents a core priority of the EnlaRed program, the participants from other programs have a more positive view than the EnlaRed respondents.

Respondents from EnlaRed also have the least positive views about the Internet's role in improving access to markets, enhancing the government's transparency, and improving the community well-being of indigenous peoples.

Their view of the Internet's limited economic impact (PPI = 0.94) is consistent with the results for the overall sample (PPI = 0.99) as well as with those for the OMAK sample (PPI = 1.00), where respondents also expressed doubts about whether the Internet has had a major impact on existing markets. The second finding, reflecting perceptions about the low positive impact of the Internet on the government's transparency (with a PII = 0.94), stands in contrast to the program's objectives. It is, nevertheless, comparable to the perception of both the overall respondents and the OMAK participants, who, respectively, gave a low value for perceived impact of 0.99 and 1.00. It is particularly noteworthy since, unlike the other ICT programs, the EnlaRed program defined enhanced government transparency through ICTs as a core objective.

Finally, the finding that EnlaRed participants have a less favorable view about the Internet's impact on community well-being (PII = 0.96) is not unexpected, since the program did not give priority to this area.

Impact on Informational Capabilities

To what extent did the program participants attain advanced informational capabilities—in particular, information literacy skills, communication capabilities, and ability to share local knowledge and experiences with others. Furthermore, to what extent did the program help to raise awareness not only among local government officials but also among indigenous communities of the ways that improved access to, and sharing of, information can improve the transparency and accountability of local government?

As mentioned, the program conducted an information needs assessment as a baseline study to evaluate the ICT readiness of each municipality, to gauge their level of ICT exposure to, use of, and proficiency of use. The assessment then focused on issues related to people's specific use of ICTs (e-mail, telephone, and community radio), their knowledge of how to use these tools, and the problems they have encountered in using them. The assessment specifically gauged the use of ICTs and not—as in the OMAK program—the role of information per se in local political processes and of ICT programs in improving existing flows of information and communication between local governments and communities.

This difference between the two programs is significant, since it demonstrates that EnlaRed, in contrast to OMAK, emphasized technology itself and failed to take a holistic approach to ICTs, which would have warranted studying the local socioeconomic, political, and cultural context *before* promoting ICT use. The program suffered from its techno-centric approach, which was too specialized to meet its ambitious agenda of improving local governance.

During my fieldwork, I had the opportunity to attend three of EnlaRed's training workshops. During the first workshop in February 2005, I discussed the needs assessment with several participants. Bernal Patty, the Batallas representative, expressed doubts about its usefulness:

>> When the project team first came to Batallas, they raised a lot of expectations. During the meetings with the project team, all the main decision makers from the municipality and surrounding communities were present. Most people were very excited to hear about the program. From the beginning, however, a serious concern for us was what the project would concretely do to improve the living conditions of our communities. I had the impression that the program team first did not know themselves what it was they really wanted to achieve, besides promoting the use of ICTs.[11]   <<

While the needs assessment reached out to a broad range of stakeholders (councilmembers and representatives of indigenous organizations and women's organizations), it failed to engage participants in a broader reflection on the role of information and communication in their communities' social, economic, and political development.[12] It did not help to raise awareness among indigenous peoples and local government officials of the ways that improved access to, and sharing of, information can ameliorate governance, transparency, and local accountability. Instead, it gathered information about people's uses of, and the municipalities' readiness to use, ICTs. This particular focus generated a great deal of statistical data but did not provide participants with a space for broader reflection on the role that ICTs can play in their communities' social, economic, and political development. Furthermore, it failed to engender a sense of ownership over the program among local stakeholders or to establish a relationship of trust between local communities and the La Paz–based project team.

As a result, the EnlaRed participants attained lower informational capabilities than participants from any other ICT program. Only 17 percent of participants reached an advanced level of informational capabilities, compared with 46 percent of the overall sample, while approximately one-third acquired only basic informational capabilities (figure 9.15).

FIGURE 9.15   Perceived impact on informational capabilities, by program

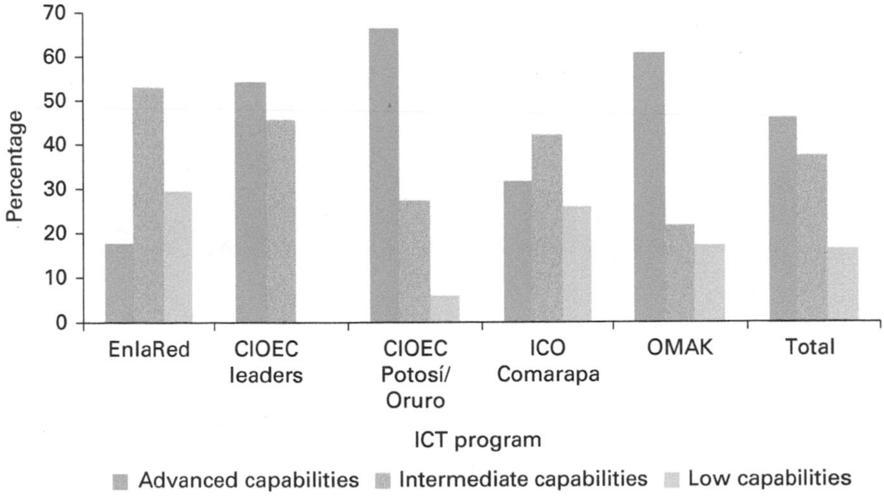

ICT program

■ Advanced capabilities   ■ Intermediate capabilities   ■ Low capabilities

When comparing these data with the results from the grassroots programs studied—OMAK and the Coordinadora de Integración de Organizaciones Económicas Campesinas Indígenas y Originarias (CIOEC)—the differences become apparent. For instance, among participants from the CIOEC Potosí and Oruro and OMAK programs, only 6 and 17 percent, respectively, said they have acquired only basic informational capabilities, while 67 and 61 percent, respectively, said they have become proficient (figure 9.15). This is particularly noteworthy since EnlaRed operated in a more conducive socioeconomic context than the other two grassroots programs; thus one would expect much more advanced uses of ICTs in the EnlaRed communities than in the other communities.

First, the overwhelming majority of EnlaRed participants (84 percent) indicated that they have either an advanced technical or a university education, compared with only 57 percent of OMAK participants (figure 9.16).

Second, only 5 percent of EnlaRed participants live in municipalities with extreme levels of poverty (with unsatisfied basic needs in categories IV and V), compared with 65 percent of OMAK participants (figure 9.17). Moreover, 35 percent of EnlaRed participants live in municipalities with satisfied basic needs.

The data demonstrate the limited positive impact of the EnlaRed program on its participants' informational capabilities in spite of a significantly better socioeconomic context. These findings provide evidence for the contention that the program's training program overemphasized the role of technology and almost completely neglected the broader issues of information literacy, information management, and communication capabilities.

In order to analyze in more depth the process of program implementation, this section describes in more detail the qualitative data gathered during the fieldwork through participatory observation and in-depth interviews with the program's participants. The program's centralized structure reinforced

FIGURE 9.16   Education level of participants, by program

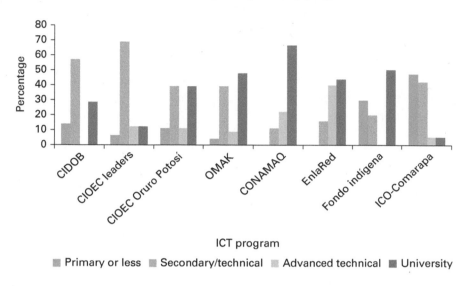

ICT program

■ Primary or less   ■ Secondary/technical   ■ Advanced technical   ■ University

FIGURE 9.17   Range of municipal poverty rates, by program

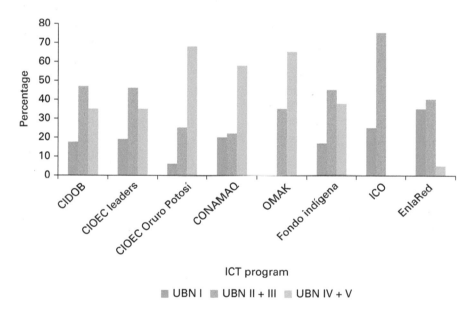

ICT program

■ UBN I   ■ UBN II + III   ■ UBN IV + V

the perception held by mayors and senior local government officials that the program was not relevant to their daily work. Most municipalities did not consider the project to be a real priority and delegated only very junior staff from their technology or communications teams to attend the training workshops. The mayors of Uyuni and Machacamarca refused to send anyone to the first workshop, and the EnlaRed program director had to travel to both municipalities to explain, in person, the project's rationale to local authorities.[13] The greatest shortcoming of the training component, then, was its centralized nature, which disenfranchised local stakeholders and impeded any meaningful discussion about the role of ICTs at the local level.

Many participants also criticized the discontinuity of the training program, mentioning, in particular, that all of the training materials were covered during the two workshops and that no follow-up was provided between or after the workshops. This problem was compounded by FAM's failure to select municipalities with adequate infrastructure to implement the program, basing selection instead on political considerations. Participants could not access the Internet at their workplace and had to access it instead in a telecenter; participants from four municipalities could not use any Internet at all once they returned to their communities, since the entire municipality was lacking access to the Internet. According to a participant from Tarabuco,

>> For us in Tarabuco it is very hard to access the Internet. Every time I want to read my e-mails, I need to travel to Sucre, which is one hour away from our community. My colleague and I were very excited to be able to participate in this workshop, and we have learned a lot of useful things here, but what do we do with this once we return to Tarabuco? We will never be able to apply this back home, since nobody is using the Internet in our municipality.[14]   <<

Another serious weakness of the training program was that it did not encourage broader, critical reflection on the role that ICTs might play in public policy making or in promoting local economic and social development; it focused instead on offering practical training in the use of different ICT tools and applications. This narrow, technology-driven view of ICTs resulted in the uncritical use of these instruments by the participants, who had no knowledge about how ICTs might support the broader agendas of democratization and decentralization. Eduardo Rojas, from Fundación Redes, who was contracted by the EnlaRed team to coordinate the second training workshop, criticized this techno-centrism:

>> For me, this course should allow people to reflect about their own roles in the municipalities and to analyze the underlying reasons why many mayors are not willing to make information publicly available. For me, this is a political issue and not one of technology. But the program team insisted that we teach the participants practical tools and avoid any major discussions. This overly techno-centric approach will not have any meaningful results. Then people will know how to use e-mails and to build websites, but they won't be able to apply this to the political context in which they work. I think for many people this is a waste of their time.[15]   <<

During the second training workshop, I observed a small group of six participants with advanced ICT skills learning how to develop websites using languages like html and flash. The teaching required a high level of technical sophistication but was indifferent to local political realities. Marcelo Carbone, the communication specialist from Comarapa and one of the most enthusiastic participants, explained the purpose of his website:

>> I am very excited that now I have finally learned how to build a website for our municipality. I have spoken several times with the mayor about this, and he seemed to be quite interested in this project. For me, the most important aspect

of the site is to show the beauty of Comarapa and its surrounding communities. I hope to use it as a major tool to promote tourism in our municipality. In this workshop I have learned how to use many new tools. For instance, one of the most useful tools is how to develop better graphics for the site. The new designs make the site really much more attractive.[16]   <<

The course improved participants' software skills and helped them to construct their own websites. However, none of the sites included any meaningful content on their municipality. The sites, while visually appealing and easy to use, omitted even the most basic information, such as contact information or socioeconomic data about the municipality. Simply put, the course failed to place the use of ICT tools into the local political or socioeconomic context.

Focusing the training exclusively on ICT tools like e-mail and website design, with little mention of local socioeconomic and political topics, alienated the participants, who found the training inapplicable to their municipalities. Eduardo Rojas sharply criticized the program:

>> The main problem of the program as of so many other ICT projects in Bolivia is that they are just replicating the experiences of other sectors, such as water or agriculture. Fundamentally, the emphasis is a purely technocratic approach, which does not consider the rich cultural and social diversity in our country, but centralizes everything in La Paz and imposes the same approach in each and every community.[17]   <<

Finally, the empirical evidence shows that the program did not significantly enhance participants' ICT capabilities. This finding is remarkable since the program placed such a strong emphasis on technology, and the capacity-building component stressed the importance of ICT tools in its curriculum. Only about 24 percent of the program's participants said they have advanced ICT capabilities, compared with 30 and 28 percent of participants from OMAK and CIOEC Potosí and Oruro, respectively (figure 9.18). Moreover almost half of the EnlaRed participants (44 percent) indicated that they have attained only basic ICT skills, even after completing the two training courses, compared with 26 and 11 percent of participants from OMAK and Potosí and Oruro, respectively. This number indicates the ineffectiveness of the training program even in an area that was deemed an important priority.

Impact on the Political Dimension of Well-Being

This section investigates the underlying reasons that the program had only a limited effect on the political dimension. This issue is central to the evaluation of e-government programs, and specifically EnlaRed, since its main objectives were political in nature.

The analysis of survey results and interview responses in the previous section shows that EnlaRed failed to improve participants' informational capabilities significantly. This section investigates the project's failure to promote ICT use as a way to improve governmental responsiveness, transparency, and accountability. A critical result from the impact survey is that the majority of the EnlaRed respondents believe that the Internet has had a minor

FIGURE 9.18 **Perceived impact on people's ICT capabilities, by program**

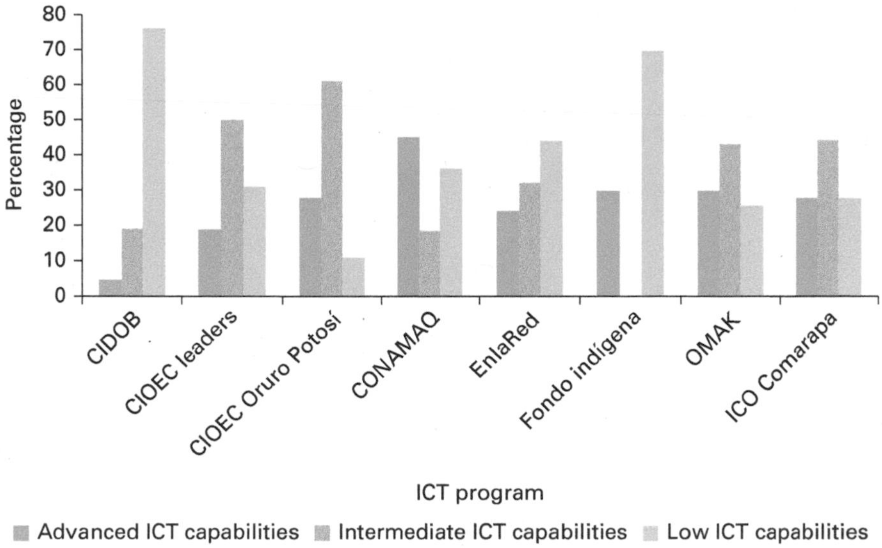

ICT program

■ Advanced ICT capabilities ■ Intermediate ICT capabilities ■ Low ICT capabilities

effect on the political dimension of their lives, with a low score for perceived impact of 0.94, and 35 percent said that the Internet has had *no effect at all* on this dimension. The overall negative perception of EnlaRed participants in this dimension becomes apparent when their responses are compared with those from the overall sample, where participants believe that the Internet has had at least a moderate impact on their lives, with the value for perceived impact reaching 1.16, which falls at the low end of the intermediate-impact category. Again, EnlaRed is the only program studied that specifically prioritized the use of ICTs to enhance peoples' political well-being.

What are the underlying causes that explain this finding? Which factors explain why the program was so unsuccessful in reaching its main objectives? What were the effects of the program on local governance structures and which barriers did it encounter in trying to enhance indigenous peoples' political well-being? Based on Pratchett's classification, which was mentioned in chapter 1, I divided the political dimension into two areas: (a) enhanced transparency and accountability and (b) improved local governance.

Impact on Transparency and Accountability

The first finding is that EnlaRed had only limited effects on the transparency and vertical accountability of local governments and thus could not significantly strengthen local democratic processes. Figure 9.19 shows the effect of the program on four components of accountability. The disaggregated results show that the program helped to improve people's access to information about their own rights and national government policies, but failed to enhance the transparency of local governments. The scores for perceived impact on both transparency (PII = 0.68) and information about government programs (PII = 0.59) are very low, indicating only a marginal positive impact on these core aspects of political well-being.

FIGURE 9.19   **Perceived lack of government transparency and accountability among EnlaRed respondents**

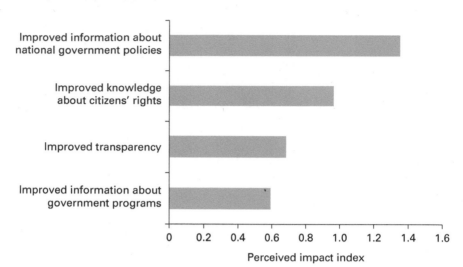

The results point to a serious incongruity in the program: while it had some positive impact on improving the access to information related to municipal development at the national level, it did not improve information flows between municipalities and local communities. The program had an overly abstract national agenda and only improved the access to information for a small elite of national policy makers, international donors, and the media; it failed to enhance in any significant manner the flow of information between local governments and citizens at the community level.

This finding is confirmed by the results of the semi-structured interviews with local government officials, NGO representatives, and community leaders in Batallas and Comarapa. In an interview, the "official mayor"[18] of Comarapa expressed his negative views about the project and the role of FAM in supporting local governments in general:

>> We have a very diverse population in Comarapa and the surrounding communities; about half the population are *mestizos*, and the other half are indigenous migrants from the highlands. Thus for us to improve our communications with all the communities is very important, but I don't see how this program can help us to better deal with this problem. The program has developed a website. How does this help us to improve our communications with more remote indigenous communities? How do these communities benefit from this? The majority of them do not even have access to electricity. How can they access a website? I really can't see the value of this type of program for us.[19]   <<

This comment needs to be understood in the context of Comarapa, a municipality that, in spite of relatively favorable socioeconomic indicators, must contend with significant social tensions between its primarily rural indigenous population and its urban *mestizo* population. Thus the senior adviser quoted above stressed the challenge of improving his municipality's capacity to communicate with local communities in order to enhance the participation of a broad range of people in municipal planning processes. This is a serious shortcoming that implementation of the LPP has confronted in most rural communities throughout Bolivia. While the law has succeeded in significantly enhancing the participation of communities in the identification of municipal development projects through broad-based consultations, in most municipalities it has been unable to improve the participation of local communities in municipal decision-making processes about critical issues of local development such as, for instance, the approval of planned public investment projects (Goudsmit and Blackburn 2001).

In Comarapa the local government frequently sided with the *mestizo* population and attempted to suppress the influence of indigenous peoples in local politics. The president of the local agricultural producer association and a Quechua leader described the exclusivity of local politics:

>> For us it is very clear that the municipality is quite biased against our communities. It is always the same issue. We can wait for hours in the municipality with our demands, and the mayor and the councilmember won't listen to our concerns. They are too busy with their political games and with representing the interests of the rich people of the city—we just don't have any voice in their decisions, and we have no information at all about what type of programs and activities they are planning for this year.[20]   <<

Claudia Camacho, coordinator of an ICT program for the Instituto de Capacitación del Oriente (ICO), a local NGO supporting small-scale agricultural producers, agreed:

>> I have been working here in Comarapa for the last four years and have been able to observe the way the municipality operates quite closely. The mayor is making his own decisions based on his political considerations, and the majority of the population has no access to any information at all. It is a rather undemocratic, closed process, whereby the indigenous farmers don't participate. For them, the only way of being heard is to organize themselves through their producer organizations and to organize a street blockage. Then the mayor usually listens to their concerns, since there is a crisis. Most of the people only pay attention when there is a crisis and when there is the threat that social tensions might explode.[21]   <<

Local authorities lacked the will to improve access to information for the general population and instead allowed a small circle of mostly urban elites to dominate the local political decision-making process. What was the role of EnlaRed in this difficult sociopolitical environment? Did the program have any influence at all on the local political process? Why did the program fail to promote the participation of indigenous communities in the decision-making process of local government?

One reason was the program's inability to react flexibly to the given sociopolitical context. Instead of working with local government officials to make the municipal planning processes more transparent and participatory, the project continued to promote its original concept: to assist the municipality in the use of technology. The program provided technical assistance in developing a central website for the municipality and stressed the training of government officials in ICT skills. These activities did not match the local sociopolitical context, nor were they appropriate for addressing the existing gaps in transparency and accountability. The program did not attempt to alter the elitist culture of local government. In brief, the program failed (a) to improve the transparency and accountability of local government and (b) to enhance people's political capabilities and help them to participate effectively in the decision-making processes. The empirical evidence presented above demonstrates the perception of local stakeholders that the EnlaRed program was too removed from communities and did not address their own realities. This illustrates the program's inability to improve local democratic processes meaningfully.

Impact on Local Governance

Another component of political capabilities is the capacity of municipalities to conduct public policy making in local communities. Within the increasingly fragmented and disaggregated structure of local communities, many local governments face considerable challenges in trying to promote local governance structures that are conducive to economic and social development (Pratchett 1999, 735). The following analysis uses the concept of "good local governance," developed by Judith Tendler, which refers to a healthy "three-way dynamics among local government, civil society, and an active central government" (Tendler 1997, 145).

At the core of the concept is the notion that a good structure of local governance is essential for effective public policy making, including collaboration between local governments, civil society organizations, and local communities, as well as collaboration between local and central governments. Local government is an effective facilitator of partnerships between local actors and stakeholders rather than an implementer of development programs.

What role can ICTs play in this system of local governance? To what extent was EnlaRed able to strengthen the capacity of local governments to assume the role of a "facilitator" of local development? To what extent was the project successful in enhancing healthy local governance in local communities?

The principal finding from both the municipalities of Comarapa and Batallas is that the EnlaRed project was unable to play a central role in enhancing the capabilities of local governments in this dimension, primarily because it did not effectively promote an institutional change within local governments toward a more democratic and inclusive approach to public policy making. Instead, local governments continued to manage local development programs in their traditional bureaucratic styles.

Asked about the role that EnlaRed played in local politics, Claudia Camacho from ICO cited the program's inability to facilitate a partnership between the local government and her NGO to support the construction of a telecenter for Comarapa:

>> When we first heard about the EnlaRed Municipal Program and that Comarapa was selected to participate in its pilot phase, we were very excited and thought that we could closely collaborate with the municipality on promoting the use of

ICTs, a key objective of our own program. Our idea was that we could share the connectivity costs with several institutions working in Comarapa and asked the mayor for his support. Unfortunately, the mayor did not see any value in this program and instead promised that the municipality would build its own telecenter for the community. Now, two years later, Comarapa still does not have the promised telecenter, and we have never received an official response from the municipality on the proposal we submitted in writing to the mayor's office.[22]   <<

This lack of collaboration among local municipalities, local NGOs, and the central government is all too common. Local governments often have the official political responsibility for but lack the institutional capacity and technical knowledge needed to implement development programs like EnlaRed. According to the mayor of Comarapa,

>> For me it is very clear. We in the municipality are the executive branch of government at the local level. It is our mandate to plan and implement all programs in our municipality. I honestly don't understand how these NGOs can operate here without coordinating their activities with us. They just don't have the mandate to do so; also since they are not elected by the population I wonder who is holding them accountable? In the case of the planned telecenter, I had several meetings with the staff from ICO and explained to them that it is our role to coordinate this project. I still think that it is a good idea to build a telecenter here in Comarapa, particularly for our youth, but we need to oversee all activities related to this project.[23]   <<

This statement demonstrates that the mayor considered it the prerogative of local government to coordinate and oversee all development programs in his municipality. While local government officials frequently insist that the municipality needs to spearhead the development programs at the local level, the severe crisis of the state has led to a situation where national and international NGOs have frequently had to fill the institutional vacuum and are de facto implementing many programs, such as in health, education, or infrastructure, that normally would be carried out by the state. This frequently leads to important tensions between local governments, NGOs, and citizens. Another reason for the poor local governance structure in Comarapa is the role that international donors frequently play in creating dependencies in both the public sector and civil society organizations at the local level. The EnlaRed program, for instance, defined its priorities and operational methods through a highly donor-centric approach. It did not allow local actors to adjust the program's planned activities to their particular social, political, and cultural context or to collaborate on its implementation. The provision of Internet services and ICT training workshops would have lent itself to such a collaborative approach, since many organizations within the municipality had expressed their strong interest in sharing the connectivity services. An important consequence of such a donor-driven approach is that it provides important disincentives for local actors to work together and to coordinate their activities with each other, since an organization is accountable primarily to its international donor and not to local stakeholders (Nicod 1996).

The inability of the municipality to react to the proposal by ICO to share the costs of the Internet connectivity demonstrates that programs like the EnlaRed program often focus solely on achieving their centrally defined development objectives rather than on facilitating partnerships among the different local actors.

Furthermore, the collaboration between the local government and the departmental and central governments is often weak. In an interview, the official mayor criticized centralized institutions such as FAM:

>> For us here in Comarapa, La Paz is very far away. We barely communicate with anybody in the central government. For us it is much more important to focus on the implementation of our municipal development plan. Sometimes people from the *prefectura* or from the central government call us, but usually only to ask us to send them some data or a report. This is not very useful to us; it only means more work for us. Also, you asked me about the FAM, well, I don't exactly know what they do in La Paz. What they say is that they need to represent our interests there. So far we have not received anything concrete from them.[24]   <<

Moreover, the results from the baseline survey indicate that only a minority of local government officials consider enhanced communication between colleagues from central government agencies as being important. More than half (56 percent) said that communication should be enhanced with international organizations rather than with the central government, with only one-third (31 percent) indicating that improving communication between the local and national governments is important.

In sum, EnlaRed failed to enhance the local governance structure at the community level, since its program had an overly abstract national agenda and its centralized approach prevented it from effecting significant institutional changes at the local level. Furthermore, the program did not facilitate effective partnerships between local actors and instead tried to implement its preconceived objectives irrespective of the local sociopolitical context. In this sense, it illustrates the shortcomings of many donor-centric programs in Bolivia. Finally, the program lacked institutional credibility with local government officials. It had no significant impact on enhancing the coordination between local government, civil society organizations, and local communities; neither did it enhance the relationship between local and central governments.

Impact on the Organizational Dimension

As mentioned, EnlaRed participants said that they have a relatively positive view about the Internet's effects on the organizational dimension, with an intermediate score for perceived impact of 1.35 (figure 9.20). This finding is noteworthy, since the data indicate that they have a more favorable view in this dimension than the overall sample, with a slightly lower score for perceived impact of 1.33.

This section disaggregates these overall results in order to investigate the extent to which EnlaRed did indeed lead to organizational changes in local governments and thus had a positive impact on their institutional capability.

FIGURE 9.20 **Perceived impact of the Internet on social capital among EnlaRed respondents**

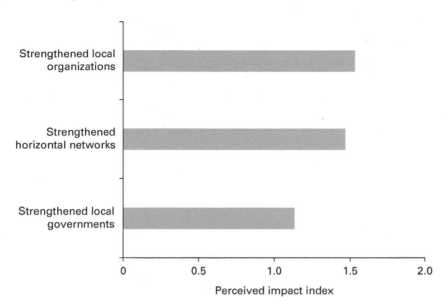

This issue is critical for evaluating the overall impact of the program, since one of its core objectives was to enhance the efficiency and efficacy of local governments by introducing ICTs into local public administrations.

First, the analysis uses the social capital index presented in chapter 8 in order to carry out a deeper analysis of the impact of the Internet on the organizational dimension. Figure 9.20 highlights that while a majority of participants perceive that the Internet has had a substantial positive impact on strengthening their horizontal networks with colleagues in other municipalities (PII = 1.53), they have a less positive view about the Internet's impact on strengthening the organizational capabilities of local government (PII = 1.14).

It is, however, noteworthy that they have a favorable view of the positive effects of the Internet on local organizations outside of government, such as community-based organizations (PII = 1.47), in spite of their negative views about its ability to bring organizational change within government. This seems to indicate that the participants believe that there are major barriers to introducing ICTs into public administrations, but consider this to be easier to achieve outside of government.

The program helped to establish a strong social network among its participants, but did not institute institutional changes within local bureaucracies. Furthermore, the observed strengthening of social capital and horizontal communication networks was the result of the ad hoc formation of informal networks between the program's participants rather than improved institutional linkages between municipalities. This more nuanced finding is confirmed by several interviews with program participants. Marcelo Carbone, from Comarapa, expressed the importance of being part of a network of like-minded professionals with whom he could share his experiences:

>> One of the most positive experiences of the workshop has been to meet my colleagues from the other municipalities. I feel that we have a lot in common since we are all IT [information technology] or communications specialists. We all face similar challenges in our municipalities. I am planning to stay in touch with several people I met here; this will be very useful when I have a question for which I can't find a quick answer. Also, I can't share these things with any of my colleagues in Comarapa. They just are not interested in the Internet or don't understand the things I am working on. There are times when this can be quite frustrating, particularly when I don't get the necessary support for my work from the mayor. So it is good to know people who face a similar situation.[25]   <<

To delve deeper into the impact on organizational well-being, I now draw on the empirical results from the two user surveys[26] carried out in October 2004 and September 2005 with local government officials in the 15 participating municipalities. The user surveys were carried out with a broad range of local officials in each municipality,[27] while the impact survey focused only on the IT and communication specialists who participated in the program's training workshops.

The first finding from the empirical evidence is that the program failed to institutionalize the use of the Internet in the administrative processes of local governments. By September 2005—a year into the second phase of the project—only about one-third (36 percent) of local government officials said that they were using the Internet for work. Although about two-thirds of respondents indicated that they had some experience in using the Internet, most said that they used the Internet only for personal, rather than professional, communication (figure 9.21). In fact, the survey results show that only

FIGURE 9.21   **Internet use by local government officials among EnlaRed respondents**

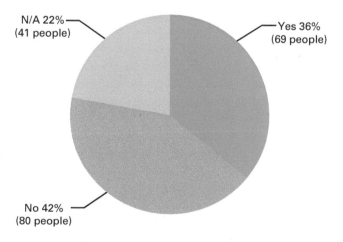

N/A 22%
(41 people)

Yes 36%
(69 people)

No 42%
(80 people)

Note: n = 190.

one-third of the respondents who use the Internet (29 percent) do so at their workplace; almost half (49 percent) said that they have to go to a telecenter or an Internet café.

Only 40 percent of government officials said they have a personal e-mail account (such as with Yahoo or Google), and none has an institutional e-mail account. The large majority (62 percent) of the respondents who have an e-mail account said that they use it only once every three days.

Moreover, key decision makers are apathetic to these technologies (figure 9.22). The large majority of mayors and councilmembers (82 percent) said that they do not consider the Internet to be important and do not use it for their work. The data indicate that only among technical staff did the majority (58.2 percent) use the Internet for their work. Technical staff also have higher rates of e-mail use: 54 percent of them have personal e-mail accounts, compared with just 18 percent of mayors and councilmembers.

The very low local appropriation of ICTs in local government is confirmed by comparing the results from the survey in October 2004 with those from the survey in September 2005: the percentage of respondents who use the Internet at work was unchanged, at about 36 percent in both years. The use of e-mail accounts actually *decreased* from 47 percent in 2004 to only 40 percent in 2005. The use of ICTs and the Internet continued to be low priorities in the 15 municipalities, in spite of their nominal participation in the program.

One reason for this low level of ICT use is the extremely high turnover of local government officials. The municipal elections in December 2004, four months into project implementation, resulted in profound political changes in a large majority of local governments and negatively affected the project: 68 percent of the respondents from the 2005 survey indicated that they had been working for less than one year in the municipality, and just 8 of the training participants, out of a total of 94, had participated in a training workshop during the first phase of the project.

This finding is confirmed by the fact that only technical staff (IT specialists and communication officers) participated in the program's training workshops

FIGURE 9.22 **Uptake of the Internet by government officials among EnlaRed respondents**

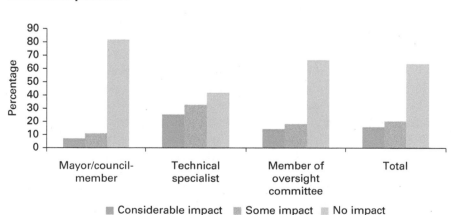

■ Considerable impact ■ Some impact ■ No impact

and that key decision makers (senior manager or councilmembers) did not show any interest in participating.

In sum, the program failed to reach one of its main objectives—to promote the use of ICTs in local public administrations in order to strengthen the organizational capabilities of local government.

Why did the project fail to bring about any significant institutional and culture change within local governments? The analysis draws on the theoretical framework developed by Stephen Peterson (Peterson 1998). According to Peterson, three tasks are critical for developing an information system and applying it in public administrations: (a) the commitment of senior managers, (b) the establishment of a process for solving organizational problems, and (c) the formation of an appropriate design team that implements a participatory systems design (Peterson 1998, 43). EnlaRed had important shortcomings in all three of these dimensions.

The findings from the qualitative research, in contrast to some of the quantitative data, clearly point to failures of the program. First, in interviews with senior government officials in both Comarapa and Batallas, it became apparent that the senior manager was not committed to the program's activities. For instance, in an interview, the official mayor of Comarapa was skeptical about the project and the role of FAM in supporting local governments in general:

>> This project is not important for us here in Comarapa. We have many administrative challenges we are facing in our daily work, and I don't see how this program has helped us with our problems. They just want to build a website. How is this useful to us? How does this improve our administration? We have many other concrete problems to solve that are much more important than to put some materials on a website.[28]   <<

The mayor of Batallas expressed a similar view:

>> We face many challenges in our daily work here in Batallas; however, it is not clear to me how the use of the Internet and the EnlaRed program specifically can help us with our daily work. It might be a good idea for someone in La Paz to have better information about issues related to municipal development, but in concrete terms I find this information of very limited use for the difficult tasks we have to deal with on a day-to-day basis.[29]   <<

These quotes indicate that the program was quite removed from the specific local realities of its participating rural municipalities and was perceived by many senior government officials as having very little value for their daily work. Furthermore, the fact that both the mayors of Uyuni and Machacamarca declined to send a staff member from their municipality to attend the training workshops also supports the finding that many senior government officials did not perceive the program to be valuable for their daily work at the local level.

Despite being highly critical of EnlaRed, the mayor of Batallas held a generally positive attitude toward ICTs and indicated several areas where the program could have been more useful for local governments:

>> For me, the issue is not whether ICTs are useful or not. I believe that these technologies can be quite useful for our work. For example, one of our key

administrative tasks is to coordinate the planning and execution of public investment projects in our municipality. Every year we elaborate an annual operating
budget that defines the projects for the coming year. For us it is quite difficult
to oversee all the different activities and expenses of the municipalities. Here,
I think the project could have really helped us. Also, since we receive about
20 percent of our budget directly from the central government, we always have
to prepare reports for either the *prefectura* or the central government. This continues to be a real headache for us, since we don't have a way to send them the
required documents, and instead the mayor or I have to travel to Santa Cruz and
La Paz to deal with these administrative processes.[30]   <<

This senior government official was not generally opposed to e-government
initiatives, but he criticized the program's failure to respond to the specific
needs of local governments and its implementation, which was based on preconceived concepts that were not a priority for the municipalities. Indeed, he
felt that the program could have helped to strengthen organizational capacities
by introducing an information system for monitoring the municipal budget
and the status of municipal public investment projects.

The National Planning System (Sistema Nacional de Planificación), which
was introduced jointly with the LPP, bases the planning processes of public
investments on the close coordination of all activities among the three principal layers of government: at the national, departmental, and municipal levels
(Molina Saucedo, Cuéllar Rivero, and Gandarillas Mansillas 2000). The effective coordination and flow of information among the different levels of government are critical, since the central government has delegated through the
LPP many responsibilities to the municipalities, including the delivery of basic
services such as infrastructure, education, and health, while it monitors the use
of funds and the effectiveness of projects. EnlaRed failed to help municipalities
to develop and implement an information system that could have improved
the ability of local officials to manage the municipal planning process. During
my fieldwork, I had the opportunity to visit the offices of the Vice Ministry
of Public Investment and External Financing (VIPFE), which is in charge of
overseeing municipal public investment projects at the central level. During my
visit, I interviewed the director of VIPFE's Information Systems Department,
who showed me the existing system for monitoring public investments (SISIN).
None of the municipalities had the institutional capacity to use the system;
instead, they submitted paper reports to the central government. According to
Clemencia Vargas,

>> Well, to be honest, the information flow between most municipalities and our
ministry is very scarce. While the municipalities are required by law to update us
regularly about the status of their different public investment programs, in praxis
most of our data are quite outdated since most municipalities don't have access
or know how to use our information system (SISIN). Instead, local government
officials usually travel to La Paz every couple of months and carry the data with
them either as paper reports or as an Excel spreadsheet on a floppy disk. This
way to share data is very problematic and causes many mistakes in data entry
and processing. Also, it causes a lot of administrative delays in the disbursement
of funds to the municipalities, since we need to monitor the use of funds and

cannot authorize payments before we have an idea about the financial records of the different investments.[31]   <<

As this account shows, the scarce flow of information between municipalities and the central government is a major reason why many public investment programs encounter serious delays in Bolivia. The large majority of rural municipalities lack the institutional capacity to monitor the progress of their programs and to report on their status to the central government. This is a strategic area in which EnlaRed could have improved the organizational capability of rural municipalities. However, the program failed to identify it as a priority and did not support municipalities in either designing or introducing an information system for monitoring their public investment programs.

While the introduction of a financial management system might have benefited municipalities like Comarapa with a relatively advanced administration, rural municipalities such as Batallas, with even lower institutional capacity, might have very different organizational and technical priorities. As the following quote from the senior adviser for Batallas demonstrates, the needs of this municipality were much more basic:

>> If you asked me about the EnlaRed program, I would have to tell you that I am not sure about its objectives. It seems to me that it uses the same approach in all municipalities, no matter what the local realities are. While for some richer municipalities the Internet might be useful, we have much more basic needs. In our administration we are still using typewriters for most of our paperwork and would need some support to change the way we are organized. For example, all birth certificates are still typed, and thus the process takes quite some time and our staff sometimes make mistakes, which we could avoid if we would use computers.[32]   <<

Thus the technology needs of most of the participating rural municipalities, particularly in the highlands, are very basic, and the Internet is not the appropriate technology for improving their organizational capacity. This finding is confirmed by the long lines of people that form outside of government offices in Batallas seeking to conduct official business. The main reason for the long lines is the inefficiency with which local government officials deal with the various administrative processes requested by the citizens. Very few officials use computers for their work, and they frequently spend a great deal of time attempting either to locate documents in their paper files or to correct mistakes in typewritten documents. The program's inability to respond to this concrete need provides a good example of the "reality-design gap" of many development programs. It is also a major reason for the program's overall low impact. The evidence from my fieldwork clearly demonstrates that the program's technological choice of the Internet was inappropriate for poor rural municipalities, such as Batallas, which face much more basic technical challenges. Instead of first assessing the institutional capacity and local context of each municipality, the program used a standard approach of technical assistance and training in all municipalities. EnlaRed failed to enhance the organizational capabilities of local governments because of its inability to respond flexibly to the very different local institutional contexts of the participating municipalities.

Applying the theoretical framework of Peterson to the case study illustrates that the principal reason for the failure of the program in the organizational dimension was not the lack of commitment by senior government officials, but the inability of the program team to respond adequately to the real priorities of local governments. It is noteworthy that senior government officials recognized the need for institutional reforms in their local administration. The program team, however, failed to embed the e-government program into a much broader local sociopolitical context and to address the existing struggle for enhanced accountability and government responsiveness at the local level. The main reasons for the relatively low positive impact of the program on enhancing the organizational capacities of local governments are that (a) it ignored the existing organizational problems in local public administrations and instead imposed a predefined, uniform technological solution (in the form of municipal websites) on the municipalities and (b) its approach was overly centralized and failed to work with local governments and citizens in a much more participatory manner.

In sum, the program had a positive impact on the social capital and horizontal communication networks of its participants but did not enhance the organizational capabilities of local governments. Its overly centralized approach and overemphasis on use of the Internet prevented it from responding adequately to local priorities and from effectively integrating ICTs in existing administrative processes. In promoting use of the Internet in local governments as a stand-alone project, the program failed to bring about any meaningful institutional change at the local level.

Impact on the Social Dimension

The empirical findings reveal that the EnlaRed participants believe that the Internet has had no substantial impact on the social dimension of people's well-being, with a low score of 0.88 for social impact and an even lower score of 0.53 for economic impact. This finding is unexpected, since the social dimension was the one area in which participants in the overall survey expressed the most positive views, with the majority saying that the Internet has had an intermediate impact on their lives (PII = 1.44).

These data need to be seen in the context of the third major objective of the EnlaRed program, which was to improve the delivery of social services by strengthening the institutional capabilities of local governments. This aspect of municipal development is particularly important for indigenous peoples, since the large majority of their communities lack access to basic social services, such as water and sanitation, education, and health. Furthermore, one of the principal objectives of the decentralization program in Bolivia was to improve substantially the delivery of social services to rural, marginalized communities. In fact, an argument in favor of the LPP has been that local municipalities should have a much closer relationship to communities due to their geographic proximity. However, the major objective of the LPP has not been realized due to the capture of municipal planning processes by local elites (Blackburn 2000). EnlaRed respondents said that the Internet has not improved the delivery of social services to indigenous communities.

This finding is consistent with the failure of the program to enhance the organizational capabilities of local government. In other words, since the program did not strengthen the institutional capacity of municipalities, it is to be expected that it also failed to have a positive impact on people's social well-being, which depends on the capacity of government to deliver social services to local communities.

To examine the various components of social well-being, the impact survey asked three questions: To what extent has the Internet improved your community's access to education? To what extent has the Internet enhanced your community's access to health care? To what extent has the Internet improved government services?

An in-depth data analysis reveals that the EnlaRed respondents have a negative perception about the Internet's impact on improving the social dimension in all three of these areas. The large majority of respondents believe that the Internet has had only a minimal impact on improving government services, as indicated by a very low score for perceived impact of 0.50 (figure 9.23). Furthermore, they have a much more negative view than the overall participants about the Internet's impact on improving rural communities' access to education (PII = 1.00 and 1.53, respectively). In health, the difference in perceptions between the EnlaRed and the overall respondents is even greater, with the perceived impact index reaching a value of 1.32 for the overall respondents and 0.63 for the EnlaRed participants.

Why did the program fail to achieve its third major objective and to improve indigenous communities' capabilities in the social dimensions of their lives? During my first visit to Batallas, I accompanied the EnlaRed team to a meeting with a senior government official of the municipality and his staff to

FIGURE 9.23   Role of the Internet in the social dimension of well-being among EnlaRed respondents

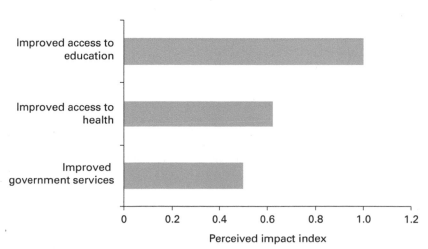

Perceived impact index

discuss the status of the program. The key actors in the municipality were not prepared for the visit or for the team's presentation, even though the visit had been officially announced two weeks before and confirmed the day before. The program team also had difficulty relating to the government official and his staff. The team was composed exclusively of urban *mestizos,* and the local staff was primarily of Aymara origin. Gradually both sides engaged in a more open conversation, but they were lacking the necessary mutual trust to work together effectively. Their wariness reflects the overly centralized nature of the government-led project; its institutional set-up was complicated from the beginning by its isolation from the social, political, economic, and cultural realities of the participating municipalities.

I was also able to observe the extremely low awareness and limited use of ICTs and the Internet in the municipality, which seemed to face many more immediate challenges than participating in a nationwide e-government program. The municipality—with about 67 percent of the population living below the poverty line—faced serious environmental degradation in the form of soil erosion that severely threatened the livelihoods of subsistence farmers. These conditions beg the question: What is the underlying rationale for an e-government program? Clearly, the opportunity costs of ICT programs need to be addressed. Was Batallas the best place to implement an ambitious e-government program? There was clearly a disjunction between the immediate, material needs of the population and the elusive benefits of the EnlaRed program. The situation in Batallas was especially complex, since the previous mayor had invested heavily in a pilot project to establish a virtual library, with little demonstrable success. That project had raised expectations in the communities, which hoped that the project would improve the education of their children. According to a community leader,

>> When the mayor first told us about this project we were very skeptical. We thought that the Internet was only something for people in the city and not for us. But our children explained to us what it was all about, and they were very excited about the project. We all thought this might be a way to improve their education. Unfortunately the computers and the library only worked for several weeks. Most of the young people were very disappointed about this.[33]   <<

Bernal, the main contact for the EnlaRed program, gave us an extensive tour of the municipality, and within a short period of time, it became apparent that the municipality faced many institutional challenges and that the project team was not well equipped to help the municipality to address them. He expressed the people's general frustration:

>> For us it is unbelievable to see this satellite dish on the roof of our municipality and still not be able to use Internet in our offices. I know that the mayor wanted to build a virtual library in the municipality and thus contracted a private sector firm to install the system [a V-sat-satellite dish] in the municipality. At first the system worked very well, and particularly the young people were very excited about the Internet. But then after several weeks, the system had some technical difficulties, I think related to the heavy dust, and suddenly everything stopped working. The situation has been like this for at least the last one and a half years.[34]   <<

This account from the local EnlaRed coordinator illustrates the many challenges that rural communities face in their efforts to use the Internet as a way to improve the delivery of social services. In the absence of a national rural connectivity program, the only way to get Internet services in remote areas is to contract private satellite companies directly. Satellite technology is unreliable in the extreme weather conditions of rural Bolivia. It is also expensive and difficult to maintain. Most small, poor municipalities are unable to pay the US$400 monthly service costs (a figure cited by the mayor) or the costs of repair. Once the technology fails, as it did in Batallas, many systems are simply abandoned. EnlaRed did not offer the technical expertise or funding needed to support the virtual library project. In the eyes of many, the program was too abstract in nature and irrelevant to their immediate needs:

>> I understand from you that your project cannot help us with our technical problems, but that you would be very interested in training our staff in the use of ICTs and to publish any local news story about our municipality on your website. It seems that your project has some good ideas, but the services you offer are so specialized that they are basically useless for us. Based on your information, there is another government project that focuses specifically on infrastructure and connectivity issues. This might make sense for a person in La Paz or the donors who fund these projects, but from our perspective we need one project that helps us in all aspects of technology and not just in parts of it.[35]   <<

This sentiment highlights a serious design-reality gap. The program did not account for the difficulty of implementing an e-government program in rural indigenous municipalities, which operate under major financial, personal, and institutional constraints. The municipality of Batallas would require comprehensive, long-term technical and financial support in order to implement an e-government project capable of enhancing its delivery of social services to its communities. While it may make sense from a donor's perspective (in this case, USAID) to limit intervention to a specific area of e-government, from the municipality's perspective such an approach is too techno-centric and centralized. In other words, the program suffered from taking on its donor's priorities. The objectives and program activities were all predefined and agreed between the donor and FAM, the executing government agency, based on the advice of a technical specialist from the donor agency and technocrats from the central government. The program's failure to implement its activities successfully at the local level revealed the urban biases of the program team and the donor's technical specialist. Once the program design was confronted with the challenging realities of rural municipalities in Bolivia, the limitations of its centralized design were revealed. In the end, the program failed to achieve its major objective of strengthening the institutional capacity of local government to provide improved services to its citizens. The problems encountered by EnlaRed are encountered by many development interventions in Bolivia, which have created expectations among indigenous communities without being able to fulfill them. In the meantime indigenous people continue to lack access to the most basic services, such as education and health services.

Conclusions

This case study has demonstrated that the EnlaRed program failed to have a significant impact on the accountability and performance of local government and did not enhance indigenous peoples' human capabilities and social well-being. According to its participants, the Internet has had only a minor impact on all of the principal dimensions of well-being of their lives, including the political, social, and economic dimensions.

A critical reason for the program's low impact is its overemphasis on technology itself and its failure to adopt a more strategic approach to the role of information per se in promoting socioeconomic development at the local level. Based on the program's overly techno-centric approach, its participants failed to attain advanced informational capabilities and thus were not able to enhance significantly their human and social capabilities.

The case study has investigated the critical factors that have resulted in such a low perceived impact of the Internet on people's well-being and the organizational capabilities of local governments. Based on the empirical evidence presented in this chapter, the program's low impact seems to have been due to both internal and external factors.

There were three main internal factors. First, the institutional weaknesses of the program's intermediary—FAM—as well as the overly centralized approach to implementation impeded the program's ability to support local municipalities and indigenous communities in the use of ICTs at the local level. Second, using the same approach in all 15 municipalities meant that the program was unable to adapt the use of ICTs to the local socioeconomic and cultural context. Third, overemphasis on the technology resulted in an inadequate focus on the role of information in socioeconomic development.

Several external factors also played a role. First, the lack of a comprehensive national strategy for developing telecommunications and ICT infrastructure in Bolivia, particularly in rural areas, meant that the program was implemented in technically unprepared regions. Second, the dependence on international donor funding weakened its effectiveness and sustainability. For example, USAID—the program's main donor—highly influenced the program's design, which resulted in an overly technocratic approach to providing specific services (like training in the use of ICTs) to local governments. Third, many local governments have limited institutional and technical capacity to develop public policies and to implement development projects.

In sum, the program was too centralized and did not address broader issues related to governance. The resulting design-reality gap undermined its ability to enhance the transparency, accountability, and performance of local governments or to have an impact on indigenous peoples' human and social capabilities. To have a more significant effect on poor people's well-being, programs have to be tailored much more precisely to the specific socioeconomic, political, and cultural realities of local communities.

Notes

1. Interview with Bernal Patty, technology specialist, Batallas, February, 25 2005.
2. Interview with Marcelo Carbone, communications specialist and main contact person for EnlaRed in Comarapa, February 25, 2005.
3. The ETIC developed a National Strategy for the Information Society for Bolivia through a participatory process that included all sectors of Bolivian society (www.etic.org.bo).
4. Carlos Soria, EnlaRed program director, stressed this point several times in my interview with him on August 25, 2005. He considered this issue to be a major reason for the important difficulties the program faced during its implementation. Furthermore, this point was raised by Sonia Paredez, communication specialist of the Tarabuco municipality, in my interview with her on February 25, 2005.
5. A *mancomunidad* is a voluntary association of several municipalities within a region that is frequently based on a common cultural identity. The Mancomunidad de Chiquitania is an association of six municipalities in the Chiquitania region of the eastern lowlands of Bolivia.
6. Interview with Justo Seoane, mayor of Concepción and former vice minister of indigenous affairs, April 22, 2005.
7. FAM is the national association of all the Bolivian municipalities with the exception of the major cities of La Paz, El Alto, Cochabamba, and Santa Cruz.
8. PADEM is coordinated by the Swiss NGO Ayuda Obrera Suiza, which works to strengthen local democratic processes and to empower poor communities throughout Latin America.
9. Interview with Carlos Soria, coordinator of EnlaRed, August 25, 2005.
10. The 15 municipalities are Cobija (Pando), Trinidad (Beni), Montero and Comarapa (Santa Cruz), Sipe Sipe and Villa Tunari (Cochabamba), Monteagudo and Tarabuco (Chuquisaca), El Puente (Tarija), Batallas and Coroico (La Paz), Machacamarca and Pampa Aullagas (Oruro), and Uyuni and Tupiza (Postosí).
11. Interview with Bernal Patty, information technology (IT) specialist, Batallas Municipality, February 25 2005.
12. Interview with Lena Calla, deputy director of EnlaRed, July 2005.
13. Interview with Lena Calla, deputy director of EnlaRed, July 2005.
14. Interview with Sonia Paredez, communication specialist, Tarabuco, February 25, 2005.
15. Interview with Eduardo Rojas, coordinator of Fundación Redes, a foundation dedicated to promoting the use of ICTs for development in Bolivia, Cochabamba, June 10, 2005.
16. Interview by author with Marcelo Carbone, Cochabamba, August 2005.
17. Interview with Eduardo Rojas, coordinator of Fundación Redes, Cochabamba, August 2005.

18. The official mayor, the most senior official in local government, is in charge of all administrative processes.

19. Interview with José Luis Alvarez, official mayor of Comarapa, July 15, 2005.

20. Interview with Julio Sánchez, president of the agricultural producer association, Comarapa, May 24, 2005.

21. Interview with Claudia Camacho, ICT program coordinator of ICO in Comarapa, May 23, 2005.

22. Interview with Claudia Camacho, ICT program coordinator of ICO in Comarapa, May 23, 2005.

23. Interview with Juan Delgadillo mayor of Comarapa, May 23, 2005.

24. Interview with José Luis Alvarez, official mayor of Comarapa, July 15, 2005.

25. Interview with Marcelo Carbone, communications specialist and main contact person for EnlaRed in Comarapa, February 25, 2005.

26. Both of the user surveys focused exclusively on issues related to the awareness of, readiness for, and use of ICTs in local governments (EnlaRed project files).

27. The survey conducted in September 2004 was based on a sample of 190 respondents, and the survey conducted in September 2005 was based on a sample of 186 participants.

28. Interview with José Luis Alvarez, official mayor of Comarapa, July 15, 2005.

29. Interview with Manuel Quispe, official mayor of Batallas, July 30, 2005.

30. Interview with Manuel Quispe, official mayor of Batallas, July 30, 2005.

31. Interview with Clemencia Vargas, director of the Information Systems Department, VIPFE, August 12, 2005.

32. Interview with Manuel Quispe, official mayor of Batallas, July 30, 2005.

33. Interview with Ivan Salinas, community leader from Batallas, June 30, 2005.

34. Interview with Bernal Patty, IT specialist, Batallas, July 30, 2005.

35. Interview with Juan Delgado, mayor of Batallas, July 30, 2005.

References

Ayo, Diego. 2003. "Evaluando la ley de participación popular: Once puntos en debate." *Umbrales* 12: 157–79.

Blackburn, James. 2000. *Popular Participation in a Prebendal Society: A Case Study of Participatory Municipal Planning in Sucre, Bolivia*. Ph.D. diss., University of Sussex.

Blair, Harry. 2000. "Participation and Accountability at the Periphery: Democratic Local Governance in Six Countries." *World Development* 28 (1): 21–39.

Booth, David, Suzanne Clisby, and Charlotta Widmark. 1996. "Empowering the Poor through Institutional Reform? An Initial Appraisal of the Bolivian Experience." Stockholm University, Department of Social Anthropology.

Calla, Lena. 2004. "EnlaRed Project Baseline Study." Federación de Asociaciones Municipales de Bolivia, La Paz.

Ciborra, Claudio. 2002. "Unveiling E-Government and Development: Governing at a Distance in the New War." Public lecture at the London School of Economics, London.

Ciborra, Claudio, and Diego D. Navarra. 2005. "Good Governance, Development Theory, and Aid Policy: Risks and Challenges of E-Government in Jordan." *Journal of Information Technology for International Development* 11 (2): 141–59.

Faguet, Jean-Paul. 2003. "Decentralization and Local Government in Bolivia: An Overview from the Bottom-up." Working Paper 29, Crisis State Programme London School of Economics and Political Science, London, UK.

Goudsmit, Into A., and James Blackburn. 2001. "Participatory Municipal Planning in Bolivia: An Ambiguous Experience." *Development in Practice* 11 (5, November): 587–96.

Healy, Kevin. 2004. "Towards an Andean Rural Development Paradigm?" *NACLA Report on the Americas* 38 (3): 28–33.

Heeks, Richard. 1999. "Information and Communication Technologies, Poverty, and Development." Development Informatics Working Paper 5, University of Manchester, Institute for Development Policy and Management, Manchester.

———. 2002. "I-Development Not E-Development: Special Issue on ICTs and Development." *Journal of International Development* 14 (1): 1–11.

———. 2003. "Most E-Government-for-Development Projects Fail: How Can Risks Be Reduced?" Institute for Development Policy and Management, University of Manchester.

INE (Instituto Nacional de Estadísticas). 2001. *National Census of Housing and Population 2001.* La Paz: Ministry of Planning and Coordination, INE.

Kay, Cristóbal, and Miguel Urioste. 2007. "Bolivia's Unfinished Agrarian Reform: Rural Poverty and Development Policies." In *Land, Poverty, and Livelihoods in the Era of Globalization: Perspectives from Developing and Transition Countries,* edited by A. Haroon Akram-Lodhi, Saturnino M. Borras, and Cristóbal Kay. London: Routledge.

Kohl, Benjamin. 2003. "Democratizing Decentralization in Bolivia: The Law of Popular Participation." *Journal of Planning Education and Research* 23 (2): 153–64.

Madon, Shirin. 1993. "Introducing Administrative Reform through the Application of Computer-Based Information Systems: A Case Study in India." *Public Administration and Development* 13 (1): 37–48.

———. 2004. "Evaluating the Developmental Impact of E-Governance Initiatives: An Exploratory Framework." *Electronic Journal on Information Systems in Developing Countries* 20 (5): 1–13.

Molina Saucedo, Hugo, Ruddy Rolando Cuéllar Rivero, and Evans Gandarillas Mansillas. 2000. "Mancomunidad municipal: Visión estratégica para una política de estado." Documento de Trabajo 6, Ministerio de Desarrollo Sostenibile y Planificación, Vice Ministerio de Planificación Estratégica y Participación Popular, La Paz.

Nicod, Chantal. 1996. "Seguimiento al proceso de participación popular en los municipios de Chuquisaca." In *Los problemas de representatividad del sistema democrático boliviano*, edited by Gonzalo Rojas Ortuste. Washington, DC: Friedrich-Ebart-Stiftung. https://openlibrary.org/works /OL2212076W/Los_problemas_de_representatividad_del_sistema _democrat%C3%ADco_boliviano.

Oyungi, Walter O. 2000. "Decentralization for Good Governance and Development: The Unending Debate." *Regional Dialogue* 21 (1): 3–22.

Peterson, Stephen B. 1998. "Saints, Demons, Wizards, and Systems: Why Information Technology Reforms Fail or Underperform in Public Bureaucracies in Africa." *Public Administration and Development* 18 (1): 37–60.

Pettigrew, Andrew M. 1985. "Contextualist Research and the Study of Organisational Change Processes." In *Research Methods in Information Systems*, edited by Enid Mumford, Rudi Hirschheim, Guy Fitzgerald, and Trevor Wood-Harper, 53–78. Amsterdam: North-Holland.

———. 1998. "Success and Failure in Corporate Transformation Initiatives." In *Information Technology and Organizational Transformation*, edited by R. D. Galliers and W. R. Baets. Chichester: Wiley.

Pratchett, Lawrence. 1999. "New Technologies and the Modernization of Local Government: An Analysis of Biases and Constraints." *Public Administration* 77 (4): 731–50.

Samoff, Joel. 1979. "The Bureaucracy and the Bourgeoisie: Decentralization and Class Structure in Tanzania." *Comparative Study of Society and History* 21 (1): 30–62.

Tendler, Judith. 1997. *Good Governance in the Tropics*. Baltimore, MD: Johns Hopkins University Press.

UNDP (United Nations Development Programme), INE (National Institute of Statistics), and UDAPE (Social and Economic Policy Analysis Unit). 2004. "Indice de desarrollo humano en los municípios de Bolivia." UNDP, INE, and UDAPE, La Paz.

UNDP and INE. 2005. "Bolivia atlas estadístico municipal 2005." UNDP and INE, La Paz.

———. 2006. "Bolivia atlas estadístico municipal 2006." UNDP and INE, La Paz.

Urioste, Miguel. 1992. *Fortalecer las comunidades: Una utopía subversiva, democrática … y possible.* La Paz: AIPE/PROCOM/TIERRA.

Walsham, Geoffrey. 1993. "The Emergence of Interpretivism in IS Research." *Information Systems Research* 6 (4): 376–94.

World Bank. 1999. *Decentralization and Accountability of the Public Sector: Proceedings of the Annual World Bank Conference on Development in Latin America and the Caribbean.* Washington, DC: World Bank.

Zoomers, Annelies. 1999. *Linking Livelihood Strategies to Development: Experiences from the Bolivian Andes.* Amsterdam: Royal Tropical Institute.

Part IV

Conclusions

CHAPTER 10

Conclusions: Toward an ICT Impact Chain

The study has demonstrated that, under certain conditions, information and communication technologies (ICTs) can significantly enhance poor people's human and social capabilities and thus have a positive impact on their well-being. At the core of the process of introducing ICTs into marginalized communities stands the notion that ICTs can (a) enhance poor people's individual and collective agency, (b) strengthen their existing individual or community assets, and (c) enhance their informational capabilities. The study has addressed two research questions: Can ICTs enhance poor people's human and social capabilities to achieve the lifestyle they value and thus improve their well-being? Under what conditions can they do so?

Principal Conclusions

Enhancing people's informational capabilities is the most critical factor determining the extent to which ICTs can enhance their well-being. That is, the expansion of people's informational capabilities not only has intrinsic value for their well-being but also, and even more important, an essential role to play in strengthening their capabilities in multiple dimensions of their lives. The analysis has shown that *informational capabilities*, similar to literacy, play a catalytic role in expanding poor people's human and social capabilities and thus enhancing their ability to engage with the formal institutions in the economic, social, and cultural spheres of their lives. In this sense, ICTs can act as important agents for social change, both for individuals and for local communities.

The analysis has, however, also demonstrated that there are important differences in the extent to which informational capabilities expand people's human and collective capabilities in the political, economic, organizational, and social dimensions of their lives. There is no direct, causal relationship between ICTs and poverty reduction. The relationship is much more complex and indirect in nature, whereby the impact on people's well-being depends to a large extent on a dynamic and iterative process between people and technology within a specific local, cultural, social, and political context.

Therefore, a critical condition for ICTs to enhance poor people's well-being is to go beyond simply providing access and promoting the use of ICTs and instead to enhance their meaningful use. ICTs receive meaning only if people use and enact them for a specific purpose and if local communities can exert control over their use by interpreting and appropriating them for their own specific sociocultural realities.

The study has also illustrated that the most immediate and direct impact of ICT programs on people's well-being is the personal empowerment of the most marginalized groups, such as indigenous women, whereby the newly acquired ICT capabilities provide women with a sense of achievement and pride, significantly strengthening their self-esteem. Marginalized groups perceive that the Internet plays a critical role in enhancing the social capabilities of their communities, but they consider its positive impact on individual human capabilities to be less significant. Thus the Internet is seen to have the strongest impact on the social and organizational dimensions of their lives. With regard to both the political and economic dimensions, there exists only a very limited relationship between the enhancement of a person's informational and human capabilities. In both dimensions, the role that ICTs play in enhancing people's well-being is significantly limited by the broader structural barriers of extreme poverty and social exclusion of poor communities in Bolivia.

Furthermore, using quantitative and qualitative information, the study confirms the two hypotheses proposed in chapter 1. First, intermediary organizations that introduce ICTs into local communities play a crucial role in determining the effects of ICTs on poor people's well-being. Thus a key recommendation for evaluating the impact of ICTs on well-being is to place the process of how ICTs are being introduced at the center of the analysis. The case studies have demonstrated that it is at least as important to analyze the process by which ICTs are being introduced as it is to analyze the type of activities that are being undertaken in local communities. For ICT programs, it is essential that community members not only gain the technical skills to make meaningful use of technology but also gradually take ownership of these technologies. Local communities need to control the process of introducing and appropriating ICTs into their communities.

The research also has confirmed the second hypothesis posed in chapter 1: ICTs have to be appropriated by local communities in order to enhance their well-being. However, tension exists between the first and second hypotheses, as the central role that intermediaries play in the process of introducing ICTs can create dependency for marginalized groups. As the case study of the Organización de Mujeres Aymaras del Kollasuyo (OMAK), the subject of chapter 8, has shown, it is particularly important for grassroots organizations that have established a relationship of trust with local communities to take the lead in managing the design and implementation of technology programs.

Toward an ICT Impact Chain

Based on the study's main findings, it is now possible to develop an impact chain that describes both the principal factors and the process by which ICTs can significantly enhance people's individual and collective well-being across multiple dimensions of their lives.

The impact chain unpacks the overall impact of ICTs on people's well-being into a five-step process that explains the conditions under which the access to and use of ICTs (a) become meaningful for poor communities, (b) are translated into enhanced informational capabilities, and (c) result in improvements in people's human and social capabilities.

Step One: Conducting an Information and Communication Needs Assessment

The first step in the impact chain is to conduct an information and communication needs assessment. This stage is critical since ICTs are not being introduced into communities in isolation from existing information and communication ecologies. They need to be embedded into the existing structures in order to (a) strengthen the communities' informational capital, (b) be accepted by the communities' principal stakeholders, and (c) be sustainable in the long term. It is thus essential to analyze the existing information ecology of communities *before* providing specific ICT services (such as access to Internet connectivity). As visualized in figure 10.1, the assessment should analyze the communities' existing information and communication needs; identify key local stakeholders, such as elders, who frequently are the traditional information brokers in local communities; assess the existing informational capital; and identify the existing channels of communication. Finally, the information needs assessment should identify critical barriers and bottlenecks that have caused information and communication gaps between local communities and national policy makers and identify mechanisms through which ICTs could promote an improved two-way flow of information and communications between these two actors.

This step is needed to ensure that ICT programs are not supply driven or push a specific technology on communities, but instead respond to real priorities and needs. It is critical to recognize the cultural and social diversity of communities.

Furthermore, most local communities rely heavily on traditional forms of communications (such as community meetings, kinship-based networks of family and friends) to obtain their information. The results from the non-user survey, presented in chapter 4, have shown that poor people's exposure to new forms of ICTs—in particular, the Internet—is very low, while their expectations about its potential positive impact on their individual and collective well-being are very high. This is particularly true for the most excluded and marginalized groups, such as indigenous women and youth, who have very positive views of the Internet's potential impact.

The case studies presented in chapters 8 and 9 show the importance not only of carrying out a needs assessment but also the manner in which it is conducted. The participatory nature of the assessment is critical for gaining

FIGURE 10.1   ICT impact chain: A 5-step process

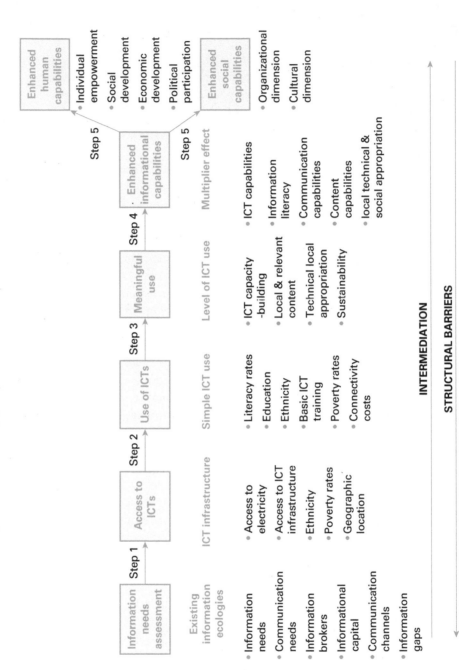

Source: Developed by author.

support from the communities for the program. The government-led program EnlaRed carried out a needs assessment, but it gauged only the use of ICTs and the e-readiness of local communities and not—as in the case of the grassroots OMAK program—the role of information per se in local political processes and the possibilities for an ICT program to improve the existing flow of information and communications between government organizations and local communities. The EnlaRed program thus overemphasized the role of technology and failed to engage local communities in a dialogue about the role that information could play in their development more generally and the extent to which ICTs could support them in their efforts to strengthen their existing information and communication systems.

Moreover, as in the case of EnlaRed, many ICT programs fail because they lack the flexibility needed to adjust planned activities to the specific local, social, cultural, and political context. Although the program identified the rich cultural and social differences of its 15 participating municipalities, it implemented its program through an overly centralized approach prescribing the same technological solution (that is, a municipal website) in all of the municipalities, irrespective of the important differences in their local, social, economic, and political context. EnlaRed's program did not respond to the specific information and communication needs of local communities and failed to strengthen participants' capabilities to communicate and share information with local governments. The case study is a good example of an ICT program that did not base its activities on the results of its information needs assessment, pursuing instead a supply-driven approach to implementation. Although access to ICTs improved, communities did not have better access to information than before. This can be attributed largely to the unwillingness of local governments to improve their transparency and vertical accountability. Thus, using the logic of the impact chain, the failure of the program can be ascribed to its inability to meet a necessary condition in the initial step of the chain—discerning the information brokers, identifying the existing informational capital, and understanding the complexity of the overall information ecology across the different municipalities.

In contrast, the grassroots-led program OMAK carried out an impact assessment that enabled poor people to engage in a broader discussion and reflection about the role of information for local development more generally, raised awareness, and instilled a sense of ownership of the program. The OMAK program met the critical factors necessary for the initial step in the impact chain, even though it faced critical constraints in enhancing poor people's uses of ICTs.

Step Two: Moving from Access to ICT Use

The second step of the impact chain addresses issues related to people's ability to use ICTs. The study has shown that in order to assess the impact of ICTs on people's well-being, it is critical to move beyond the concept of ICT access and to study the multiple factors that enable people's use of ICTs within their socioeconomic, political, and cultural context. Access to ICTs is not sufficient for enhancing poor people's actual use of ICTs. In fact, as shown in chapter 6, a substantial majority of indigenous peoples (58 percent) who have access to the Internet do not use it. Significant barriers prevent marginalized groups from using ICTs even when the technology is readily available. The central finding of chapter 6 is that ethnicity (belonging to a specific indigenous group) and socioeconomic factors (literacy rates and poverty levels) rather than individual factors (age, gender, and education) are critical in explaining the marked inequalities in ICT use within Bolivia.

The empirical evidence has shown that, while education matters to ICT use, important differences between the indigenous and non-indigenous populations persist even when controlling for educational attainment. Even when indigenous peoples reach the same level of tertiary education as the non-indigenous population, they continue to be significantly disadvantaged in their Internet use, at a ratio of almost 2 to 1.

Furthermore, there is a strong correlation between poverty, ethnicity, and ICT use. Among the non-indigenous population, 33 percent live above the poverty line and 40 percent use ICTs. Among the indigenous population, 26 percent live above the poverty line and only 14 percent use ICTs. There also are significant differences among indigenous groups—an Aymara is 2.8 times less likely than a Quechua and 4.5 times less likely than a non-indigenous person to use the Internet. Thus digital inequalities in Bolivia reflect the deep structural inequalities within society, whereby ethnicity is the critical factor determining whether a person is poor and lacks the human capabilities and social opportunities to reach the lifestyle she or he values.

Furthermore, the high cost of Internet use in rural areas is a critical impediment to use. Indigenous peoples frequently cannot afford to use the Internet, especially if the service is provided by private entrepreneurs whose priority is to ensure their own financial sustainability.

Finally, the study has demonstrated that the presence of an intermediary organization is the most significant factor explaining poor people's ICT uses. The intervention of an intermediary organization enables local communities to acquire the basic capabilities needed to use ICTs, even if they do not have access to these technologies within their communities. The majority of participants in the OMAK program use the Internet in intermediary cities or towns and combine their weekly visits to regional markets with a visit to a public Internet access point. This intriguing finding highlights that the knowledge of how to use the Internet is more important for poor people's Internet uses than having access to the technology within their communities. Consequently, for people to reach a stage in which they not only have access to but also use the Internet, it is essential for ICTs to have an impact on their well-being. The study has highlighted that, in spite of existing structural constraints, the intermediation through a local and effective ICT program can significantly enhance poor people's Internet use even if communities continue to lack access to Internet connectivity in their communities. The study thus has found that there is significant room for reducing digital inequalities within Bolivia through targeted interventions that promote the use of ICTs among rural indigenous communities, the most disadvantaged group in Bolivia.

In sum, the second step of the impact chain emphasizes that poor people's actual use of the Internet, rather than access to the technology, is a critical precondition for ICTs to have an impact on their well-being. The study, however, takes the analysis a step further. It argues that it is essential to investigate people's capabilities to make "meaningful" uses of ICTs and that it is not sufficient to evaluate the factors influencing whether or not poor people have the opportunities to use ICTs. The impact chain thus differentiates between simple and meaningful use of ICTs. Simple use of the Internet does not require proficiency, while meaningful use captures the depth, usefulness, and level of expertise, gauging Internet use in terms of how efficient, informed, and beneficial it is.

Step Three: Enabling Meaningful ICT Use

The third step in the impact chain highlights the conditions under which simple ICT use is converted into meaningful use. Applying this framework helps to identify the factors that impede people from giving their ICT use meaning and from deriving real benefits from it for their individual and collective well-being. The study has shown that four conditions have to be met to enable people to realize meaningful use of ICTs: (a) enhancement of their ICT capabilities, (b) availability of local and relevant content, (c) local appropriation of ICTs, and (d) financial and social sustainability of ICT programs.

First, poor people face multiple barriers to improving their ICT capabilities. For indigenous peoples in rural areas, the expansion of their ICT capabilities is independent of their individual sociodemographic characteristics, including age, gender, and education, but it is heavily dependent on broader economic, social, political, and cultural factors. Thus ethnicity and not individual agency is the determining factor influencing the degree to which people can enhance their ICT capabilities. A major unexpected result is that indigenous peoples' ICT capabilities are independent of educational attainment. As such, indigenous peoples with only basic education can nevertheless reach advanced levels of ICT capabilities if an effective local intermediary supports them in adapting ICTs to their local, cultural, and social context. In fact, the statistical data have demonstrated that more than half of the indigenous participants who took part in a grassroots-level ICT program targeting communities with extremely high poverty and illiteracy rates were nevertheless able to attain advanced levels of ICT capabilities.

Consequently, chapter 6 presents two major findings: (a) the ICT program, similar to the simple use of ICTs, is the most important variable determining people's ICT capabilities, and (b) grassroots organizations are significantly more successful than programs led by government or nongovernmental organizations (NGOs) in enhancing people's ICT capabilities. Both of these major findings indicate that marginalized groups with the support of an effective and local intermediary indeed can overcome the multiple socioeconomic constraints and significantly enhance their ICT capabilities, meeting one of the principal conditions for converting simple use into meaningful use.

Second, the OMAK case study has demonstrated that the lack of local and relevant content is a major limitation for poor people's ability to make meaningful use of the Internet. Indigenous peoples perceive that the Internet is an urban medium that propagates the values and norms of Western, non-indigenous societies. The women participating in the OMAK program expressed their frustration about the content that they found on the Internet, which is presented in an academic style and is too abstract to be directly applicable to their lives. Language, too, remains a prohibitive barrier, as the large majority of the Internet's content is in English. As the OMAK case study has emphasized, the intermediary organization plays a central role in this dimension by supporting local users to identify content that is relevant to their needs and that represents their own cultural and social norms. For example, OMAK staff referred participants to several useful sites that were developed by indigenous peoples in the Andean region. Nevertheless, the continuous lack of local and relevant content is a major barrier to meaningful Internet use among indigenous peoples.

Third, the local technical appropriation of ICTs by indigenous communities is critical to attaining meaningful use. This concept assumes that people are provided the necessary space in which to explore and interpret technologies on their own terms, to define which tools and applications they consider suitable for their needs, and to adapt these technologies to their local economic, social, and cultural context. As shown in chapter 9, programs such as EnlaRed that are implemented in an overly centralized manner deny people the opportunity to adapt ICTs to their own circumstances and frequently impose preconceived technical solutions (in this case, municipal websites) on local communities. The approach of EnlaRed failed because the technical solutions did not correspond to the local priorities in most municipalities. The technical solutions were not based on real information and communication needs of local governments, were chosen in a top-down manner by centralized technocrats, and were then rolled out to local municipalities. This important design-reality gap resulted in the extremely low appropriation of the promoted technological applications in all of the participating municipalities. In the context of the extremely low institutional capacity of the municipality of Batallas, for example, local government officials considered the use of word processing and Excel spreadsheets to be a much higher priority for their work than the Internet. The program nevertheless insisted on promoting the development of a municipal website, even though the municipality did not have access to the Internet.

In contrast, the OMAK program provided participants with ample space in which to explore and interpret technologies on their own, to discuss their meaning with other participants, and to facilitate a process in which indigenous women were empowered to appropriate the technologies to their own needs. Based on the expressed needs of its participants, the program provided computer training in basic computer skills, including word processing and basic spreadsheet use. Participants considered this approach to be highly empowering and pointed out that these types of ICT uses were the most rewarding and meaningful to them since they were able to apply the newly acquired ICT capabilities directly to their daily lives.

Finally, ICT programs need to reach financial and social sustainability in order to provide people with the opportunity to use ICTs in the long term, enabling them to attain a meaningful level of use. Frequently, ICT programs fail shortly after installing the ICT infrastructure or carrying out their initial phase of training activities due to the lack of community ownership. For instance, the EnlaRed program faced significant challenges with regard to both financial and social sustainability. It failed to base its activities on the priorities of local stakeholders or to develop strong partnerships with the existing social organizations in the communities or NGOs working in the project area. Thus it could not raise funds to support the program in the long term and was neither financially nor socially sustainable. Once the U.S. Agency for International Development, its principal donor, decided to cancel any funding, the project was forced to suspend its activities. Thus there was no long-term continuity, which would have been an important requirement for enabling participants to derive meaningful ICT use.

In contrast, while the OMAK program also suffered from a lack of financial sustainability, it nevertheless built a strong base for the long-term use of ICTs in the participating communities. Based on its highly participatory approach, local communities gradually gained a sense of ownership over the program and continued to use ICTs, even when the program had to suspend some of its main activities. In the municipality of Curahuara de Carangas, located in Oruro, community members succeeded in convincing the mayor to finance the installation of a community-based telecenter, creating a permanent space in which to explore and use ICTs.

Step Four: Enhancing Informational Capabilities

The fourth step in the impact chain analyzes the conditions that have to be met so that people's meaningful use of ICT also enhances their informational capabilities. This step is essential, since the extent to which ICT programs succeed in enhancing a person's informational capabilities is the most critical factor determining the impact of ICTs on poor people's well-being. The study has illustrated that there are significant differences between expanding people's ICT capabilities and expanding their informational capabilities. The concept of ICT capabilities encapsulates a person's ability to make efficient use of computer hardware, software, and ICT tools; the concept of informational capabilities is an information-centric approach that deemphasizes the role of technology per se. It instead includes four components: ICT capability, information literacy, communication capabilities, and content capabilities.

Transforming people's meaningful use of ICTs into enhanced informational capabilities depends on the extent to which individuals can (a) enhance their capabilities in all four dimensions of informational capabilities, (b) strengthen their existing informational capital, and (c) enhance their individual and collective agency regarding the use of information. A critical factor in reaching this step is the process of local appropriation of ICTs by local communities, as facilitated by an effective and local intermediary.

In fact, a principal finding of the study is that the intermediary organization (ICT program) is the variable that exerts the strongest influence on people's informational capabilities. Programs conducted by grassroots organizations (Coordinadora de Integración de Organizaciones Económicas

Campesinas Indígenas y Originarias [CIOEC] and OMAK), in particular, have been very successful in enhancing poor people's informational capabilities, while programs led by government (EnlaRed) and NGOs (Instituto de Capacitación del Oriente [ICO]) have been relatively unsuccessful in reaching this objective.

To what extent have different ICT programs reached the four dimensions of informational capabilities? The empirical findings have demonstrated that a critical success factor for ICT programs is that they not only enhance people's ICT capabilities but also take an information-centered approach and enhance people's informational capabilities.

An essential aspect of informational capabilities is information literacy, which emphasizes a person's ability to collect, process, evaluate, use, and share information with others within her or his own sociocultural context. One of the key lessons learned from OMAK's experience is that most of the difficulties that indigenous peoples encounter in using the Internet are related to the analysis and interpretation of information rather than the use of technology itself. However, the case study has also shown that even poor people with relatively limited formal education can enhance their information literacy skills if they receive hands-on support, guidance, and specific capacity-building activities related to the interpretation of information.

In contrast, the EnlaRed program overly emphasized technology itself and provided little guidance on issues related to the use, processing, and evaluation of information. It failed to place the use of ICTs within the broader local social, cultural, economic, and political context and thus failed to improve the information literacy skills of its participants.

A good example of the critical differences between ICT capabilities, meaningful use, and enhanced informational capabilities is the use of ICTs (Internet and community radio) to improve small-scale farmers' access to market prices. The evaluation of an ICT program coordinated by ICO in the Santa Cruz Valley emphasized that while it was highly successful in enhancing the "CT" capabilities of participants to use the Internet to find market price information, it failed to enhance their "I" capabilities to interpret, evaluate, process, and share information with others. While this type of use can be considered meaningful, many indigenous farmers were not able to understand how local and regional markets work and why market prices for their agricultural goods fluctuate so widely. Thus improved CT capabilities and enhanced access to the raw market data, without the necessary information literacy skills to interpret the data, failed to enable farmers to apply the information directly to their daily challenge of selling agricultural products in local and regional markets.

Enhancing people's communication capabilities is also an essential aspect of improved informational capabilities. The OMAK program significantly improved the ability of its participants to communicate with family members, friends, and professional contacts. For poor people—due to their strong social networks—the strengthening of their communication capabilities enhanced primarily their horizontal communications with other communities and to a much lesser degree their vertical communications with policy makers and the urban population.

The EnlaRed program succeeded only in enhancing people's communication capabilities and failed to strengthen participants' capabilities in the other three dimensions of informational capabilities. The program significantly strengthened the informal networking between local government officials who participated in the program, but had no positive impact on their ICT capabilities, information literacy, or content capabilities. In fact, the program's failure to enhance the participants' informational capabilities constituted a major reason for its overall low impact.

In contrast, the OMAK program enhanced people's capabilities not only to "consume" but also to produce their own local content and to share it with others. The case study has shown that the issue of "content capabilities" is particularly important for indigenous peoples due to (a) their strong demand for local information and (b) the lack of local Internet content in most rural communities. For this reason, OMAK developed a partnership with Aymaranet—an indigenous-led website—which hosted a site for the organization and provided OMAK and program participants with a space in which they could create and disseminate their own content and share their experiences with other indigenous peoples and the public in general.

Moreover, expanding people's existing informational capital through the use of the Internet also plays a central role in determining whether or not they have enhanced their informational capabilities. Only those ICT interventions that enable communities to appropriate ICTs both technically and socially are successful in enhancing people's informational capital. The technical aspect of the local appropriation process focuses on providing people with opportunities to select and adapt communication tools based on their own information needs, while the social aspect of the process highlights the ability of the community to adapt technologies so that they are rooted in its own social, economic, and cultural processes.

The OMAK case study has shown that this grassroots organization was successful in assuming simultaneously the role of a technical intermediary and that of a social intermediary. With regard to technical appropriation, it enabled indigenous peoples to explore, use, and adapt technologies under their own terms and conditions by facilitating an open, secure, and collective learning environment. Moreover, it provided the necessary technical support (that is, ICT training, local content) for the technical appropriation of ICTs and thus enabled participants to make meaningful use of ICTs.

In relation to the social appropriation of the Internet, OMAK was instrumental in providing the social space for participants, particularly indigenous women, to come together and explore the meaning of technologies and their applicability to their individual and collective well-being. A central aspect of this social process was that OMAK embedded these technologies in the existing social and organizational community structures by fully integrating them into the local women's organizations. The ICT program played a critical social role in providing indigenous women with a social space in which to come together and discuss the multiple constraints they were facing in their daily struggle to improve their own and their families' well-being. By stressing the social dimension, the program went far beyond simply promoting the technical appropriation of technology. Furthermore, OMAK perceived its ICT program to be an instrument for strengthening the rich indigenous knowledge systems of local communities. Different from NGO or government programs, OMAK built its activities on a deep understanding of poor people's world-views and indigenous knowledge and strengthened the communities' existing informational capital instead of weakening it or even attempting to replace it, as is the case with many conventional ICT interventions. A critical factor for success of the project as a whole was that this process led to the gradual trans-fer of "ownership" to the participating communities, which was evidenced by the proactive role that many participants assumed in the management and preparation of specific program activities.

Finally, enhancing participants' individual and collective agency with regard to their use of information is also essential. This concept stresses the politi-cal dimension of information and places ICTs into the broader sociopolitical and economic context, particularly in view of the underlying structural factors that have caused the persistently high poverty rates and lack of economic and social opportunities for indigenous peoples in Bolivia (described in chapter 3). It emphasizes the need to engage marginalized groups in a process of self-reflection about the role that the lack of equal access to information and com-munication in their relations with the formal institutions of the state, civil society, and the market plays in their systematic social exclusion and discrimi-nation within Bolivian society. Several grassroots programs (CIOEC, OMAK) have facilitated such a reflective process for indigenous peoples. For instance, the OMAK program has used Freire's approach of "conscientization" and thus enabled indigenous participants to gain an increased awareness and under-standing of the political dimension of information and the way it forms an integral part of their systematic social exclusion and lack of equal economic opportunities within Bolivian society. Such a process was instrumental for the individual and collective empowerment of the participants in the sense that it enabled them to define the meaning, motivation, and purpose of their ICT use and to gain a "sense of agency" related to the use of information. The program particularly engaged the communities in defining the value and meaning of information and ICTs for both their individual human and social capabilities and thus strengthened both their individual and their collective agency.

A critical aspect of the expansion of people's individual and collective agency is that poor people gain the necessary knowledge and human capabili-ties to use, manipulate, and control ICTs based on their own cultural values and worldviews. The use of technologies has to be rooted in the local commu-nities, and their introduction has to be facilitated by a grassroots organization that originates from within the communities. In this sense, poor people's own-ership of and control over the use and management of ICTs and the resulting enhancement of their informational capabilities can act as a critical source for their individual and collective empowerment.

In sum, for ICTs to have an impact on people's well-being, it is critical to have an intermediary organization that supports people in such a manner that their meaningful use of ICTs also enhances their informational capabilities. If poor people are enabled to take this critical step based on their own cultural and social values, enhanced informational capabilities similar to literacy can enhance their human capabilities to make strategic life choices and to interact more effectively with the formal institutions of the state and the market.

Step Five: Enhancing Human and Social Capabilities

The final step in the impact chain investigates the extent to which advanced informational capabilities can enhance people's human and social capabilities and identifies the dimensions in which the meaningful use of ICTs can play a transformative role in their lives. According to the alternative evaluation framework developed in chapter 1, whether or not the use of ICTs expands poor people's human and social capabilities depends on the extent to which informational capabilities (a) enhance their individual and collective agency, (b) strengthen their existing capital (human, financial), and (c) have a positive multiplier effect on the other capabilities. Consequently, the impact of ICTs on people's human and social capabilities is the strongest the more robust the effects of enhanced informational capabilities are on these three aspects of people's lives.

The last step of the impact chain aims to unpack the indirect effects that the enhanced informational capabilities might have on the multiple dimensions (economic, social, and political) of a person's life and the factors that determine the existence and strength of the ICT multiplier effect on people's individual and collective well-being. It is, however, also important to consider the principal socioeconomic and political barriers that marginalized groups face in their daily lives and thus to analyze the major impediments encountered for such a transformative process to take place.

The first dimension in the last step of the impact chain is the personal dimension and the extent to which people's uses of ICTs can result in their individual empowerment. The study finds that personal empowerment is the only dimension in which the use of ICTs can directly enhance people's human well-being. As indicated by both the quantitative and qualitative evidence, enhanced proficiency in ICT use can have a significant and direct impact on people's psychological well-being, particularly for indigenous women and youth, who belong to the most vulnerable and excluded groups within Bolivia. In fact, the OMAK case study has illustrated that indigenous women expressed a sense of pride and accomplishment as a result of their newly acquired knowledge of how to use the Internet and other forms of ICTs. The analysis indicates that enhanced ICT capabilities can be the source of improved individual agency and significantly strengthen the participants' self-esteem.

The social dimension of people's well-being is the second dimension in which enhanced informational capabilities can play an important role in improving people's human capabilities. Indigenous peoples perceive the Internet to have had the strongest impact on their social well-being, with three-quarters (74 percent) of survey respondents with advanced informational capabilities indicating that the Internet can significantly enhance their communities' access to education. Indigenous peoples perceive the Internet to have had the strongest impact on the collective and social capabilities of their communities and a smaller impact on their individual human capabilities.

Investigating the three processes by which enhanced informational capabilities can be converted into enhanced human and social capabilities, it becomes apparent that the Internet has the biggest potential to enhance individual rather than collective capabilities in this dimension of people's lives. With respect to education, for instance, advanced informational capabilities can strengthen people's individual agency by raising their awareness of educational opportunities and by improving their access to nonformal education. Second, enhanced informational capabilities can significantly strengthen an individual's human capital—in particular, if the program focuses on building capacity in information literacy, for example. Third, enhanced informational capabilities in the area of education have a strong multiplier effect, in the sense that, similar to literacy, they can significantly enhance a person's human capabilities and enable her or him to reach higher levels of education. The OMAK case study has demonstrated that ICT capacity-building programs can play a critical role as part of adult education and vocational training programs. As such, ICTs can play an important role in improving poor people's access to nonformal education.

The empirical data, however, have also revealed that the opportunities for ICTs to enhance poor communities' access to education and thus expand their social capabilities are much more limited. While the use of ICTs can enhance the collective agency of poor people in this dimension, the positive spillover effects of informational capabilities are rather limited. Improved informational capabilities and enhanced ability to channel and voice their demand for better education do not translate into improved delivery of government services. Indigenous peoples continue to be excluded from equal access to education and suffer from very low-quality education in their communities. This can be attributed largely to structural challenges within Bolivia's education system, including lack of qualified teachers, scarcity of teaching materials, and poor infrastructure. These issues affect most rural schools and cannot be addressed by ICTs. Thus, while indigenous peoples have very high expectations about the positive impact of the Internet on education—particularly on the access to and quality of education—the structural inequalities between the indigenous and non-indigenous populations, particularly in rural areas, are too significant to be overcome by the use of ICTs.

In the economic dimension of people's well-being, the study concludes that the use of ICTs has somewhat limited positive effects on human well-being. The empirical evidence has indicated that enhanced informational capabilities have weak effects on strengthening people's individual and collective economic agency, do not improve their existing economic or financial capital, and have only limited positive multiplier effects on their economic well-being.

The main reason for this observation is that while ICTs can help to enhance people's access to market *prices*, they cannot meaningfully alter existing market inequalities or reduce the structural barriers that impede participation in the market. Information asymmetry is only one of many factors (such as high transportation costs, limited production capacity) that have led to the economic marginalization of indigenous peoples, and improving access to market information is not, by itself, sufficient to enhance their economic well-being significantly. As the empirical evidence has shown, the use of ICTs does not improve the "negotiating power" of indigenous peoples in local and regional markets, fails to reduce the high "transaction costs" they face when bringing their products to the markets, and does not have any positive impact on incomes.

Moreover, ICTs have the lowest impact on the political dimension of poor people's well-being. As shown in both the survey results and the case studies, many barriers exclude marginalized groups from participating in the political system at the local and central levels of government, and these barriers are too significant to be overcome by the use of ICTs. While ICTs can help indigenous peoples to enhance their individual and collective political agency by, for instance, exerting their right to information, they cannot enhance the transparency of government institutions in the absence of a significant cultural change within government institutions themselves. In fact, the data show that there are no positive multiplier effects between the enhancement of a person's informational capabilities and improvements in her or his human capabilities in political terms. Bolivia's long history of weak "vertical accountability" is the main reason for the lack of indigenous peoples' political participation in the government's decision-making processes. A fundamental change of behaviors and attitudes by politicians and government officials alike is necessary to enhance the accountability and transparency of government institutions. Central to this issue is the lack of "information accessibility," not the lack of access to ICTs.

In relation to social capabilities, the study examines the impact of ICTs on the organizational and cultural dimensions of indigenous peoples' well-being. The expansion of informational capabilities is strongly correlated with organizational capabilities, indicating that indigenous peoples value the Internet primarily as a means to enhance their collective or social capabilities and less as a means to enhance their individual human capabilities.

As shown in chapter 6 and both of the case studies, enhanced informational capabilities can significantly enhance poor people's collective agency, in the sense that the meaningful use of ICTs can strengthen local organizations and regional networks. In particular for grassroots organizations, the Internet and enhanced informational capabilities can significantly strengthen their institutional capabilities to communicate with international donors and to form networks among like-minded organizations (citizenship rights network), which can be instrumental in achieving their overall objectives.

Furthermore, the empirical evidence has shown that the meaningful use of ICTs can significantly strengthen poor people's social capital. In fact, the large majority (61 percent) of indigenous peoples with advanced informational capabilities indicated that the Internet has had a strong influence on their social capital. The data also have indicated that the use of the Internet primarily enhances the existing horizontal channels of communication and information networks at the local level. Thus the positive multiplier effect of informational capabilities is limited to a specific form of social capital—bonding social capital—and does not have a significant positive impact on bridging social capital. Thus the meaningful use of ICTs significantly strengthens existing horizontal networks, but cannot improve vertical communications with the formal institutions of the state, markets, and civil society. The social exclusion of indigenous peoples has led to the existence of parallel channels of communication, with indigenous peoples relying largely on more traditional forms of ICTs, including community and amateur radio, and the non-indigenous population primarily using television, commercial radio stations, and cell phones to obtain information and to communicate with family members and friends. Thus enhanced use of the Internet does not alter the existing "segregation" of communications and information exchanges, and the lack of effective intercultural communications between the indigenous and non-indigenous populations remains unresolved.

Finally, the study has shown mixed results concerning the impact of ICTs on strengthening poor people's well-being in the cultural dimension. A surprising finding is that only a very small minority of indigenous peoples (15 percent) consider the Internet to be a threat to their cultural identity. Due to past negative encounters that indigenous peoples in Bolivia have had with technology, this finding is quite unexpected. For instance, the Green Revolution in the 1970s, which promoted the introduction of new agricultural farming technologies and the transfer of Western-based scientific knowledge in rural communities, caused many problems in indigenous communities, including soil erosion and environmental degradation. Nevertheless, the large majority of indigenous peoples (80 percent) indicated that they believe that the Internet has the potential to improve their well-being either significantly (46 percent) or somewhat (34 percent). The qualitative data from the case studies have shown that indigenous peoples consider the Internet to be relatively "alien" to their culture and thus do not see it as an important instrument to enhance their individual or collective agency in cultural expression. Thus the enhancement of informational capabilities has only a modest multiplier effect in the cultural dimension, with 54 percent of indigenous peoples with advanced informational capabilities indicating that they believe that the Internet can significantly enhance their well-being in this dimension.

The Critical Role of Effective and Local Intermediaries

Furthermore, the impact chain illustrates that the study has confirmed its principal hypothesis—that the presence of an effective and local intermediary organization is the essential factor for enabling poor people to enhance their well-being through the use of ICTs. They help local communities to interpret, appropriate, and enact ICTs to their local sociocultural context; to make the use of ICT meaningful to their everyday lives; and to enable people to enhance their informational capabilities, which ultimately translate into improvements in their human and social capabilities. The ICT impact chain illustrates this critical finding by tracing the path of an ICT program from its initial stage of the information and needs assessment to the enhancement of people's human and social capabilities.

Enabling Environment or Structural Barriers

Finally, the impact chain emphasizes the multiple structural barriers within Bolivian society that severely restrict the ability of marginalized groups to make meaningful use of ICTs. The study has shown that indigenous peoples' lack of access to ICTs and low ICT capabilities reflect the broader socioeconomic, political, and cultural mechanisms that have led to their systematic social exclusion, poverty, and discrimination within Bolivian society. It is noteworthy that indigenous peoples are bound to be significantly disadvantaged in each and every step of the ICT impact chain and that only a small minority of indigenous peoples—in particular, indigenous leaders—have managed to overcome the multiple barriers encountered in the process of accessing, using, and appropriating ICTs. Grassroots organizations play a critical role in assisting marginalized communities to overcome these structural barriers, and only individuals who succeed in enhancing their individual and collective agency, strengthening their existing informational capital, and acquiring advanced informational capabilities can enhance their human and social capabilities through their use of ICTs.

Principal Policy Implications and Recommendations

The following policy implications and recommendations are intended to address the manner in which ICT programs should be designed in order to be most effective in enhancing the well-being of poor communities.

First, *ICT programs should be fully integrated into the overall objectives of a sectoral program instead of being implemented as a stand-alone ICT program.* In fact, the study has shown that the potential benefits are largest when ICTs are fully integrated into other sectoral development programs (in education or health). As the OMAK case study has demonstrated, ICTs can make a significant contribution toward reaching the core objectives of a program—in the OMAK case, the empowerment of indigenous women—when they are made an integral part of the overall program.

Second, *prior to initiating any project activities, it is critical to conduct a detailed assessment of existing information flows and needs.* The assessment should focus on how the new technologies can strengthen existing communication and information exchanges within and between communities. The assessment should also identify key information intermediaries in the community and analyze existing power relationships as they relate to the transfer of knowledge within communities. It is essential for programs to ensure that indigenous communities first identify and define their own needs and development priorities and only then decide whether and how ICTs can support the community's development goals. As shown in the EnlaRed case study, if such a process is not undertaken and the objectives of the ICT project are not well defined, ICT programs frequently fail.

Third, *it is important for evaluating the impact of ICTs on the well-being of poor communities to analyze the process of how ICTs are being introduced.* Outside agents or intermediaries are playing a key role in supporting communities in appropriating the technologies to meet their own local and cultural needs. The case studies have shown that capacity-building activities, the provision of local content, and the gradual transfer of management to local communities by the intermediary are the most important factors influencing whether or not an ICT program will strengthen poor people's human and social capabilities and thus enhance their well-being.

Fourth, *ICTs are most effective when combined with traditional media.* The study thus recommends the use of various forms of ICTs. As the analysis has demonstrated, the use of two different technologies—the Internet and community radio stations—combines the advantages of both media. While the Internet is a powerful tool for connecting networks and exchanging large amounts of information across long distances, community radio has a very broad reach in Bolivia and is the most accessible and inclusive technology for poor remote communities. Due to the strong oral tradition of indigenous communities in Bolivia, this is of particular importance, considering that its use does not require literacy.

Fifth, *ICT programs should proactively address social, political, and cultural factors during implementation of their activities at the community level.* The study has demonstrated that the most important factors influencing whether an ICT program has positive or negative outcomes are social, political, and cultural in nature; technical issues involved in the provision of ICTs often do not play a key role. ICT programs frequently do not respond to a concrete need expressed by communities but are designed through a top-down, supply-driven approach. To avoid the potential negative social effects of such an approach, it is crucial to frame any ICT intervention around the communities' existing social structures. Doing so helps to strengthen traditional information systems by building on indigenous knowledge and existing information channels without undermining the community structures.

Finally, *it is essential to place the human development of people, rather than technology itself, at the center of the design, implementation, and evaluation of ICT programs.* The advantage of using the capability approach as the basis for evaluating technology programs is its emphasis on studying the effects of ICTs on improving the human living conditions of poor and marginalized groups, in contrast to more conventional approaches, which overemphasize the significance of technology for social change. The capability approach thus enables the researcher to evaluate the impact of ICT programs from the vantage point of poor people themselves, rather than from the perspective of an outsider.

APPENDIX A

Map of Indigenous Peoples in Bolivia

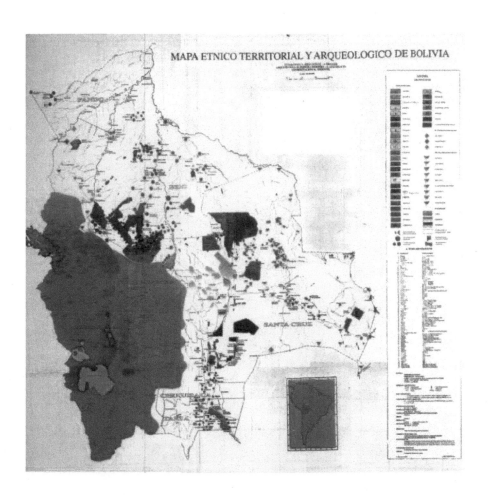

APPENDIX B

Indigenous Peoples' Development Plan

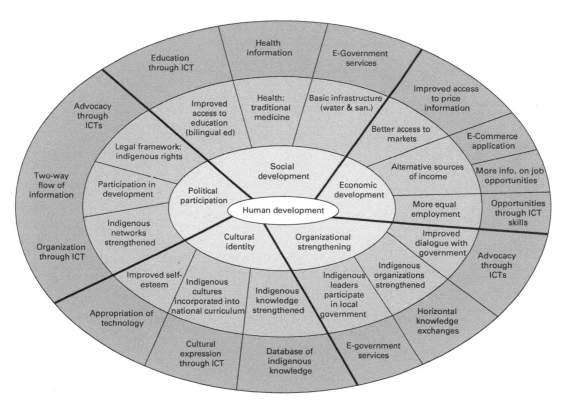

Source: Results from national consultations with indigenous peoples in Bolivia in 1998 (VAIPO, 1999, La Paz).

APPENDIX C

Instrumental and Substantive Social Capabilities

The research distinguishes between the following two types of social capabilities. The first type is instrumental in nature and serves as a catalyst in the enhancement of individual human capabilities. An example of this type of social capability is a women's savings group, where women come together and act collectively so that they can receive access to credit. In this example, participants in the group have enhanced their capabilities in a way that they could not have accomplished alone. At the same time, the incentives for forming the group are based on the individual need to access credit; therefore, this type of group or social capability has an instrumental character.

The second type of social capabilities not only plays an instrumental role in enhancing the human capabilities of individuals, it also plays a *substantive* role of its own. These capabilities are especially relevant for indigenous peoples, who frequently define their own well-being through the well-being of their community. Thus substantive social capabilities are keys to understanding their vision of an autonomous indigenous development based on their identity and worldview. An example of a social capability that is substantive rather than instrumental in nature is a cultural or religious ritual or festival. Festivals strengthen the communities' identity and traditional institutions and thus are substantive for the community; often they do not have the objective of enhancing the human capability of any one of their members. In fact, they can often be detrimental to the well-being of the individual.[1]

Another example of a substantive social capability can be found in the traditional information systems of indigenous communities, which are based on the notion of indigenous knowledge as a collective good. In each community there are individuals (frequently elders) who are preserving and passing on the community's indigenous knowledge, and the way the knowledge is passed on and applied is purely collective in nature. Due to their oral tradition, the preservation and passing on of knowledge is a process that can only be enacted through the very nature of a group. In addition, a large part of traditional knowledge is applied through collective action. For instance, the traditional Andean agricultural technique of "raised fields" is only possible through collective work (*minka*), and the benefits of applying this traditional knowledge are shared by the community. In this sense, indigenous knowledge represents an important "social capability" for indigenous communities, and its substantive character allows for more than just the instrumental enhancement of individual human capabilities.

Note

1. For instance, the cargo system in the Andean highlands requires the individual to spend his or her life savings for the community (Allen 1988).

Reference

Allen, Catherine. 1988. "The Hold Life Has: Coca and Cultural Identity in an Andean Community." Ethnographic Inquiry 12, Smithsonian Institution, Washington, DC.

APPENDIX D

The Ethnicity Index by Xavier Albó

The study draws on the ethnicity index developed by Xavier Albó, which emphasizes that the language criteria frequently used in national censuses throughout Latin America are insufficient to differentiate appropriately between indigenous and non-indigenous peoples. As shown in table D.1, Albó has developed a more nuanced index that uses three criteria to differentiate between high, medium, and low degrees of ethnicity: (a) principal language spoken as a child was an indigenous language, (b) born in an indigenous community (rural or semi-urban area), and (c) self-identification as being indigenous (table D.1; Molina Barrios and Albó 2006; Albó and Quispe 2004).

A high level of ethnicity refers to a person who meets all of these criteria. An intermediate level of ethnicity indicates a person who self-identifies as indigenous, was born in an indigenous community, but did not speak an indigenous language as a child. A low level of ethnicity indicates a person who does not self-identify as indigenous, whose principal language spoken as a child was not an indigenous language, but who was born in an indigenous community. A person who does not meet any of the three criteria is considered non-indigenous.

TABLE D.1 Criteria for the construction of Albó's ethnicity index

Principal language spoken as a child is an indigenous language	Born in an indigenous community (rural and semi-urban area)	Self-identification as indigenous	Ethnicity index
Yes	Yes	Yes	High
Yes	No	Yes	Intermediate
No	Yes	No	Low
No	No	No	Non-indigenous

References

Albó, Xavier, and Víctor Quispe. 2004. "¿Quiénes son indígenas en los gobiernos municipales?" Cuadernos de Investigación CIPCA 55, Centro de Investigación y Promoción del Campesinado, Plural (CIPCA), La Paz.

Molina Barrios, Ramiro, and Xavier Albó. 2006. *Gama étnico y linguística de la población Boliviana.* La Paz: Sistema de las Naciones Unidas en Bolivia.

APPENDIX E

Statistical Data

TABLE E.1 Information sources, by cluster (data for figures 4.2 and 4.4)

Source	Indigenous organizations (national)	Indigenous organizations (international)	Grassroots (highlands)	Grassroots (lowlands)	OMAK youth	NGO	Total	Chi-square[a]	df	Asymp. sig.[a]
Meetings	1.67	1.59	1.48	1.30	1.05	1.03	1.30	18.51	5	0.00
Family or friends	1.40	1.12	1.21	1.59	1.31	0.87	1.24	16.17	5	0.01
Community radio	1.29	1.73	1.40	1.48	1.60	1.45	1.51	9.98	5	0.08
Television	0.60	1.45	0.98	1.30	1.58	1.17	1.19	34.99	5	0.00
Telephone	0.38	1.00	0.58	0.00	0.93	0.29	0.74	24.73	5	0.00
Letters	1.21	0.75	0.73	0.00	0.74	0.60	0.78	6.62	5	0.25
Newspapers	0.67	0.65	0.44	0.61	0.98	0.46	0.67	10.57	5	0.06
Internet	0.30	0.33	0.05	0.36	0.46	0.18	0.32	8.04	5	0.15

a. Kruskal-Wallis significance test.

TABLE E.2 Most frequently used information sources, by program (data for figure 4.3)

Source	CIOEC	Fondo Indígena	Omak leaders	CIOEC Potosí	CIOEC Oruro	APCOB	CIDOB	OMAK youth	ICO	Total	Chi-square[a]	df	Asymp. sig.[a]
Meetings	1.67	1.59	1.50	1.50	1.42	1.33	1.25	1.05	1.03	1.30	18.89	8	0.02
Family or friends	1.40	1.12	1.43	1.10	0.93	1.73	1.43	1.03	0.87	1.24	22.39	8	0.00
Community radio	1.29	1.73	1.73	1.54	1.35	1.67	1.20	1.60	1.45	1.51	13.02	8	0.11

a. Kruskal-Wallis significance test.

TABLE E.3   Most frequently used information sources, by ethnicity (data for figure 5.5)

Source	Spanish	Spanish and indigenous language	Indigenous language	Total	Chi-square[a]	df	Asymp. sig.[a]
Community radio	1.59	1.49	1.47	1.51	1.04	2	0.60
Meetings	1.03	1.41	1.23	1.30	7.62	2	0.02
Family or friends	1.18	1.23	1.35	1.24	1.64	2	0.44
Television	1.41	1.06	1.22	1.19	7.79	2	0.02
Letters	1.06	0.73	0.73	0.78	3.09	2	0.21
Telephone	0.93	0.72	0.71	0.74	0.69	2	0.71
Newspapers	0.57	0.80	0.60	0.67	4.55	2	0.10
Internet	0.48	0.24	0.31	0.32	4.43	2	0.11

a. Kruskal-Wallis significance test.

TABLE E.4   Frequently used communications channels, by program

Channel	CIOEC	Fondo Indígena	Omak leaders	CIOEC Potosí	CIOEC Oruro	APCOB	CIDOB	CONAMAQ	OMAK youth	ICO	Total	Chi-square[a]	df	Asymp. sig.[a]
Personal visits	1.14	1.50	1.44	1.40	1.50	1.65	1.62	1.57	1.34	0.81	1.39	22.68	9	0.01
Amateur radio	1.25	1.24	1.19	1.30	0.63	1.67	1.00	1.29	1.38	0.80	1.24	15.21	9	0.09

a. Kruskal-Wallis significance test.

TABLE E.5A   Frequently used communications channel, by profession

Channel	Indigenous leader	Councilmember	Farmer/artisan	Student	Total	Chi-square[a]	Df	Asymp. sig.[a]
Personal visits	1.50	1.40	0.84	1.35	1.39	17.14	3	0.00
Amateur radio	1.27	1.00	0.73	1.25	1.24	7.55	3	0.06

a. Kruskal-Wallis significance test.

TABLE E.5B   Cross-tabulation for communications channel, by profession

Indicator	Indigenous leaders	Councilmembers	Farmers	Students	Total
How frequently do you use personal visits? (%)					
A lot	61.8	60.0	25.8	56.1	55.6
Somewhat	26.5	20.0	32.2	22.7	25.9
None	11.8	20.0	41.9	21.2	18.5
Total (%)	100	100	100	100	100
Total (number)	136	10	31	66	243

TABLE E.5C   Statistical significance test for personal visits, by profession

Test	Value	df	Asymp. sig.
Chi-square	19.611	6	0.03
Cramer V	0.201		0.03

TABLE E.6   Information gap analysis (data for figure 4.17)

Indicator	Festivals and religious events	Job opportunities	Government policies	Market prices	Health information	Citizenship rights	Local news	Agricultural practices	Educational materials
Information needs	0.92	1.15	1.22	1.42	1.42	1.43	1.53	1.53	1.57
Information received	0.76	0.70	0.95	1.03	0.91	0.92	1.21	0.98	0.98
Information gap	0.16	0.45	0.27	0.39	0.51	0.51	0.32	0.55	0.59
Wilcoxon (Z)	−2,283	−5,815	−3,836	−5,510	−7,160	−6,556	−3,496	−7,128	−8,751
Stat. significance	0.022	0.000	0.000	0.000	0.000	0.000	0.000	0.000	0.000

TABLE E.7   Information gap analysis for OMAK youth program (data for figure 4.18)

Indicator	Festivals or religious events	Job opportunities	Government policies	Market prices	Health information	Citizenship rights	Local news	Agricultural practices	Educational materials
Information needs	0.92	0.90	0.75	0.92	0.80	0.90	—	0.97	0.94
Information received	0.93	1.12	1.00	0.95	0.88	0.83	—	1.02	0.69
Information gap	−0.01	−0.22	−0.25	−0.03	−0.08	0.07	—	−0.05	0.25
Wilcoxon (Z)	−0,350	−1,798	−2,011	−0,342	−1,436	−0,029	—	−0,850	−2,507
Stat. significance	0.726	0.072	0.044	0.732	0.151	0.977	—	0.396	0.012

Note: — = not available.

TABLE E.8 Information gap analysis without OMAK youth program (data for figure 4.20)

Indicator	Festivals, religious events	Job opportunities	Government policies	Market prices	Health information	Citizenship rights	Local news	Agricultural practices	Educational materials
Information needs	0.91	1.25	1.42	1.59	1.67	1.65	1.53	1.72	1.82
Information received	0.67	0.51	0.93	1.06	0.92	0.96	1.21	0.97	1.10
Information gap	0.24	0.74	0.49	0.53	0.75	0.69	0.32	0.75	0.72
Wilcoxon (Z)	−3,14	−7,07	−6,05	−6,18	−8,37	−7,47	−3,5	−8,4	−8,47
Stat. significance	0.002	0.00	0.00	0.00	0.00	0.00	0.00	0.00	0.00

TABLE E.9 Information gap analysis for educational materials, by program (data for figure 4.21)

Indicator	OMAK youth	CIOEC	OMAK leader	APCOB	Fondo Indígena	CIDOB	ICO	CIOEC Potosí	CIOEC Oruro	Total	Chi-square[a]	df	Asymp. sig.[a]
Information needs	0.94	1.88	1.81	1.89	1.73	1.82	1.78	1.87	1.87	1.57	77.37	8	0.00
Information received	0.69	1.50	1.42	1.25	1.03	1.06	0.93	1.00	0.79	0.98	26.34	8	0.00
Information gap	0.25	0.38	0.39	0.64	0.70	0.76	0.85	0.87	1.08	0.59			

a. Kruskal-Wallis significance test.

TABLE E.10A Information needs for educational materials, by ethnicity for the full sample (data for figure 4.22)

Indicator	Spanish	Spanish and indigenous language	Indigenous language	Total
How much information do you need about educational materials? (%)				
A lot	76.4	81.5	34.4	68.9
Somewhat	18.2	13.3	31.1	18.7
None	5.5	5.2	34.4	12.4
Total (%)	100.0	100.0	100.0	100.0
Total (number)	55	135	61	251

TABLE E.10B Statistical significance test for information needs for educational materials, by ethnicity for the full sample

Test	Value	df	Asymp. sig.
Chi-square	53.0	4	0.00
Cramer V	0.32		0.00

TABLE E.11A   Information gap analysis for educational materials, by ethnicity, for the full sample (data for figure 4.23)

Indicator	Spanish	Spanish and indigenous language	Indigenous language	Total	Chi-square[a]	df	Asymp. sig.[a]
Information needs	1.71	1.76	1.00	1.57	50.26	2	0.00
Information received	1.04	1.05	0.76	0.98	6.07	2	0.04
Information gap	0.67	0.71	0.24	0.59			

a. Kruskal-Wallis significance test.

TABLE E.11B   Information gap analysis for educational materials, by ethnicity, for Aymara and Quechua separately

Indicator	Spanish	Spanish and indigenous language	Aymara	Quechua	Total	Chi-square[a]	df	Asymp. sig.[a]
Information needs	1.71	1.76	0.96	1.08	1.57	50.27	3	0.00
Information received	1.04	1.05	0.64	1.18	0.98	9.97	3	0.02
Information gap	0.67	0.71	0.31	−0.10	0.59			

a. Kruskal-Wallis significance test.

TABLE E.12A   Information gap analysis for educational materials, by ethnicity without OMAK youth sample (data for figure 4.24)

Indicator	Spanish	Spanish and indigenous language	Indigenous language	Total
How much information do you need about educational materials? (%)				
A lot	76.4	87.2	75.0	84.4
Somewhat	18.2	11.0	25.0	12.8
None	5.5	1.8		2.8
Total (%)	100	100	100	100
Total (number)	52	115	12	179

TABLE E.12B   Statistical significance test for information needs for educational material, by ethnicity (without OMAK youth)

Test	Value	df	Asymp. sig.
Chi-square	4.345	4	0.361
Cramer V	0.11		0.36

TABLE E.13   Information gap analysis for educational materials, by ethnicity, without youth sample (data for figure 4.25)

Indicator	Spanish	Spanish and indigenous language	Indigenous language	Total	Chi-square[a]	df	Asymp. sig.[a]
Information needs	1.75	1.85	1.75	1.82	1.91	2	0.38
Information received	1.06	1.13	1.00	1.10	0.62	2	0.73
Information gap	0.69	0.72	0.75	0.72			

a. Kruskal-Wallis significance test.

TABLE E.14   Information gap analysis for educational materials, by gender (for figure 4.26)

Indicator	Male	Female	Total	Chi-square[a]	df	Asymp. sig.[a]
Information needs	1.66	1.35	1.57	6.66	1	0.01
Information received	1.06	0.78	0.98	6.33	1	0.01
Information gap	0.60	0.57	0.59			

a. Kruskal-Wallis significance test.

TABLE E.15   Information gap analysis for agricultural practices, by program (data for figure 4.27)

Indicator	OMAK youth	CIOEC	OMAK leader	APCOB	Fondo Indígena	CIDOB	ICO	CIOEC Potosí	CIOEC Oruro	Total	Chi-square[a]	df	Asymp. sig.[a]
Information needs	0.97	2.00	1.08	1.82	1.63	1.73	1.80	1.88	1.56	1.53	69.02	8	0.00
Information received	1.02	1.59	0.64	1.09	0.90	0.95	1.00	1.00	0.53	0.98	20.73	8	0.01
Information gap	−0.05	0.41	0.43	0.73	0.74	0.78	0.80	0.88	1.02	0.55			

a. Kruskal-Wallis significance test.

TABLE E.16   Information gap analysis for citizenship rights, by program (data for figure 4.28)

Indicator	OMAK youth	CIOEC	OMAK leaders	APCOB	Fondo Indígena	CIDOB	ICO	CIOEC Oruro	CIOEC Potosí	Total	Chi-square[a]	df	Asymp. sig.[a]
Information needs	0.90	1.83	1.33	1.57	1.72	1.80	1.60	1.63	1.88	1.43	55.73	8	0.00
Information received	0.83	1.42	1.00	1.00	1.24	0.62	0.69	0.54	1.13	0.92	21.81	8	0.01
Information gap	0.07	0.41	0.33	0.57	0.48	1.18	0.91	1.09	0.75	0.51			

a. Kruskal-Wallis significance test.

TABLE E.17A   Cross-tabulation for information needs for citizenship rights, by education level

Indicator	Primary school and less	Secondary school	Technical schooling (intermediate level)	Technical schooling (advanced) and university	Total
How much information do you need about citizenship rights? (%)					
A lot	71.8	51.2	70.3	55.6	59.6
Somewhat	15.4	26.4	29.7	38.9	23.8
None	12.8	22.5		5.6	16.7
Total (%)	100	100	100	100	100
Total (number)	78	129	15	18	240

TABLE E.17B   Statistical significance test for information needs for citizenship rights, by education level

Test	Value	df	Asymp. sig.
Chi-square	15.612	6	0.01
Cramer V	0.180		0.01

TABLE E.18A   Cross-tabulation for information needs for citizenship rights, by age

Indicator	35 years and younger	36 years and older	Total
How much information do you need about citizenship rights? (%)			
A lot	53.2	70.5	58.9
Somewhat	24.7	23.1	24.2
None	22.2	6.4	16.9
Total (%)	100	100	100
Total (number)	158	78	236

TABLE E.18B   Statistical significance test for information needs for citizenship rights, by age

Test	Value	df	Asymp. sig.
Chi-square	10.359	2	0.006
Cramer V	0.210		0.006

TABLE E.19A   Information gap analysis for market prices, by age (data for figure 4.29)

Indicator	35 years and younger	36 years and older	Total	Chi-square[a]	df	Asymp. sig.[a]
Information needs	1.22	1.80	1.42	34.074	1	0.00
Information received	0.98	1.12	1.03	1.571	1	0.210
Information gap	0.24	0.68	0.39			

a. Kruskal-Wallis significance test.

TABLE E.19B **Cross-tabulation for information needs for market prices, by age**

Indicator	35 years and younger	36 years and older	Total
How much information do you need about market prices? (%)			
A lot	42.5	81.6	55.1
Somewhat	36.9	17.1	30.5
None	20.6	1.3	14.4
Total (%)	100	100	100
Total (number)	160	76	236

TABLE E.19C **Statistical significance test for market price information, by age**

Test	Value	df	Asymp. sig.
Chi-square	34.220	2	0.000
Cramer V	0.381		0.000

TABLE E.20 **Information gap analysis for market prices, by program (data for figure 4.30)**

Indicator	OMAK youth	CIOEC	OMAK leader	APCOB	Fondo Indígena	CIDOB	ICO	CIOEC Potosí	CIOEC Oruro	Total	Chi-square[a]	df	Asymp. sig.[a]
Information needs	0.92	1.95	0.71	1.45	1.70	1.71	1.33	1.88	1.92	1.42	77,14	8	0.00
Information received	0.95	1.75	0.50	1.18	1.40	1.20	0.63	0.79	0.60	1.03	47,26	8	0.00
Information gap	−0.03	0.20	0.21	0.27	0.31	0.51	0.70	1.10	1.32	0.39			

a. Kruskal-Wallis significance test.

TABLE E.21 **Government information gap analysis, by program (data for figure 4.31)**

Indicator	OMAK youth	CIOEC	OMAK leader	APCOB	Fondo Indígena	CIDOB	ICO	CIOEC Potosí	CIOEC Oruro	Total	Chi-square[a]	df	Asymp. sig.[a]
Information needs	0.75	1.13	1.75	1.81	1.22	1.33	1.33	1.14	1.54	1.22	53,60	8	0.00
Information received	1.00	0.77	1.33	1.35	0.68	0.79	0.78	0.58	0.86	0.95	21,22	8	0.01
Information gap	−0.25	0.36	0.42	0.45	0.54	0.55	0.56	0.56	0.68	0.27			

a. Kruskal-Wallis significance test.

TABLE E.22 Government information gap analysis, by gender for the full sample (data for figure 4.32)

Indicator	Male	Female	Total	Chi-square[a]	df	Asymp. sig.[a]
Information needs	1.31	1.00	1.22	6.95	1	0.01
Information received	0.93	1.00	0.95	0.33	1	0.568
Information gap	0.38	0.00	0.38			

a. Kruskal-Wallis significance test.

TABLE E.23 Government information gap, by ethnicity, for full sample (data for figure 4.33)

Indicator	Spanish	Spanish and indigenous language	Indigenous language	Total	Chi-square[a]	df	Asymp. sig.[a]
Information needs	1.31	1.30	0.93	1.22	8.84	2	0.01
Information received	0.79	0.98	1.02	0.95	2.47	2	0.29
Information gap	0.52	0.32	−0.09	0.27			

a. Kruskal-Wallis significance test.

TABLE E.24 Government information gap, by ethnicity, for Aymara and Quechua (without youth sample)

Indicator	Spanish	Spanish and indigenous language	Aymara	Quechua	Total	Chi-square[a]	df	Asymp. sig.[a]
Information needs	1.31	1.30	0.84	0.94	1.42	1.399	2	0.499
Information received	0.79	0.98	0.97	1.13	0.93	2.421	2	0.298
Information gap	0.52	0.32	−0.13	−0.19	0.47			

a. Kruskal-Wallis significance test.

TABLE E.25 Government information gap, by ethnicity, without youth sample (data for figure 4.34)

Indicator	Spanish	Spanish and indigenous language	Indigenous language	Total	Chi-square[a]	df	Asymp. sig.[a]
Information needs	1.67	1.43	1.35	1.42	1.399	2	0.499
Information received	1.12	0.96	0.88	0.93	2.421	2	0.298
Information gap	0.55	0.47	0.47	0.47			

a. Kruskal-Wallis significance test.

APPENDIX F

Definition of ICT Capability Variable

The definition of the variable "ICT capability" is based on question 22 of the user survey: For which activities have you used the Internet? For each sub-question, the respondents had four options: (1) a lot, (2) somewhat, (3) none, or (9) no answer. The variable "ICT capability" was calculated using the mode value—the most frequently occurring value in a set of discrete data. The mode value, instead of the median or mean value, was chosen because the mode value is the only way to indicate the typical response for nominal variables. Table F.1 identifies the mode value. The overall value 1 indicates an advanced level of use, the overall value 2 indicates an intermediate level of use, the overall value 3 indicates a basic level of use, and the overall value of 9 indicates that participants did not answer this question.

ICT capabilities are considered advanced (1) if the most common response was a lot and if more than 50 percent (5 of 9) of the questions had this value. ICT capabilities are considered intermediate (2) if the most common response was somewhat and if more than 50 percent (5 of 9) of the questions had this value. ICT capabilities are considered basic (3) if the most common response was "none" and if more than 50 percent (5 of 9) of the questions had this value. The same logic is applied to define the "no answer" category. If the result was more than one mode value, only the first mode value is counted.

TABLE F.1   Mode value for question 22 regarding ICT capacity

ICT capability	22.1. Receive and send e-mail	22.2. Search information using a search engine	22.3. Read the news	22.4. Distribute and publish information about your organization	22.5. Participate in online forum
Advanced (1)	1	1	1	2	1
Intermediate (2)	2	2	2	2	3
Basic (3)	2	2	3	3	3
No answer (9)	1	1	1	9	9

ICT capability	22.6. Coordinate activities of the indigenous movement	22.7. Present funding proposals	22.8. Take online courses	22.9. Search for a job
Advanced (1)	1	1	2	2
Intermediate (2)	2	3	3	3
Basic (3)	2	2	3	3
No answer (9)	9	9	2	9

APPENDIX G

Definition of Informational Capability Variable

The variable "informational capability" has two components. The first component is the variable "ICT capability" (see appendix F). The second component is based on the variable "information skills," which is based on question 16 of the user survey: "To what extent has the use of the Internet helped you to (16.1) better communicate with family members and friends, (16.2) find information that you need, and (16.3) share information with others." For each subquestion, respondents had four options: (1) a lot, (2) somewhat, (3) none, and (9) no answer.

The variable "information skills" was calculated using the mode value—the most frequently occurring value in a set of discrete data. Table G.1 helps to identify the mode value: The overall value 1 indicates advanced use, the overall value 2 indicates intermediate use, the overall value 3 indicates basic use, and the overall value 9 indicates that the participant did not answer the question.

Information skills are considered advanced (1) if the most common response was a lot and if the responses to two out of the three questions had this value. Information skills are considered intermediate (2) if the most common response was somewhat and if two out of the three questions had this value. Information skills are considered basic (3) if the most common response was "none" and if two out of the three questions had this value. The same logic is applied to define the "no answer" category. If the result was more than one mode value, only the first mode value is counted.

In a second step, the variable "informational capabilities" was calculated using the weighted arithmetic mean of the ICT capability and information skills variables:

$$\bar{x} = \frac{\sum_{i=1}^{n} w_i \cdot x_i}{\sum_{i=1}^{n} w_i}. \tag{G.1}$$

TABLE G.1   Mode value for question 16 regarding information skills

Information skills	16.1. Better communicate with family members and friends	16.2. Find information that you need	16.3. Share information with others
Advanced (1)	1	1	3
Intermediate (2)	1	2	2
Basic (3)	3	3	2
No answer (9)	9	9	3

Informational capability = [ICT capability variable * (1/3) + Information skills variable * (2/3)].

I assigned the ICT capabilities variable a weight of one-third and the information skills variable a weight of two-thirds since they are composed of three subcomponents: information literacy, communications skills, and content skills. For the construction of the informational capabilities variable, informational skills outweigh people's ICT capabilities and thus are weighted higher than ICT capabilities. In order to interpret the resulting data in terms of people's differences in their ability to apply ICTs and information in a meaningful manner to the different aspects of their lives, I created three categories:

- Low informational capabilities: 0.0–0.49
- Intermediate informational capabilities: 0.5–1.49
- High informational capabilities: 1.5–2.0.

Definition of Sector and Thematic Indexes

The various sector and thematic indexes are based on the following statements or questions.

Social Dimension Index

23.1. The Internet has improved your knowledge about global issues.
23.2. The Internet has improved your access to education.
23.3. The Internet has improved your access to health.
23.4. The Internet has strengthened your leadership skills.
23.5. The Internet has strengthened your relationships with other leaders.
23.6. The Internet has improved the communications within my community.

Economic Dimension Index

25.1. The Internet has improved my access to market price information.
25.2. The Internet has reduced the need to travel to sell my products.
25.3. The Internet has helped me to find new markets for my products.
25.4. The Internet has helped me to obtain new knowledge to improve my production.
25.5. The Internet has provided me with new opportunities to find employment.
25.6. The Internet has helped me to communicate better with family members and friends who are sending me regularly money.
25.7. The Internet has helped me to increase my income.

Political Dimension Index

27.1. The Internet has improved my access to information about national policies.

27.2. The Internet has improved my participation in the decision-making process of local government.

27.3. The Internet has helped me to better participate in national policy dialogues.

27.4. The Internet has helped me to better know my rights.

27.5. The Internet has helped me to be better informed about international news.

27.6. The Internet has helped me to be in touch with others to discuss politics.

Organizational Dimension Index

36.1. The Internet has strengthened indigenous organizations.

36.2. The Internet has improved the access to information about the management of indigenous organizations.

36.3. The Internet has strengthened networks among indigenous organizations.

36.4. The Internet has improved the communications with the central government.

Personal Development

32.1. The Internet has provided me with new job opportunities.

32.2. I feel that through the use of the Internet I have many more opportunities in comparison to other members of my community.

32.3. I feel that the Internet has helped me to be less isolated from the rest of Bolivia.

32.4. The Internet has provided me with new opportunities to improve my well-being.

32.5. I feel proud to know how to use the Internet.

Cultural Dimension

34.1. The Internet has strengthened my cultural identity.

34.2. The Internet has led to a loss of cultural identity in my community.

34.3. The Internet has improved my knowledge about my indigenous language.

34.4. The Internet has helped me to better know my own culture.

34.5. The Internet has helped me to disseminate information about my culture.

Vertical Communication Index (with National Government)

27.1. The Internet has improved my access to information about national policies.

27.3. The Internet has helped me to better participate in national policy dialogues.

30.2. The Internet has improved the transparency of government institutions.

36.4. The Internet has improved the communications with central government.

Vertical Communication Index (with Local Government)

30.6. The Internet has improved the participation of indigenous peoples in local government.

26.2. The Internet has improved my knowledge about development projects of the local government.

30.3. The Internet has improved the social control over the local government's programs.

Horizontal Communication Index (between Indigenous Leaders)

23.5. The Internet has strengthened my relationship with other leaders.

30.1. The Internet has strengthened networks among indigenous organizations.

Vertical Communication Index (between Indigenous Leaders and Communities)

23.6. The Internet has helped me to improve the communication with my community.

36.2. The Internet has improved the access to information about the management of indigenous organizations.

31. How do you perceive the information exchanges between indigenous organizations and local communities: (i) good, (ii) regular, (iii) bad, and (iv) there is no information exchange.

Transparency Index

27.1. The Internet has improved my access to information about national policies.

30.2. The Internet has improved the transparency of government institutions.

30.3. The Internet has improved the social control over the local government's programs.

Social Capital Index

26.1. The Internet has strengthened the community-based cooperatives in my community.

36.1. The Internet has strengthened indigenous organizations.

36.3. The Internet has strengthened networks among indigenous organizations.

Interview List

International Indigenous Leaders

- Mateo Martínez, director, Fondo Indígena, January 15, 2005
- Juan Tarifa, program leader, Fondo Indígena, January 16, 2005

National Indigenous Leaders

- Andrea Flores, president and director, OMAK, June 5, 2005
- Jorge Flores, program manager, OMAK, May 10, 2005
- Félix Gutiérrez, director, Indigenous Communications Initiative, Red Wayra, La Paz, July 7, 2005
- Renato García, national program coordinator, CIOEC, La Paz, July 3, 2005

Local Indigenous Leaders

- Juana González, Aymara women's group leader, Achaca, Tihuanacu, April 15, 2005
- Julio Sánchez, president, Agricultural Producer Association, Comarapa, May 24, 2005
- Elvina Montenegro, Aymara women's group leader from Luribay, Tihuanacu, July 19, 2005
- Elizabeth Corneo, Aymara women's group leader, Tihuanacu, July 19, 2005
- Angelica Morales, treasurer, OMAK women's group, Tihuanacu, July 20, 2005
- Fernanda Villegas, program participant, OMAK, Tihuanacu, July 20, 2005
- Eulalia Ruiz, program participant, OMAK, Tihuanacu, July 20, 2005

- Ivan Salinas, community leader from Batallas, June 30, 2005
- Teresa Nacho, Aymara women's group leader and OMAK program participant, July 21, 2005
- Rosa Morales, president, OMAK women's center, Tihuanacu, July 21, 2005
- María Morales, Aymara youth leader, Achaca, Tihuanacu, July 22, 2005
- Julio Ximénez, indigenous leader from Concepción, April, 22, 2005

NGO Representatives

- Juergen Riesler, director, ABCOB, Santa Cruz, April 24, 2005
- Edwin Rocha, director, ICO, Santa Cruz, June 18, 2005
- Eduardo Rojas, Fundación Redes, June 10, 2005

National Government Representative

- Clemencia Vargas, director of Information Systems Department, VIPFE, La Paz, August 12, 2005

Local Government Representatives

- Manual Quispe, official mayor of Batallas municipality, July 30, 2005
- Juan Delgado, mayor of Batallas, July 30, 2005
- José Luis Alvarez, official mayor of Comarapa, Santa Cruz, August 20, 2005
- Justo Seoane, mayor of La Concepción and former vice minister of indigenous affairs, April 22, 2005
- Lena Calla, deputy director, EnlaRed Municipal, FAM, La Paz, July 6, 2005
- Carlos Soria, director, EnlaRed Municipal, FAM, La Paz, August 25, 2005
- Marcelo Carbonne, project coordinator, EnlaRed Municipal, Comarapa, La Paz, February 25, 2005
- Lino Condori, mayor of Tihuanacu, La Paz, August 15, 2005
- Juan Delgadillo, mayor of Comarapa, May 23, 2005

ICT Specialists

- Bernal Patty, information technology specialist, municipality of Batallas, La Paz, February 25, 2005
- Sonia Paredez, communications specialist, municipality of Tarabuco, La Paz, February 25, 2005
- Eliana Rojas, ICT program coordinator, CIDOB, Santa Cruz, April 23, 2005

- Juan Burgos, coordinator of ICT project, APCOB, Santa Cruz, April 24, 2005
- Claudia Camacho, ICO Comarapa, coordinator of ICT project, Comarapa, May 23, 2005

Church Representative

- Padre Claudio, Catholic priest, Universidad Católica, Tihuanacu, July 22, 2005

List of Focus Groups

TABLE J.1 Pilot and subsequent focus groups

Location	Target group	Date	Number of participants
Pilot focus group			
La Paz	Fondo Indígena, national leaders from Latin America	December 12, 2004	12
16 focus groups			
La Paz	EnlaRed, local government officials (national)	February 25, 2005	11
La Paz	CONAMAQ, national indigenous leaders	March 05, 2005	12
Tihuanacu	OMAK, grassroots women's group leaders	April 14, 2005	10
Tihuanacu	OMAK, community leaders (highlands)	April 19 2005	14
Concepción	APCOB, grassroots leaders (lowlands)	April 25, 2005	12
Comarapa	ICO, small-scale farmers (lowlands)	May 24, 2005	10
Comarapa	ICO, small-scale farmers (lowlands)	May 23, 2005	14
Tihuanacu	OMAK, women's group leaders	May 28, 2005	12
Cochabamba	EnlaRed, local government officials	June 10, 2005	15
Tihuanacu	OMAK, indigenous youth (highlands)	July 19, 2005	12
Tihuanacu	Tihuanacu, grassroots women's group leaders (highlands)	July 19, 2005	10
Mairana	ICO, small-scale farmers (lowlands)	August 19, 2005	9
Comarapa	ICO, small-scale farmers (lowlands)	August 20, 2005	14
Acha	CIOEC, national indigenous leaders	August 24, 2005	15
Challapata	CIOEC, grassroots leaders (highlands)	August 27, 2005	15
Potosí	CIOEC, grassroots leaders (highlands)	September 2, 2005	11

APPENDIX K

Survey Forms

Non-User Survey Form

Encuesta sobre el rol de las Tecnologías de Información y Comunicación para el desarrollo indígena (para no-usarios de Internet)

> **Instrucción de llenado de boleta:** Por favor lea atentamente las preguntas y encierre en un círculo su respuesta.

I. INFORMACIÓN PERSONAL

1. Departamento: _____
2. Municipio: _____
3. Comunidad: _____
4. Organización: _____
5. Tel: _____
6. Correo Electrónico: _____
7. Sexo
    1. Hombre
    2. Mujer
8. Edad_____
9. Cuál de los siguientes idiomas es el que más utiliza para comunicarse en su hogar?
    1. Castellano
    2. Castellano y alguna lengua originaria/indígena
    3. Aymará
    4. Quechua
    5. Guaraní
    6. Otra lengua originaria/indígena

10. Grado de instrucción

1	Menos de 5 años de primaria
2	Primaria
3	Secundaria
4	Técnico medio
5	Técnico superior
6	Universidad
7	Otros

11. ¿Qué cargo ocupa en su comunidad o municipio? ¿Cuál es su ocupación principal?
    1. Mallku
    2. Dirigente
    3. Concejal
    4. Técnico
    5. Agricultor/artesano
    6. Comerciante
    7. Estudiante
    8. Otro

12. ¿Existe en su comunidad servicio de electricidad?
    1. Si
    2. No

II. USOS DE MEDIOS DE COMUNICACIÓN E INFORMACIÓN

13. En su comunidad, que medios de comunicación existen?

1	Cabina de teléfono fijo
2	Celular
3	Radio emisora
4	Televisión
5	Correo y cartas
6	Internet
7	Ninguno
8	Otro

14. Para comunicarse con familiares o amigos que viven fuera de su comunidad con qué frecuencia utiliza:

	1. Mucho	2. Poco	3. Nada
1. Teléfono	1	2	3
2. Celular	1	2	3
3. Radio aficionados	1	2	3

4. Cartas	1	2	3
5. Encargo	1	2	3
6. Visita personal	1	2	3
7. Internet	1	2	3

15. Con que frecuencia recibe información a través de:

	1. Mucho	2. Poco	3. Nada
Familiares o amigos	1	2	3
Reuniones de la comunidad	1	2	3
Periódico	1	2	3
Radio emisora	1	2	3
Televisión	1	2	3
Teléfono/fax	1	2	3
Cartas	1	2	3
Internet	1	2	3

III. VALOR DE LA INFORMACIÓN Y EL INTERNET

16. ¿Cuán importante es para usted, el acceso a la información para el desarrollo indígena?
    1. Muy importante
    2. Importante
    3. Poco
    4. Nada

17. ¿Cuán útil piensa usted es el uso del Internet para el desarrollo indígena?
    1. Muy útil
    2. Util
    3. Poco
    4. Nada

IV. NECESIDADES DE COMUNICACIÓN E INFORMACIÓN

18. Qué tipo de información necesita usted?

	1. Mucho	2. Poco	3. Nada
Información educativa	1	2	3
Información sobre agricultura	1	2	3
Sobre leyes (sus derechos)	1	2	3
Sobre salud	1	2	3
Sobre empleo	1	2	3
Sobre política nacional	1	2	3
Sobre precios de productos	1	2	3

Sobre fiestas y actividades religiosas	1	2	3
Noticias y eventos locales	1	2	3

19. Qué tipo de información recibe en este momento?

	1. Mucho	2. Poco	3. Nada
Información educativa	1	2	3
Información sobre agricultura	1	2	3
Sobre leyes (sus derechos)	1	2	3
Sobre salud	1	2	3
Sobre empleo	1	2	3
Sobre política nacional	1	2	3
Sobre precios de productos	1	2	3
Sobre fiestas y actividades religiosas	1	2	3
Noticias y eventos locales	1	2	3

20. ¿Que tipo de información en cuanto a proyectos le parece más útil?

	1. Mucho	2. Poco	3. Nada
Proyectos comunitarios locales	1	2	3
Proyectos comunitarios en América Latina	1	2	3
Proyectos del gobierno central	1	2	3
Proyectos de apoyo internacionales	1	2	3
Proyectos de organizaciones no gubernamental (ONG)	1	2	3

21. ¿Qué servicios de información serian los más necesarios?

	1. Mucho	2. Poco	3. Nada
Intercambio de experiencias entre diferentes comunidades	1	2	3
Intercambio de información entre organizaciones indígenas y las comunidades	1	2	3
Información sobre organizaciones internacionales	1	2	3
Información sobre programas del gobierno local	1	2	3
Información sobre programas de organizaciones no gubernamentales (ONG)	1	2	3

Comercialización de sus productos a través del Internet	1	2	3
Información para mejorar la productividad en agricultura, artesanía y otros	1	2	3

22. ¿Que tipo de comunicación es más necesaria para Usted?

	1. Mucho	2. Poco	3. Nada
Entre comunidades de base	1	2	3
Entre organizaciones indígenas a nivel nacional	1	2	3
Entre organizaciones indígenas y las comunidades de base	1	2	3
Entre organizaciones indígenas a nivel internacional	1	2	3
Entre organizaciones indígenas y el gobierno	1	2	3
Entre organizaciones indígenas y organizaciones internacionales	1	2	3
Entre organizaciones indígenas y organizaciones no gubernamentales (ONG)	1	2	3

V. ACESSO

23. ¿Tiene acceso al Internet?
    1. Si
    2. No

VI. USO DEL INTERNET

24. ¿Usted usa o ha usado alguna vez el Internet?
    1. Si
    2. No

25. Si todavía no ha usado el Internet, lo quiere usar en un futuro próximo?
    1. Si
    2. No

26. Cuáles son las razones principales para no usar el Internet?
    1. No tengo acceso al Internet
    2. No se cómo funciona el Internet
    3. Es muy caro
    4. Problemas de idioma
    5. No veo la utilidad de usarlo
    6. Otro _____

VII. INTERCAMBIO DE INFORMACIÓN

27. ¿Actualmente cómo le parece el intercambio de información entre las comunidades y el gobierno?

1.	Muy bueno
2.	Bueno
3.	Regular
4.	Problemático
5.	No hay ningún intercambio de información

28. ¿Cómo le parece el intercambio de información entre las organizaciones indígenas y las comunidades de base?

1.	Muy bueno
2.	Bueno
3.	Regular
4.	Problemático
5.	No hay ningún intercambio de información

29. ¿Piensa que el Internet puede mejorar las relaciones entre los pueblos indígenas y el gobierno?
    1. Si
    2. No

30. ¿En qué forma piensa que puede mejorar las relaciones?

1.	Puede mejorar el intercambio de información
2.	Puede mejorar la transparencia del gobierno
3.	Puede mejorar la participación de los pueblos indígenas en el gobierno local
4.	Otro _____

31. ¿Piensa que el Internet puede mejorar los servicios del gobierno a las comunidades indígenas?
    1. Si
    2. No

32. ¿De qué manera piensa que el Internet puede mejorar estos servicios del gobierno a las comunidades indígenas?

1.	Puede aumentar el acceso a información sobre los programas sociales del gobierno
2.	Permite de estar mejor informado sobre las acciones políticas del gobierno
3.	Puede mejorar el acceso a educación
4.	Puede mejorar al acceso a servicios de salud

5.	Puede servir para hacer trámites
6.	Otro _____

33. Porque piensa que NO puede mejorar los servicios del gobierno.
    1. Desconfianza del gobierno
    2. Desconfianza de las autoridades y dirigentes
    3. Otro _____

VIII. OPORTUNIDADES Y DESAFIOS

34. Usted cree que el Internet puede traer a su comunidad:
    1. Más cosas buenas
    2. Más cosas malas
    3. Todo será igual

35. ¿Qué cosas buenas (positivas) puede traer el Internet para su comunidad?
    1. Mayor conocimiento
    2. Mejor educación
    3. Mayor participación en la política
    4. Mayor acceso a servicios públicos
    5. Mayor unidad entre comunidades
    6. Todo
    7. Ninguna cosa buena
    8. Otro _____

36. ¿Que cosas malas (negativas) puede traer el Internet en su comunidad?
    1. Conflictos entre comunidades
    2. Conflictos dentro la comunidad
    3. Mayor influencia de empresas internacionales (transnacionales)
    4. Podemos perder nuestra cultura
    5. Ninguna cosa mala
    6. Otro _____

Muchas gracias por su tiempo!!

User Survey Form

Encuesta sobre el rol de las tecnologías de información y comunicación para el desarrollo indígena (para usuarios de Internet)

Instrucción de llenado de boleta: Por favor lea atentamente las preguntas y encierre en un círculo su respuesta.

I. INFORMACIÓN PERSONAL

1. **Departamento:** _____
2. **Municipio:** _____
3. **Comunidad:** _____
4. **Nombre de organización:** _____
5. **Tel:** _____
6. **Correo electrónico:** _____
7. **Sexo**
    1. Hombre
    2. Mujer
8. **Edad** _____
9. **Cuál de los siguientes idiomas *es el principal idioma* que utiliza para comunicarse en su hogar?**
    1. Castellano
    2. Castellano y alguna lengua originaria/indígena
    3. Aymará
    4. Quechua
    5. Guaraní
    6. Otra lengua originaria/indígena
10. **Máximo grado de instrucción:**

1.	Menos de 5 años de primaria
2.	Primaria
3.	Secundaria
4.	Técnico medio
5.	Técnico superior
6.	Universidad
7.	Otros

11. **¿Qué cargo ocupa en su comunidad o municipio?**
    1. Mallku
    2. Dirigente
    3. Concejal
    4. Director o profesor de unidad educativa
    5. Miembro de la comunidad
    6. Representante de organización territorial de base
    7. Miembro del comité de vigilancia
    8. Otro _____

12. ¿A qué tipo de organización representa usted?

1.	Organización indígena/campesina
2.	Alcaldía
3.	Organización productiva (cooperativa o OECA)
4.	Organización de mujeres
5.	Unidad educativa (colegio)
6.	Otro:_____

II. CAPACIDADES INFORMACIONALES

13. ¿Usted participó en alguna capacitación sobre el uso del Internet?
    1. Si
    2. No

14. ¿Cuál cree usted que es el nivel de su conocimiento en el uso del Internet?
    1. Básico
    2. Medio
    3. Avanzado

15. ¿Cuán importante es para usted capacitarse más en el uso del Internet?
    1. Muy importante
    2. Poco importante
    3. Nada importante

16. En qué medida cree que el uso del Internet le ha ayuda a:

		1. Mucho	2. Poco	3. Nada
1.	Comunicarse mejor con familiares y amigos	1	2	3
2.	Encontrar información que necesitaba	1	2	3
3.	Compartir información con otras personas	1	2	3
4.	Difundir sus ideas y experiencias	1	2	3
5.	Otro:_____	1	2	3

17. Cuán confiable le parece la información recibida por Internet?
    1. Muy confiable
    2. Poco confiable
    3. Nada confiable

18. Cuán difícil le parece:

		1. Mucho	2. Poco	3. Nada
1.	Evaluar la calidad de la información que recibe	1	2	3
2.	Usar la información en mi trabajo	1	2	3

3.	Aplicar la información del Internet a mi realidad	1	2	3
4.	Compartir la información con otros miembros de mi comunidad	1	2	3
5.	Encontrar soluciones a problemas de mi comunidad	1	2	
6.	Difundir mis ideas y conocimientos a través del Internet	1	2	3

III. USO DEL INTERNET

19. **¿Hace cuanto tiempo utiliza el Internet?**

1.	Menos de un año
2.	Entre un año y tres
3.	Entre tres y cinco
4.	Más de cinco años

20. **¿Dónde está utilizando el Internet?**

1.	En un telecentro
2.	En mi colegio
3.	En mi organización
4.	En la alcaldía
5.	Otro: _____

21. **Con qué frecuencia se comunica a través del correo electrónico con una:**

		1. Mucho	2. Poco	3. Nada
1.	Organización indígena nacional	1	2	3
2.	Organización indígena en América Latina	1	2	3
3.	Organización indígena a nivel mundial	1	2	3
4.	Agencia de ayuda internacional	1	2	3
5.	Organización del gobierno	1	2	3
6.	Organización no-gubernamental (ONG)	1	2	3
7.	Otro: _____	1	2	3

22. **Cuán difícil le parece realizar las siguientes actividades en el Internet?**

		1. Muy difícil	2. Poco	3. Nada
1.	Mandar y recibir correos electrónicos	1	2	3
2.	Buscar información a través de un buscador	1	2	3

3.	Leer noticias	1	2	3
4.	Difundir/publicar información sobre su organización	1	2	3
5.	Participar en foros de discusión	1	2	3
6.	Coordinar acciones del movimiento indígena	1	2	3
7.	Presentar propuestas	1	2	3
8.	Tomar cursos por Internet	1	2	3
9.	Buscar trabajo	1	2	3
10.	Otro:_____	1	2	3

IV. DESARROLLO SOCIAL (dimensión)

23. **En qué medida el Internet me ha ayudado a:**

		1. Mucho	2. Poco	3. Nada
1.	Conocer mejor lo que pasa en el mundo	1	2	3
2.	Tener mayores posibilidades de educación	1	2	3
3.	Realizar mejor mi trabajo	1	2	3
4.	Fortalecer mi capacidad de liderazgo	1	2	3
5.	Fortalecer mi relación con otros dirigentes	1	2	3
6.	Mejorar la comunicación con mi comunidad	1	2	3

24. **En qué medida el Internet le ha ayudado a su comunidad a mejorar:**

		1. Mucho	2. Poco	3. Nada
1.	El acceso a educación	1	2	3
2.	El acceso a los servicios de salud	1	2	3
3.	El acceso a información sobre proyectos sociales	1	2	3
4.	La información sobre programas de beneficio social del gobierno (Ej. Bonosol)			
5.	Los servicios del gobierno a las comunidades indígenas			

V. DESARROLLO ECONÓMICO

25. **En qué medida el Internet me ha ayudado a:**

		1. Mucho	2. Poco	3. Nada
1.	Tener mayor acceso a información sobre el mercado y precios	1	2	3
2.	Tener menos necesidad de realizar viajes(menor costo de transporte y pérdida de tiempo	1	2	3
3.	Encontrar nuevos mercados para sus productos	1	2	3
4.	Recibir nuevos conocimientos para mejorar la producción	1	2	3
5.	Más oportunidades de encontrar trabajo	1	2	3
6.	Comunicarse mejor con familiares y amigos que le envían dinero regularmente	1	2	3
7.	Mejorar sus ingresos (venta de sus productos)	1	2	3

26. **¿En qué medida el Internet le ha ayudado al desarrollo económico de su comunidad?**

		1. Mucho	2. Poco	3. Nada
1	Mejor acceso a financiamiento internacional	1	2	3
2	Mejor conocimiento de proyectos productivos de la alcaldía	1	2	3
3	Fortalecer a las organizaciones productivas de mi comunidad	1	2	3
4	Promover el turismo en mi comunidad	1	2	3
5	Promover el comercio entre comunidades (ferias)	1	2	3

VI. DESARROLLO POLÍTICO (Participación política)

27. **En qué medida el Internet me ha ayudado a:**

		1. Mucho	2. Poco	3. Nada
1.	Tener mayor información sobre políticas nacionales	1	2	3
2.	Mejorar la participación en la toma de decisiones a nivel del gobierno local	1	2	3

3.	Participar en discusiones políticas nacionales Asamblea Constituyente, nacionalización y otros)	1	2	3
4.	Conocer mejor mis derechos.	1	2	3
5.	Tener mayor información sobre noticias internacionales	1	2	3
6.	Hacer contactos con otras personas para compartir ideas políticas	1	2	3

28. **Piensa que el Internet ha mejorado las relaciones entre los pueblos indígenas y el gobierno?**
   1. Si
   2. No

29. **En qué medida el Internet le ha ayudado a:**

		1. Mucho	2. Poco	3. Nada
1.	Fortalecer redes entre organizaciones indígenas/campesinas	1	2	3
2.	Mayor transparencia de instituciones gubernamentales	1	2	3
3.	Mayor control sobre las actividades del gobierno local	1	2	3
4.	Mayor participación de organizaciones indígenas en eventos internacionales	1	2	3
5.	Promover los derechos indígenas a nivel nacional e internacional	1	2	3
6.	Mejorar la participación de los pueblos indígenas en gobiernos locales	1	2	3

30. **¿Actualmente cómo califica usted el intercambio de información entre las comunidades y el gobierno?**

		1. Mucho	2. Poco	3. Nada
1.	Bueno	1	2	3
2.	Regular	1	2	3
3.	Mala	1	2	3
4.	No hay ningún intercambio de información	1	2	3

31. ¿Cómo califica el intercambio de información entre las organizaciones indígenas y las comunidades de base?

		1. Mucho	2. Poco	3. Nada
1.	Bueno	1	2	3
2.	Regular	1	2	3
3.	Mala	1	2	3
4.	No hay ningún intercambio de información	1	2	3

VII. DESARROLLO PERSONAL

32. En qué medida el Internet me ha ayudado:

		1. Mucho	2. Poco	3. Nada
1.	De tener mayor seguridad para conseguir un mejor trabajo	1	2	3
2.	De sentirme con más posibilidades respecto a otras personas de mi comunidad	1	2	3
3.	De sentirme menos aislado de lo que pasa en Bolivia	1	2	3
4.	Me abre un nuevo mundo para mejorar mi calidad de vida	1	2	3
5.	De sentirme orgulloso de saber manejar el Internet	1	2	3

33. En qué medida el Internet le ha ayudado a:

		1. Mucho	2. Poco	3. Nada
1.	Que su comunidad sea conocida en el mundo	1	2	3
2.	Que su comunidad avance más rápido que las comunidades vecinas que no tienen Internet	1	2	3
3.	Que su comunidad tenga posibilidades de mejorar la calidad de vida	1	2	3
4.	Le hace sentir orgulloso de su comunidad	1	2	3

VIII. CULTURAL

34. En qué medida el Internet me ha ayudado a:

		1. Mucho	2. Poco	3. Nada
1.	Fortalecer su identidad cultural	1	2	3
2.	Podría resultar en la pérdida de su identidad cultural	1	2	3

3.	Mejorar el conocimiento de su lengua originaria	1	2	3
4.	Conocer mejor su cultura	1	2	3
5.	Difundir información sobre mi cultura	1	2	3

35. **En qué medida usted cree que el Internet:**

		1. Mucho	2. Poco	3. Nada
1.	Es un instrumento para difundir sus valores culturales	1	2	3
2.	Ha fortalecido la identificación cultural de su comunidad	1	2	3
3.	Ha ayudado a que su comunidad se pueda relacionar mejor con otras culturas indígenas	1	2	3
4.	Ha ayudado a rescatar conocimientos indígenas tradicionales de su comunidad	1	2	3
5.	Puede ser usado por personas extrajeras para robar conocimientos indígenas tradicionales	1	2	3

IX. ORGANIZACIONAL

36. **En qué medida el Internet ha ayudado a:**

		1. Mucho	2. Poco	3. Nada
1.	Fortalecer las organizaciones indígenas	1	2	3
2.	Mejorar el acceso a información sobre la gestión de organizaciones indígenas	1	2	3
3.	Fortalecer redes entre organizaciones indígenas	1	2	3
4.	Mejorar la comunicación con el gobierno central	1	2	3

37. **Cuán útil piensa usted que es el Internet para el trabajo de su organización?**
    1. Muy útil
    2. Poco útil
    3. Nada útil
    4. No se/no respondo

X. OPORTUNIDADES Y DESAFIOS

38. **Usted cree que el Internet ha traído a su comunidad:**
    1. Más cosas buenas
    2. Más cosas malas
    3. Todo es igual

39. **¿Qué cosas buenas (positivas) ha traído el Internet para su comunidad?**
    1. Mayor conocimiento
    2. Mejor educación
    3. Mayor participación en la política
    4. Mayor acceso a servicios públicos
    5. Mayor unidad entre comunidades
    6. Mayores oportunidades económicas
    7. Fortalecimiento de nuestra cultura
    8. Todo
    9. Ninguna cosa buena
    10. Otro: _____

40. **¿Qué cosas malas (negativas) ha traído el Internet a su comunidad?**
    1. Conflictos entre comunidades
    2. Conflictos dentro la comunidad
    3. Conflictos dentro las organizaciones indígenas
    4. Mayor influencia de empresas internacionales (transnacionales)
    5. Estamos perdiendo nuestra cultura
    6. Ninguna cosa mala
    7. Otro: _____

Muchas gracias por su tiempo!!

Index

Boxes, figures, maps, notes, and tables are indicated by b, f, m, n, and t, respectively.